ESO ASTROPHYSICS SYMPOSIA
European Southern Observatory

Series Editor: Philippe Crane

Springer
Berlin
Heidelberg
New York
Barcelona
Budapest
Hong Kong
London
Milan
Paris
Santa Clara
Singapore
Tokyo

D. Minniti H.-W. Rix (Eds.)

Spiral Galaxies in the Near-IR

Proceedings of the ESO/MPA Workshop
Held at Garching, Germany,
7–9 June 1995

Springer

Volume Editors

Dante Minniti
European Southern Observatory
Karl-Schwarzschild-Strasse 2
D-85748 Garching, Germany
and
Lawrence Livermore National Laboratory
MS L 401, P.O. Box 808
Livermore, CA 94550, USA

Hans-Walter Rix
Max-Planck-Institut für Astrophysik
Karl-Schwarzschild-Strasse 1
D-85748 Garching, Germany
and
Steward Observatory
University of Arizona
Tucson AZ 85721, USA

Series Editor

Philippe Crane
European Southern Observatory
Karl-Schwarzschild-Strasse 2
D-85748 Garching, Germany

Cataloging-in-Publication Data applied for.

Die Deutsche Bibliothek - CIP-Einheitsaufnahme

Spiral galaxies in the near IR : proceedings of the ESO workshop, held at Garching, Germany, 7 - 9 June 1995 / D. Minniti ; H.-W. Rix (ed.). - Berlin ; Heidelberg ; New York ; Barcelona ; Budapest ; Hong Kong ; London ; Milan ; Paris ; Santa Clara ; Singapore ; Tokyo : Springer, 1996
(ESO astrophysics symposia)
ISBN 3-540-60937-7
NE: Minniti, Dante [Hrsg.]; European Southern Observatory

ISBN 3-540-60937-7 Springer-Verlag Berlin Heidelberg New York

Printed in Germany

Typesetting: Camera ready by author/editor
SPIN: 10517724 42/3142-543210 - Printed on acid-free paper

Preface

From June 7–9, 1995, the European Southern Observatory (ESO) and the Max Plank Institut für Astrophysik (MPA) jointly held the Workshop on Spiral Galaxies in the Near-IR. This meeting took place at the ESO headquarters in Garching bei München, Germany. The weather was changing, with the *biergarten* closed, but that did not stop 85 people from all over the world from attending the meeting. The three days were intensive, with talks and coffee and posters from 9 am to 6 pm, and very productive indeed for everyone. The topics covered the stellar populations of the Milky Way and other more distant spirals, the role of dust, the dynamics of spiral galaxies, and the nuclear activity seen at near-IR wavelengths. This volume presents the original contributions from the participants, including several papers that review the state-of-the-art knowledge in these various subjects.

The editors would like to thank first and foremost Christina Stoffer, for she took care of everything. The meeting would not have been so successful without her expertise and efficiency.

We are deeply indebted to the directors of MPA and ESO Science, Simon White and Jacqueline Bergeron, for their support and encouragement.

We would also like to thank the other members of the scientific organizing committee: R. Genzel, K. Freeman, A. Moorwood, S. White, M. Rieke and E. Athannasoula, for their advice with the organization of the program.

We also thank G. Rieke, R. Genzel, L. Athannasoula, A. Renzini and R. Terlevich for accepting, without previous notice, to give a concluding summary of the meeting, discussing the most important results presented and trying to predict the future.

We thank Alicia and Mary Beth, our wives, for their help always. We are very grateful to Alicia for the help in editing and putting these proceedings in final form. We thank S. Courteau for comments about the proceedings.

We are very grateful to the ESO fellows and students for their assistance throughout the meeting: D. Clements, F. Courbin, P. Goudfroig, M. Villar, L. Kaper, M. Kissler-Patig, F. Patat, J. van Loon, G. Dudziak, and J. Rönback. We are grateful to P. Crane (editor of this series), I. J. Danziger, J. Spyromilio, R. Fosbury, G. Meylan, C. Tinney, and M. H. Ulrich for their advice. We also thank Cornelia Rickl for the organization of the banquet.

Finally, we would like to thank all the people who attended the meeting for a very educating and stimulating meeting.

Garching, December 1995

Dante Minniti
Hans-Walter Rix

Contents

I. Stellar Populations in the IR

II. IR Observations of Local Group Galaxies

I

STELLAR POPULATIONS IN THE IR

When Do Near-IR Colors Help in Studying Stellar Populations?

David R. Silva

NOAO/KPNO, P.O. Box 26732, Tucson, AZ, USA, 85725-6732

Abstract. Integrated near-IR colors can provide useful constraints on the coolest constituents of composite stellar populations. This is particularly true in quiescent systems such as globular clusters and E/S0 galaxies. However, the presence of dust and active star formation in Irr/BCD and spiral galaxies can significantly reduce the precision of such constraints. At present, an empirical approach to interpreting near-IR colors is favored over a theoretical approach since current evolutionary population synthesis models do not correctly reproduce the near-IR colors of real galaxies.

1 Introduction

Conventional wisdom holds the following truths about integrated near-IR colors to be self-evident. First, small bursts of star formation cause large excursions in optical color but small changes in near-IR colors. For example, there is a much higher dispersion in the integrated optical colors of star forming galaxies than in their integrated near-IR colors. Thus, near-IR colors are considered to be less sensitive to recent star formation history. Second, if dust is present, optical colors are affected more than near-IR colors. Thus, near-IR colors are a more direct probe of the underlying stellar population. On the other hand, optical and near-IR colors are sensitive to different sub-populations so they also provide different perspectives. Third, optical colors are very age/metallicity degenerate – dissimilar changes in age and metallicity lead to similar integrated colors. In contrast, it is commonly held that near-IR colors are mostly sensitive to mean metallicity and relatively insensitive to mean age. Taken together, these self-evident truths lead to a model where a combination of optical and near-IR colors can lead to definitive constraints on age and metallicity, even in the presence of dust.

In what follows, this framework is discussed in the context of interpreting the integrated light of nearby stellar systems. While near-IR colors do provide interesting stellar population constraints, in many cases, especially spiral disks, they do not provide definitive answers unless they are coupled with additional information, such as measurements of the depth of the CO 2.36μm bandhead, a sensitive luminosity indicator in M stars. I make no pretense that this review is complete or unbiased – these are the tools and caveats I keep in mind as I pursue my own projects.

2 The Basic Frameworks

2.1 Empirical Building Blocks

In any area of stellar populations, it is often easier to interpret empirical observations of complex stellar systems (a.k.a. "galaxies") in terms of empirical observations of more simple systems (a.k.a. "stars" and "clusters"). The chief drawback to this approach is that it is often impossible to find a set of simple systems which adequately span age-metallicity parameter space. The colors of Galactic globular clusters, Magellanic Cloud clusters, and E/S0 galaxies are most often used as color templates in the near-IR.

It was first shown that the integrated near-IR colors of globular clusters are dominated by giant stars and driven by metallicity by Aaronson, Cohen, Matthews, & Malkan 1978 (hereafter ACMM78). Broadly, this color-metallicity relationship holds because near-IR colors measure mean first-ascent giant branch (FGB) temperature. In turn, FGB temperatures are driven by envelope opacity which is dominated by H–. Since metals are the primary electron donors to H–, as metallicity increases, FGB envelope opacity increases, FGB effective temperatures decrease, and mean FGB colors become redder. It is commonly argued (e.g. Burstein 1985) that Fe is the dominant electron donor and therefore J–K color traces relative Fe abundance almost directly. If true, J–K would make an excellent tracer of SN Type I enrichment. Unfortunately, the combination of Mg and Si contribute the same fractional amount of electrons as Fe, weakening the efficacy of J–K in tracing Type I Sn enrichment (Renzini 1977).

Frogel, Persson, Aaronson, & Matthews (1978) (FPAM78) extended this result to E/S0 galaxies and argued that their large aperture near-IR colors were dominated by old, FGB stars. However, this interpretation required measurements of the 2.36μm CO bandhead. Independent of the CO data, constraining the relative dwarf and giant light contributions by JHK color alone is not possible. Nevertheless, it has become common practice to conclude that if the integrated light of some extragalactic stellar population has similar JHK colors to elliptical galaxies, that population must be giant dominated in the near-IR. In the absence of additional information, such as CO measurements, this practice is clearly suspect.

The ACMM78 and FPAM78 results led to the common assumption that redder J–K colors imply higher metallicity. But the work of Persson *et al.* (1983) on Magellanic Cloud clusters demonstrated that the integrated colors of intermediate-age clusters are very red, particularly in H–K, due to the presence of C/M stars on the intermediate-age AGB (see also Frogel, Mould, & Blanco 1990). This complicates the interpretation of near-IR colors of other systems because red J–K colors could either imply high (i.e. near solar) metallicity (if H–K ≤ 0.2) or a significant intermediate-age component (if H–K ≥ 0.2). J–K color alone cannot be used to choose between these possibilities.

Persson *et al.* also found that the integrated near-IR colors of the youngest Magellanic clusters were contaminated by the light of hot, young main sequence stars, i.e. that the near-IR colors were not just probes of the cool, "old" population in these clusters.

Thus, even in "simple" systems, J–K colors alone do not provide enough information to uniquely characterize the current luminosity-weighted age and metallicity. Measurements of the H–K color are necessary to constrain the contribution of intermediate-age AGB stars while CO bandhead measurements are needed to constrain the relative contribution of dwarf, giant, and supergiant light, even in the absence of dust.

2.2 Current Evolutionary Models: Do They Work in the Near-IR?

Unfortunately, the basic near-IR color empirical framework does not completely span age-metallicity parameter space. In principle, evolutionary population synthesis models overcome this difficulty since modelers can tune their models to any specific age and metallicity, as well as accounting for such things as non-solar abundance ratios and internal reddening. This assumes, of course, that the model inputs (e.g. stellar evolutionary theory at all stellar ages, metallicities, and masses; stellar atmosphere models) are accurate enough to model real galaxies. The basic criteria of a useful model is that it reproduce the observed colors of real galaxies.[1]

In Figure 1, the frequently used models of Bruzual & Charlot (1993) and Worthey (1994) are compared to the observed colors of real elliptical galaxies. These models are thought to reproduce the optical colors and line indices of elliptical galaxies with high fidelity. Alas, as Figure 1 shows, these models do not reproduce the observed near-IR colors of elliptical galaxies very well nor do they agree with each other. Similar discrepancies also exist in the J–K vs V–K plane. This may indicate that either post-main sequence stellar evolution, particularly of mid-K and cooler FGB and AGB stars, is not being modeled properly or that the input stellar atmospheres used to calibrate the models are faulty. Whatever the cause of these discrepancies, interpreting the absolute near-IR colors of real galaxies within the framework of these

[1] To paraphrase the Hippocratic Oath: "First, do no harm."

models should be done cautiously and skeptically. Furthermore, even differential comparisons (e.g. a change in color corresponds to a specific change in metallicity or age) are suspect.

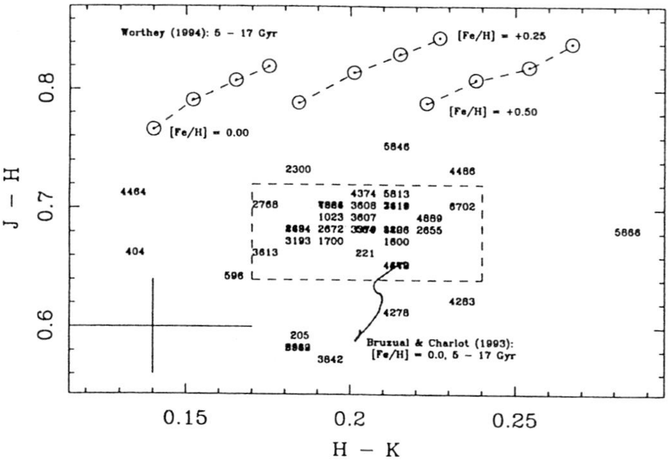

Fig. 1. Models vs. Reality: Bruzual & Charlot (1993) and Worthey (1994) models for ages 5 - 17 Gyrs compared to FPAM78 E/S0 large aperture near-IR colors (denoted by NGC numbers). Typical FPAM78 error bars shown in lower left corner. Models get redder as they get older.

3 Lessons from E/S0 Galaxies

The main conclusions of FPAM78 and related contemporaneous work (e.g. ACMM78; Frogel, Persson, & Aaronson 1979) were that the infrared light of E/S0 galaxies were dominated by old, metal-rich K/M AGB and FGB stars.

It has been argued recently that some ellipticals may have formed a significant fraction of their stars very recently, i.e. within the last 5 – 7 Gyr. In Figure 2, near-IR colors of elliptical galaxies determined to be "young" by either Schweizer & Seitzer (1992) or González (1995) are compared to the colors of Galactic globulars clusters, Magellanic Cloud clusters, and FPAM78 ellipticals. Not all "young" ellipticals have available near-IR colors, but those that do are clearly not dissimilar in color from the much larger FPAM78 sample. The same conclusion can be drawn from the CO vs H–K plane. These large aperture near-IR colors constrain the relative mass contribution of intermediate-age stars, i.e. the bulk of the mass in these galaxies cannot be

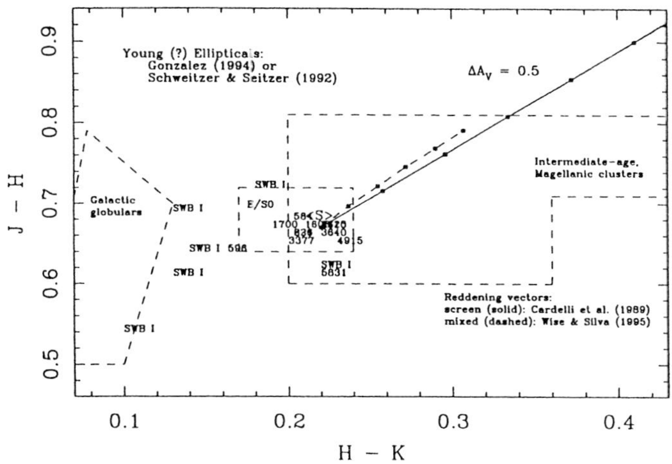

Fig. 2. "Young" Ellipticals in Near-IR: Putatively "young" ellipticals (denoted by NGC number) compared to distributions of FPAM78 E/SO large aperture near-IR colors (central dashed box), Galactic globular clusters, and intermediate-age Magellanic clusters. Colors of youngest Magellanic clusters denoted by "SWB I". Clearly, these ellipticals have similar colors to putatively "old" ellipticals. They do not have colors like younger clusters nor do they show the signature of significant internal reddening.

as young as 5 – 7 Gyr or these galaxies would have much redder H–K colors (Bothun & Silva 1996).[2]

Bothun & Gregg (1990) provides another illustrative example of how near-IR colors can provide interesting age constraints in "old" populations. Using small aperture photometry, they concluded that that some S0 disks had redder H–K colors at a given J–H color than other S0 disks (e.g. see their Figure 8) and probably contain significant contributions from intermediate-age AGB stars. If only J–K colors were available, it would not be possible to conclude whether this was an age or metallicity effect.

---

[2] M32 illustrates this point: despite the existence of a resolvable, near-IR bright, extended giant branch, thought to be the intermediate-age AGB (Freedman 1992; Elston & Silva 1992), the integrated global near-IR colors of M32 are not inconsistent with its absolute magnitude (FPAM78).

4 Spiral Galaxies: Probing the Maelstrom

4.1 Spiral Bulges: Dusty or Bursty?

In one of the earliest large near-IR surveys of spirals, Aaronson (1977) found that the near-IR colors of his spiral sample were the same as the FPAM78 ellipticals (cf. Griersmith, Hyland, & Jones 1982). This is not that surprising since within the large apertures used, bulge light dominated disk light.

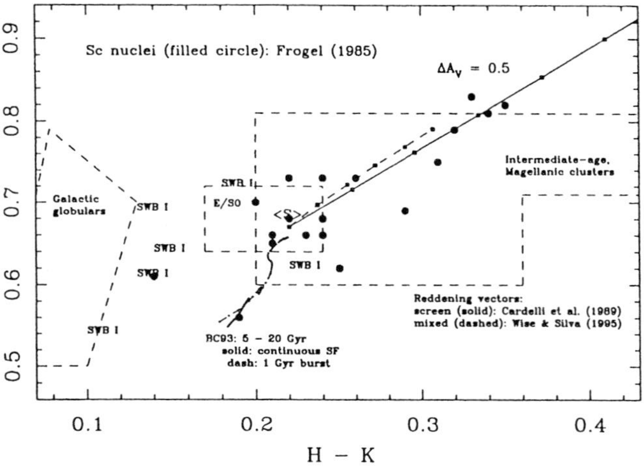

Fig. 3. Frogel (1985) Sc Nuclei Study: Sc nuclei colors compared to color distributions of Galactic globular clusters, FPAM78 ellipticals, and young (SWB I) and intermediate-age Magellanic clusters. Tick marks along the reddening vectors have intervals of $\Delta A_V = 0.5$. Based on these colors alone, it is unclear whether the nuclei lying along the reddening vectors have significant internal reddening or a significant intermediate-age component.

The Frogel (1985) small aperture study of Sc nuclei demonstrates the real difficulty in interpreting the near-IR colors of spiral bulges. Figure 3 illustrates the situation in the J–H vs H–K color plane. Based on similar diagrams and additional CO 2.36μm bandhead data, Frogel concluded that Sc nuclei had similar stellar populations to E/S0 galaxies and that extensions to redder colors were most likely the result of internal reddening. Figure 3, however, demonstrates the difficulty of distinguishing between internal reddening and an intermediate-age component based on near-IR colors alone.

Furthermore, even under the assumption that the redder colors are caused by dust, constraining the optical depth is difficult because of dependences on geometry and dust grain distributions (see, e.g. Peletier and Witt in these proceedings). For now, a simple screen model based on the Cardelli *et al.* (1989) Galactic reddening law models and a simple radiative transfer model that mixes the stars and dust from Wise & Silva (1995) are compared. The mixed model leads to much higher optical depths and therefore much larger effective dust masses.

4.2 Irr/BCD: Disk Analogues?

Irregular and blue compact dwarf (Irr/BCD) galaxies are very interesting analogues to spiral disks. These galaxies have high enough surface brightnesses that it is relatively easy to acquire precise color measurements. They also illustrate many of the interpretative difficulties of spiral disks: they are actively forming stars and therefore contain stars with a range of age and metallicity; they often have strong localized HII continuum and line emission; and they probably contain dust. If one cannot constrain the stellar populations of these objects, how can one expect to constrain the stellar populations of spiral disks?

Choosing the right question is often the key to proper interpretation. An interesting question is whether or not Irr/BCD galaxies are forming stars for the first time or not. Based on near-IR aperture photometry, Thuan (1983) concluded that, after the effects of recent star formation and internal reddening were taken into account (see Figure 1 in Thuan 1983; also see Telesco & Gatley 1984; Joy & Lester 1988; Rieke, this conference), the colors of these galaxies were consistent with the colors of an "old", evolved population and therefore these galaxies were not forming stars for the first time (see also Hunter & Gallagher 1985; Salzer & Elston 1990). However, Arimoto & Tarrab (1990) argued that the reddest Thuan galaxies could be dominated by young, red supergiants at 2μm, not internal reddening or old stars, and that their metallicities implied that some gas preprocessing had already occurred, indicating that this is not the first generation of stars in these objects.

Delineating between giants and supergiants cannot be done by near-IR color alone; measurements of a luminosity indicator, such as the 2.36μm CO bandhead are required (see, e.g., Campbell & Terlevich 1984). One difficulty in interpreting photometric CO bandhead measurements in star forming objects is in-fill by hot stars (e.g. Joy & Lester 1988). This can be largely overcome by K-band spectroscopy. Using such spectroscopy, Silva, Bershady, & Elston (1996) have recently demonstrated that some of the reddest Thuan objects are in fact supergiant dominated at 2μm as suggested by Arimoto & Tarrab.

Higher spatial resolution data also helps in the interpretation of near-IR colors. This is so obvious, it is often overlooked. Consider, for example, the giant Irr NGC 4449. Based on a grid of 8" diameter apertures, Thronson *et al.* (1987) concluded that the near-IR colors of NGC 4449 are consistent with an old, cool stellar component. Yet, even relatively crude near-IR images of NGC 4449 (Silva & Bothun 1990) show regions which are very red in H−K, consistent with the presence of intermediate-age AGB stars or significant internal reddening. Followup spectroscopy of these regions would be very interesting.

4.3 Spiral Disks: Can Near-IR Colors Be Helpful?

Just acquiring precise color data on spiral disks is incredibly difficult. Consider that at one scale length, the typical spiral disk has a K-band surface brightness of roughly 18 – 20 (e.g. de Jong 1995) while the typical K-band background surface brightness is 13 – 13.5! And yet H–K and CO index differences of 0.05 mag reflect very significant stellar population differences (see, e.g., Campbell & Terlevich 1984), making high precision photometry necessary to delineate useful astrophysical constraints. Since measurements of this photometric precision at interesting spatial resolutions are very difficult (see, however, Rix & Rieke 1993), it is common practice to only measure azimuthally averaged color profiles in spiral disks. Such averaging undoubtedly washes out much of the inherent stellar population information given the range of star formation activity in spiral disks.

Early large aperture, bulge-dominated work on spiral galaxies lead to the conclusion that spirals had similar near-IR colors, and thus similar stellar populations, to E/S0 galaxies (Aaronson 1977; Griersmith *et al.* 1982). In contrast, Bothun *et al.* (1984) argued that a sample of disk-dominated Sc galaxies showed a trend of redder J–K color with increasing total H magnitudes through large apertures. While arguing that this was consistent with lower luminosity disks having lower mean metallicity, they could not eliminate the possibility of a trend of decreasing intermediate-age AGB light contribution with decreasing luminosity. While these conclusions were based on relatively simple models, it is not at all clear that the more "sophisticated" models readily available today are qualitatively better or have higher astrophysical fidelity.

Near-IR imaging campaigns of large samples of spiral galaxies are beginning to be published (e.g. Peletier *et al.* 1994; Terndrup *et al.* 1994; de Jong 1995). A common implicit astrophysical goal is to compare the bulge and disk populations and ask whether they were coeval or not.[3] Is a consistent picture emerging? Peletier *et al.* conclude that most of the color differences within spiral galaxies are driven by dust but de Jong concludes that it must be a combination of dust, age, and metallicity effects (see also Rix & Rieke 1993). Terndrup *et al.* conclude that similar near-IR colors imply similar stellar populations in the spiral bulges and disks while optical-IR colors indicated significant age or metallicity differences; yet, they concede they cannot conclusively separate dust effects from stellar population effects by rJK photometry alone. None of these studies use H-band or CO data to try to probe

[3] Is this a "good" question? Given that: (1) many spiral disks have been continuously forming stars for 10 Gyr; and (2) we can't answer this question definitively for our own galaxy, it is certainly a "hard" question!

for the presence of intermediate-age AGB and young red supergiant stars. Such stars must contribute an appreciable fraction of the near-IR light in these continuously star forming systems.

Near-IR colors have not made studying the stellar populations of spiral disks easier, it has only raised new (albeit interesting) questions. Higher spatial resolution data, H-band and CO bandhead data, and near-IR spectroscopy would help. Rhoads (this conference) discusses an interesting step in this direction.

5 Summary: *Caveat Observer*

The basic tenet of near-IR color conventional wisdom is that two stellar systems with similar colors have similar stellar populations and therefore similar star formation histories. Or in other words, if J–K ≈ 1, the underlying stellar population is "old" (T > 10 Gyr) and most of the 2μm light is coming from FGB stars. In the absence of dust and on-going star formation, this is probably true and near-IR colors have provided very useful constraints on the star formation histories of E/S0 galaxies, as discussed in section 3.

The interpretation of the near-IR colors of systems which are actively forming stars or recently formed stars is more difficult. The combination of dust effects, H II region emission, and photospheric light from "hot" young main sequence stars and "cool" young supergiant, intermediate-age asymptotic giant, and first-ascent giant stars can conspire in numerous combinations to produce colors similar to the standard benchmarks, the Galactic globular cluster sequence and E/S0 galaxies. H-band and CO data are helpful in constraining the situation but do not completely lift the inherent degeneracies.

The current evolutionary models clearly need more fine-tuning: they do not adequately reproduce the observed near-IR colors of real galaxies. Two well-known problem areas are available stellar atmosphere models of cool stars and evolutionary models of AGB stars: neither reproduce the observed reality as well as desired. It is encouraging to see the authors of the most commonly used codes collaborating to try to delineate what areas need work next (Charlot, Worthey, & Faggotto 1995). In the meantime, the empirical approach to interpreting near-IR colors, for all its pitfalls, remains the most sound.

The observational frontier appears to be higher spatial resolution, higher S/N imaging and more near-IR spectroscopy, from space if possible. Like any other situation, near-IR colors are most useful when the astrophysical questions are well-defined. If I were trying to decipher the mysteries of spiral disks, I would make JHK maps and then use IR spectroscopy to ask specific, well constrained questions like "is it dust or is it AGB stars?". As at other wavelengths, some combination of near-IR imaging and spectroscopy will undoubtedly be needed to make real progress in constraining the nature of spiral galaxy stellar populations.

References

Arimoto, N., & Tarrab, I. (1990): A&A, 228, 6.

Aaronson, M. (1977): Ph.D. thesis, Harvard Univ.

Aaronson, M., Cohen, J.G., Mould, J., & Malkan, M. (1978): ApJ, 223, 824 (ACMM78).

Armandroff, T.E. (1988): AJ, 97, 375.

Baldwin, J.R., Frogel, J.A., & Persson, S.E. (1973): ApJ, 184, 427.

Bothun, G.D., & Gregg, M.D. (1990): ApJ, 350, 73.

Bothun, G.D., & Silva, D.R. (1996): in preparation.

Bothun, G.D., Romanshin, W., Strom, S.E, & Strom, K.M. (1984): AJ, 89, 1300.

Bruzual, A., G., & Charlot, S. (1993): ApJ, 405, 538.

Burstein, D. (1985): PASP, 97, 89.

Campbell, A.W., & Terlevich, R. (1984): MNRAS, 211, 15.

Cardelli, J.A., Clayton, G.C., & Mathis, J.C. (1989): ApJ, 345, 245.

Charlot, S., Worthey, G., & Bressan, A. (1995): ApJ, in press.

de Jong, R.S. (1995): Ph.D. thesis, Groningen.

Elston, R., & Silva, D.R. (1992): AJ, 104, 1360.

Freedman, W.L. (1992): AJ, 104, 1349.

Frogel, J.A. (1985): ApJ, 298, 528.

Frogel, J.A., Mould, J., & Blanco, V.M. (1990): ApJ, 352, 96.

Persson, S.E., Frogel, J.A., & Aaronson, M. (1979): ApJSup, 39, 61.

Frogel, J.A., Persson, S.E., Aaronson, M., & Matthews, K. (1978): ApJ, 220, 75 (FPAM78).

González, J.J. (1995): Ph.D. thesis, Univ. of California, Santa Cruz.

Griersmith, D., Hyland, A.R., & Jones, T.J. (1982): AJ, 87, 1106.

Hunter, D.A., & Gallagher, J.S. (1985): AJ, 90, 1457.

Joy, M., & Lester, D.F. (1988): ApJ, 331, 145.

Malkan, M. & Aaronson, M. (1980): unpublished.

Peletier, R.F., Valentijn, E.A., Moorwood, A.F.M., & Freudling, W. (1994): A&AS, 108, 621.

Persson, S.E., *et al.* (1983): ApJ, 266, 105.

Renzini, A. (1977): in Advanced Stages of Stellar Evolution, Eds. P. Bouvier & A. Maeder (Geneva Obs., Sauverny), p. 149.

Rix, H.-W., & Rieke, M.J. (1993): ApJ, 418, 123.

Salzer, J.J., & Elston, R. (1990): in Astrophysics with IR Arrays, Ed. R. Elston (ASP Conf Series, San Francisco), p. 41.

Schweizer, F., & Seitzer, P. (1992): AJ, 104, 1039.

Silva, D.R., Bershady, M., & Elston, R. (1996): in preparation.

Silva, D.R., & Bothun (1990): in Astrophysics with IR Arrays, Ed. R. Elston (ASP Conf Series, San Francisco), p. 38.

Telesco, C.M., & Gatley, I. (1984): ApJ, 284, 557.

Terndrup, D.M., *et al.* (1994): ApJ, 432, 518.

Thronson, H.A. *et al.* (1987): ApJ, 317, 180.

Thuan, T. (1983): ApJ, 268, 667.

Wise, M. W., & Silva, D.R. (1995): ApJ, submitted.

Worthey, G. (1994): ApJSup, 95, 107.

Photometric and Spectrophotometric Models in the NIR

B. Rocca-Volmerange

Institut d'Astrophysique de Paris, 98 bis Bd Arago, F-75014 Paris

Abstract. A brief comparison of photometric and spectrophotometric models of population synthesis is presented. Two complementary modelling in the NIR are proposed. One is a refined spectral synthesis of starbursts in the $H + K$ band. Another model aims to predict synthetic spectra (continua and lines) and colors of starbursts and the Hubble Sequence from 220Å to 5μm.

1 Introduction

Star-forming galaxies and early-type galaxies have both stellar populations dominant in the near-infrared. In starbursts, bright red supergiants and AGB stars superimposed to the underlying giant population rapidly evolve in an anisotropic dusty environment(Puxley and Brand, 1994, Rieke et al, 1985). In early-type galaxies, giant populations contribute to a fraction of more 80% of the light. However up to now, modelling evolution of galaxies in the near-infrared did not get the level of availability reached in the visible. The evolutionary details of the AGB phase (early-phase, thermal pulses, the hypothetic "AGB manqué" phase) have recently been investigated and related to the previous evolutionary phases (Boothroyd and Sackmann, 1988, Groenewegen and de Jong, 1993). For early-type galaxies, the emission is highly sensitive to the calibration of the effective temperatures of the giant branch which suffer of large uncertainties. The energy distribution of these cold stars peaks in the J band and large error bars on the determination of spectral types or T_{eff} will strongly modify colors. The response of most receivers shows a low sensitivity in the J band and atmospheric bands are unfortunately located at the curvature change of the stellar energy distribution (SED) of most stars.

At last a significant simulation of the stellar emission from the *blue* to the *near-infrared* with gas and dust contributions requires a peculiar attention to the consistency of data in the two wavelength ranges. For this reason, two complementary approaches were developed. First, near-infrared spectra of starbursts were observed with the Fourier Transform Spectrograph (FTS) at the 3.60m CFHT instrument through the H and K bands. Continuum and emission lines were analysed with a spectral evolutionary synthesis of short timescale starbursts in a dusty medium (Lançon and Rocca-Volmerange,1994, 1995). Another series of works proposes a new version of our spectral evolution model (Fioc and Rocca-Volmerange, 1995) based on a new algorithm of integration and continuously extended from the far-UV to the near-IR. A new algorithm of integration on a small timestep, a strict coherency of spectral and photometric input data, a revised stellar library available on the whole wavelength range, stellar evolutionary tracks including AGB phase and thermal pulses are main caracteristics of this new model. This new model allows to interpret high-redshift samples of galaxies, by calculating cosmological and evolution (k- and e-) corrections, (Rocca-Volmerange et al, 1995). This model is specifically adapted to interpret the increasing sample of faint galaxy counts and the evolution of high redshift radiogalaxies or starbursts in a given cosmology.

2 Input data for photometric and spectrophotometric models

The first generation of evolutionary models of galaxies was based on the photometric properties of stars. Initially they were developed in the UBV bands by Tinsley, 1972, Searle et al, 1973. Their objectives were to interpret the extremely blue colors of galaxies by a current burst of star formation. Such interpretation was possible because most stars formed during the burst were still on the main sequence and so, well identified by the photometric colors as B-V and U-B. The nebular emission (continuum and lines) was later taken into account by evaluating the number of Lyman continuum photons (Huchra, 1977). At this epoch, the extension to the UV colors was made possible with the data from the far-UV satellites as Astronomical Netherland Satellite ANS and the Orbital Astronomical Observatory OAO. The satellite Aura-D2B (Maucherat-Joubert et al, 1980) observed the Magellanic Clouds which were analysed, with evolution in the UV and metallicity effects (Rocca-Volmerange et al, 1981). Other recent photometric models (Arimoto and Yoshii, 1986) were also used to study the chemical evolution.
The evolutionary models of secund generation were spectrophotometric based on spectral stellar libraries in the visible from ground-based observations (Gunn and Stryker, 1983, Jacoby et al, 1984) and in the far-UV from the IUE satellite (Wu et al, 1983, Heck et al, 1983). In the near-infrared, the models of atmosphere are not satisfactory for the coolest stars. Moreover, the best way of interpreting the evolution of a sample of bursting galaxies is to use a stellar library observed with the same instrument. When possible,

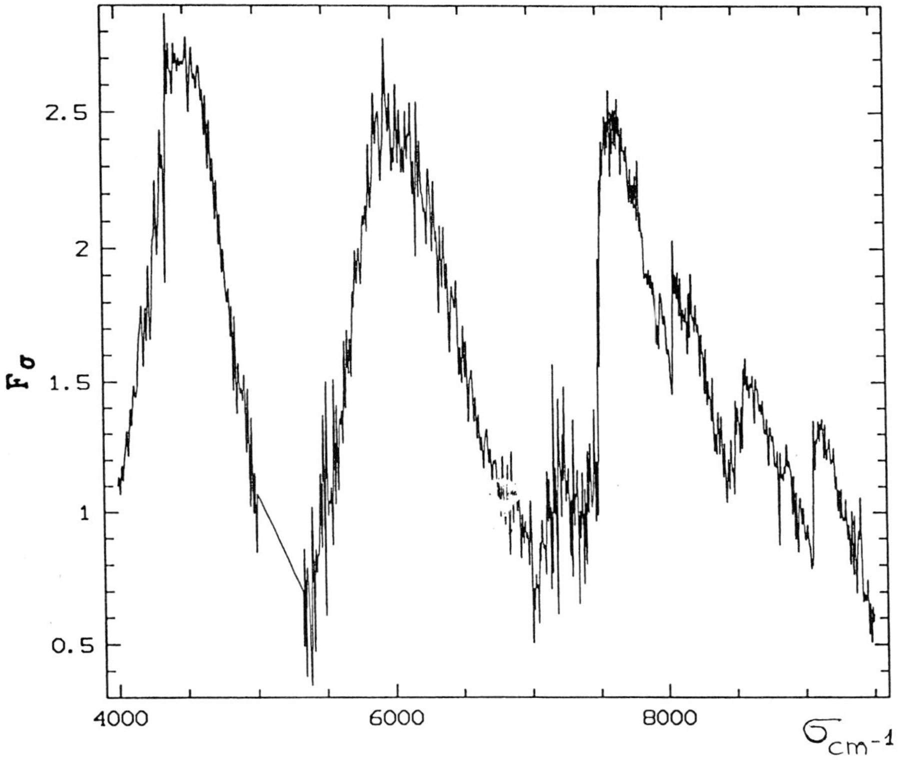

Fig. 1. The FTS/CFHT spectrum of a star M8III from 1.06μm (σ=9434cm^{-1}) to 2.5μm (σ=4000cm^{-1}), Rocca-Volmerange et al, 1995

this principle allows to avoid uncertainties due to instrumental effects. For this reason, we preferred to observe a spectral library of stars in the near-infrared H and K (Lancon and Rocca- Volmerange, 1992) with the Fourier Transform Spectrograph at the Canada France Hawaii telescope. The main caracteristics of this library is its large wavelength range in the mid-infrared (1.428 to 2.5 μm) corresponding to the H and K filters. The library has a significant coverage of the Hertzsprung-Russell diagram. The gravity range covers supergiant to dwarf stars and the effective temperature from 2000K to 45000K. The spectral resolution power (R=500) is sufficient to follow the evolution of the CO and H2O molecular bands as functions of gravity and temperature. The observations were recently extended to the J band (Rocca-Volmerange et al, 1995) and continuously connected to the visible and the ultraviolet libraries. Another example of stellar library for M stars was observed by Flucks et al, 1994. The main uncertainties of these libraries, however essential, are due to the determination of spectral types of stars.

Nevertheless the spectrophotometric models opened a new area for our understanding of galaxy evolution. They simulateneously predict an extended and continuous SED for various spectral types of galaxies. The far-UV extension allowed to model the evolution of galaxies at high- redshift (Bruzual, 1983, Guiderdoni and Rocca-Volmerange, 1987, Yoshii and Takahara, 1989). The recent coherent connection to the near-infrared (Fioc and Rocca-Volmerange, 1995, Bruzual and Charlot, 1993) complete the interpretation from the ultra-violet and blue in a so-called bolometric estimate. Moreover several significant spectral signatures can be used for constraints. Metal indices as Mg_2, Ca lines (Worthey , 1993), the hydrogen and so-called 4000A discontinuities (Hamilton, 1985). Other typical signatures are observed in the near-infrared with the 1.6 μm signature, and molecular bands as CO and H_2O. These H_2O bands are needed to distinguish between red giants from young supergiants, while CO bands are sensitive to an age effect and to dwarf effect (Lançon and Rocca-Volmerange,1992). Our spectral library includes several M giants with $3750K \leq T_{eff} \leq 2200K$ (fig. 1). They are supposed to be of normal metallicity. But the question of metallicity in relation with the evolutionary stages of these stars is not clearly solved. From a recent analysis in the low-metal abundance Magellanic Clouds (Bessel, 1995), color-magnitude diagrams show more numerous and luminous red supergiants than predicted from recent isochrones (Bertelli et al, 1994).

To summarise the comparison, spectrophotometric models are much more accurate than photometric models. The atlases of synthetic spectra of galaxies (see for example Rocca- Volmerange and Guiderdoni, 1988) give, as main outputs, a series of synthetic spectra of galaxies reddened by extinction, equivalent widths of absorption lines and the energy released in most main emission lines. Synthetic colors are calculated by convolving the SED with the passbands of filters in any photometric system, coherently calibrated on the absolute flux of Vega or of another star. The SED of various spectral types are fitted on observed templates, but due to the difficulty to get templates corrected of extinction, inclination, aperture and flux calibration, we often prefer to use synthetic colors through broad bands or color indices through narrow filters(Worthey, 1995) to test the availability of models. So, color-color diagrams or color-magnitude diagrams remain the most statistically meaningful constraints of nearby evolved galaxies of the Hubble sequence.

3 A sample of near-infrared starbursts

A sample of ten infrared luminous galaxies has been observed between 1.428 to 2.5μm with the Fourier Transform Spectrograph at the 3.60m CFHT (Lançon et al, 1994). Characterised by a high spectral resolution on a large spectral range, the spectra of NGC253 (fig.2), NGC 1614, Arp 243, UGC 5101, M 82, NGC 3690AB, Mrk 331, Arp 220, NGC 6240 and those of UGC

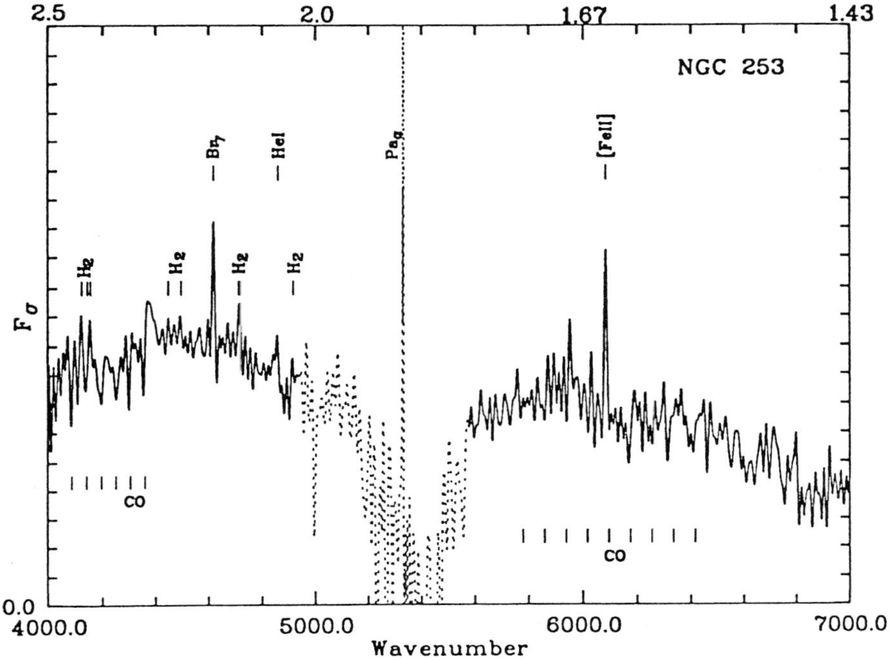

Fig. 2. The H+K spectrum of NGC253 in arbitrary units Fσ units is presented on the restframe wavenumber scale (cm^{-1})

5101 and Mrk 231 are distributed in two groups, the first one is dominated by stellar emission while the secund one shows a major non-stellar contribution to the near-IR. Only the first group has tentatively been interpreted.

The spectral synthesis proposed by Lançon and Rocca-Volmerange, 1995 is a detailed simulation of star formation bursts in the near-infrared, including continua from 1.428 to 2.5μm (H+K filter)in significant agreement with the V light and the far-infrared emission, effects of dust and the main emission lines (Brγ, HeIλ2.06μm, [FeII]λ1.64μm and rotation-vibration lines of H_2). The modelling of evolution is specifically suited to the starburst data, through the H+K bands. The integration time step of 10^5years, roughly corresponds to the lifetime duration of a 100$M_\odot$ star. Stars of the library and galaxies were observed with the same instrument FTS in the near-IR wavelength range and data processed with the same software. The adopted stellar evolutionary tracks are from the Geneva group, 1992 extended to the Asymptotic Giant Branch and the Horizontal Branch phases which are energetically dominant in the near-IR. The evolution of stellar signatures (the H-K colour, the photospheric Brackett line absorption, the equivalent widths of CO and H_2O) is used for the starburst datation and for determining main other properties. The emission lines of H and He (Brγ, Paα, HeI$\lambda 2.06\mu$m), as well as [FeII]$\lambda 1.64\mu$m and rotation-vibration lines of H_2 detected in most starburts are fitted on the current number of Lyman continuum photons and the number of supernova remnants. The ratio HeI/Brγ is no more considered as a significant indicator of the superior limit of the Initial Mass Function, as proposed by Doyon et al, 1992. Various distributions of dust and stars (dust screen, homogeneous mixing of gas or more complex structures) were considered into a multicomponent extinction modelling. The role of each parameter and a clear view of the sensitivity of results to these parameters is presented.

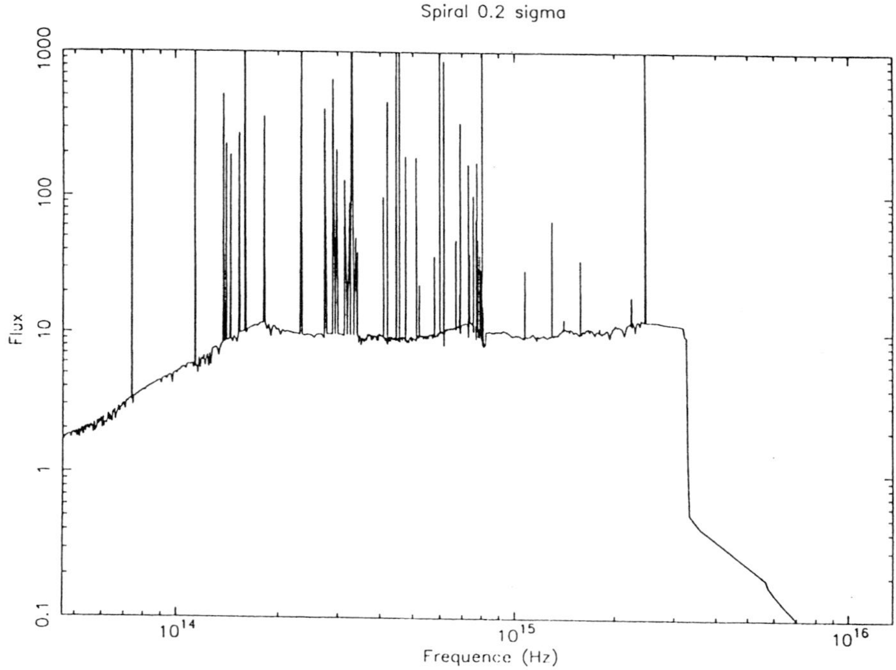

Fig. 3. Synthetic far-UV to near-IR energy distribution of a typical spiral with emission lines.

From the galaxy sample, a coherent method of analysis, based on a series of observables, is proposed and applied to the typical starburst galaxy NGC1614. The contribution of the underlying population is evaluated and inside the central 5", the burst duration is estimated to 11Myr with a truncated IMF ($60M_\odot$, $3M_\odot$). But a more standard IMF is however acceptable with constraints on the dust distribution in the HII region (Lançon and Rocca-Volmerange, 1995). Statistical results will be derived by applying the method to a complete analysis of the sample.

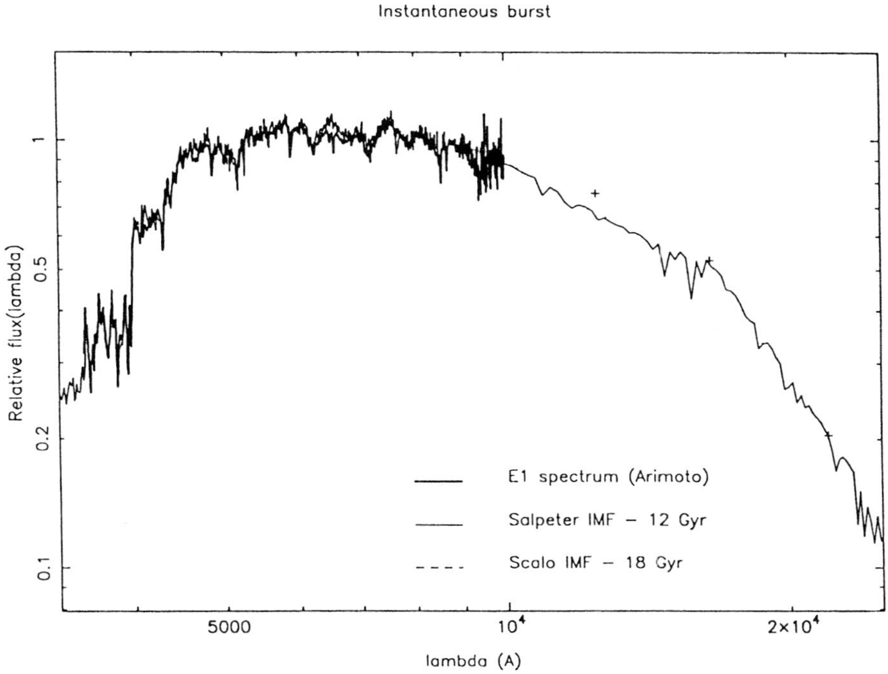

Fig. 4. The observed far-UV to near IR energy distribution of an elliptical galaxy is compared to an evolved single burst calculated with our new series of models (Fioc and Rocca-Volmerange, 1995)

4 A continuous model of spectral evolution from the UV to the near-infrared

Another approach is to predict a new spectral synthetic atlas of about 10 different spectral types from instantaneous starbursts to the old evolved elliptical galaxies ($\geq$ 15 Gyrs)(fig 3 and fig 4). The reddened SED and the nebular lines (continuum and lines) are continuously extended from the far-UV and visible to the near-infrared JHKL bands. The model (Fioc and Rocca-Volmerange, 1995) is built on a new algorithm of integration, respecting the energy released through the stellar spectra along the evolutionary tracks. Tracks are from the Padova group and have been prolongated through the AGB phase to the thermal pulses (Groenewegen and de Jong, 1988). The stellar library has been revised from Rocca-Volmerange and Guiderdoni, 1988 and extended with input data in the photometric system of Bessel and Brett, 1988. The wavelength range is 220Å to 5μm . Previous works proposed such predictions (Bruzual and Charlot, 1993) but outputs apparently are deficient in the near-infrared, missing a part of the JHK emission in elliptical galaxies. This model is adapted to interpret multispectral faint galaxy counts at various redshifts. The complete atlas of synthetic spectra and other output data as the k- and e- corrections (Rocca-Volmerange et al, 1995) will be published in CD-ROM format.

References

Arimoto, N., Yoshii, Y., 1986, Astron. Astrophys., 164, 260

Bertelli, G., Bressan, A., Chiosi, C., Fagotto, F., Nasi, E., 1994, Astron. Astrophys., 289, 665

Bessel, M.S., Brett, J.M., 1988, Pub. Astron. Soc. Pacific, 100, 1134

Boothroyd, A.I., Sackman, I.J., 1988, Astrophys. J., 328, 653

Bessel, 1995, private communication

Bessel, M.S., Brett, J.M., 1988, Publications of the Astron. Soc. of the Pacific, 100, 1134

Bruzual, G., Charlot, S., 1993, Astrophys. J.,Astrophys. J., 405, 538

Bruzual, G., Astrophys. J., 273, 105

Doyon, R., Puxley, P.J., Joseph, R.D., 1992, Astrophys. J., 397, 117

Fioc, M., Rocca-Volmerange, B., 1995, to be submitted

Flucks, M.A., Plez, B., Thé, P.S., de Winter, D., Westerlund, B.E., Steenman, H.C., Astron. Astrophys., 287, 1039

Groenewegen, M.A.T., de Jong, T., 1993, Astron. Astrophys., 267, 410

Guiderdoni, B., Rocca-Volmerange, B., 1987, Astr. Astrophys., 186, 1

Gunn, J., Stryker, L.L., 1983, Astrophys. J. Sup. Series, 52, 121

Hamilton, D., 1985, Astrophys. J., 297, 371

Heck et al, 1983, ESA/ IUE Ultraviolet spectral atlas

Huchra, J.P., 1977, Astrophys. J., 217, 928

Jacoby, G.H., Hunter, D.A., Christian, C.A., 1984, Astrophys. J. Sup. Series, 56, 257

Lançon, A., Rocca-Volmerange, B., 1992, Astron. Astrophys. Supp. S., 96, 593 (LRV92)

Lançon, A., Rocca-Volmerange, B.,Trinh X. Thuan, 1994, preprint

Lançon, A., Rocca-Volmerange, B., 1995, submitted (LRV95)

Maucherat-Joubert, M., Lequeux, J.,Rocca-Volmerange, B., 1980, Astron. Astrophys., 86, 299

Puxley, P.J., Brand, P., 1994, Mont. Not. Roy. astr. Soc., 266,431

Rieke, G.H., Cutri, R.M., Black, J.H., Kailey, W.F., McAlary, C.W., Lebofsky,M.J., Elston, J., 1985, Astrophys. J. 290, 116

Rocca-Volmerange, B. et *al*, 1995, to be submitted

Rocca-Volmerange, B., Guiderdoni, B., 1988, Astron. Astrophys. Supp. S., 75, 93

Searle, L., Sargent, W.L.W., Bagnuolo, W., 1973, Astrophys. J., 179, 427

Tinsley, B., 1972, Astron. Astrophys., 20, 383
Worthey, G., 1993, Astrophys. J., 409, 530
Wu et al, 1983, IUE Ultraviolet Spectral Atlas
Yoshii, Y., Takahara, F., 1989, Astrophys. J., 346, 28

Stellar Libraries: Spectral Features in the Near-Infrared

Beatriz Barbuy

Universidade de São Paulo, Depto. de Astronomia, C.P. 9638, São Paulo 01065-970, Brazil

Abstract. The stellar libraries available in the range 0.8 to 5 μm are reviewed. The strong features present in this range are discussed in the light of synthetic spectra results compared to observations.

1 Introduction

Progress made in recent years in infrared observational techniques made it possible to obtain spectrophotometry as, e.g. with the IRSPEC infrared spectrometer (Moorwood et al. 1991).

A number of lines present in the near-IR, in particular a variety of molecular bands, as well as a list of useful atomic lines, can be used to analyze stars and composite systems independently or complementary to the visible region. The near-IR shows the advantage of negligible reddening, although care must be taken with dust emission for wavelengths > 2 μm. In particular, the near-IR is adequate for the study of M stars in composite stellar systems, where the integrated light is dominated by them.

Considering that the region designated as near-infrared covers the wavelength range 0.8-5 μm, in this paper we review the available stellar libraries in this range, and describe the main spectral features useful as stellar populations indicators.

2 Stellar Libraries

Stellar spectral libraries are a useful and necessary tool for population synthesis. The ideal would be to have available a grid of stellar spectra covering a large wavelength range, from the UV to the far-IR, for stars of temperatures from 2000 to 50000 K and luminosity classes from dwarfs to supergiants, for metallicities in the range +0.5 < [Fe/H] < -4.0. In such a case, a galaxy evolution code whose results are fractions of stellar populations in different evolutionary stages and ages, covering a wide range of temperatures, gravities and metallicities binded to the spectral library would result in predicted galaxy spectra. Such is done by several authors (e.g. Rocca-Volmerange, this volume), but the libraries of stellar spectra are still not fully complete, in particular in what regards cool stars.

2.1 Libraries including the range 0.8-1 μm

Gunn & Stryker (1983) provide spectra in the range 3130-10800 Å, at a resolution of 40 Å in the near-IR, with 175 entries including stars from O to M, with luminosity classes from I to V. Pickles (1985) gives spectra in the range 3600-10000 Å for 200 stars at a resolution of 15 Å, from spectral types of B to M and luminosity classes V to III, with several metal-rich stars. Bica & Alloin (1987) observed 30 star clusters with ages in the range 10^6 to 1.65×10^{10} yr, and metallicities -2.0 < $[Z/Z_\odot]$ < 0.1, and 62 reference galaxy nuclei from types E to Sc, in the wavelength range 6300 - 9700 Å. Besides, Alloin & Bica (1989, hereafter AB89) observed 63 stars from F to M, and classes I to V, including most of the 62 stars studied by Jones et al. (1984), in the wavelength range 8000-9000 Å, at a resolution of 3 Å. Terndrup et al. (1990) obtained spectra of 320 K and M giants in low-absorption fields along the minor axis of the galactic bulge, plus 60 late-type giants and dwarfs in the solar neighbourhood. The spectra cover the wavelength region 6000-9000 Å. Diaz et al. (1989) obtained spectra for a sample of 106 late-type stars in the range 7900-9100 Å.

2.2 Libraries in the range 1-5 μm

Kleinmann & Hall (1986, hereafter KH86) obtained spectra in the K band at 4150-4950 cm^{-1} (2.0-2.4 μm) for a grid of 26 reference stars from F8 to M7 and ranging in luminosity from dwarfs to supergiants. Arnaud et al. (1989) observed 73 stars in the range 2.0-2.45 μm with a resolution of 0.02 μm. Their sample included solar-neighbourhood supergiants, giants, dwarfs and subdwarfs, together with several very cool and metal-rich giants in Baade's Window. Terndrup et al. (1991) have obtained spectra for 18 solar neighbourhood and 14 Baade Window stars in the range 0.45-2.45 μm, at a resolution of $\sim$ 1000. Origlia et al. (1993, hereafter OMO93) obtained spectra in the H-band at 1.5-1.7 μm of 13 A-M supergiants, 2 G-K dwarfs, and 33 K-M giants. Lines of ^{12}CO, ^{13}CO, OH, Mg, Al, Si, Ca and Fe are available and measureable in the region. Lançon & Rocca-Volmerange (1992, LR92) observed 56 stellar spectra in the wavelength range 1.427 - 2.5 μm, using the Fourier Transform Spectrometer at the 3.6m CFHT telescope; their sample contains stars of temperatures $2000 < T_{eff} < 45000$, with luminosity types I, III and V. Lázaro et al. (1994) observed 15 carbon stars in the range 1-4.2 μm.

3 Model atmospheres and synthetic spectra

A grid of model atmospheres and synthetic spectra was computed by Kurucz (1992) covering the spectrum from the ultraviolet up to the far infrared. His theoretical spectra are very useful for flux distribution, however detailed reproduction of observed spectra is not reached, as explained in Kurucz (1995a,b): "For opacity and model atmosphere calculations where only statistical accuracy is needed, the current list of 58 million lines, most with predicted wavelengths, works reasonably well. However, when computing spectra using only lines with good wavelengths, the quality is very poor. One half the lines are missing. Most of the gf values need to be corrected and even the laboratory wavelengths are not always reliable." Kurucz's line list contains a large number of molecular lines (see Kurucz 1995a,b), but yet it is not complete: in the near-IR CO (vibration-rotation) and OH are included, but H_2O is missing. In any case, this is the most complete grid of models and synthetic spectra available and it must be said that the work accomplished by R. Kurucz is not less than herculean.

Concerning model atmospheres, the grid of models by Gustafsson et al. (1975) and further unpublished sets of models made available by Gustafsson and collaborators together with Plez et al. (1992) models provide an important grid of model atmospheres. In this case the synthetic spectra are not available but it is possible to compute them with a spectrum synthesis code plus a line list.

A problem with theoretical spectra concerns model atmospheres for cool stars ($T_{eff} \leq 3500$ K), for which fully adequate grids of models are not available:

Bessell et al. (1989) produced models and presented synthetic spectra for Mira variables; their synthesized spectra show TiO and VO bands stronger than observed, but are otherwise satisfactory.

Plez (1992), Plez et al. (1992) have built spherically symmetric models of giants in the temperature range $3000 < T_{eff} < 4000$ K, having obtained a good match to observed spectra (see Fig. 7 of Plez et al. 1992). Using Plez et al. code, Fluks et al. (1994) computed models and spectra for a list of stars in the temperature range $3900 < T_{eff} < 2500$ K.

OMO93 computed LTE synthetic spectra of the 1.5-1.7 μm region. They used model atmospheres by Johnson et al. (1980). The input line list included

6000 lines of Mg, Al, Si, Ca, Fe, Ti, ^{12}CO, ^{13}CO, OH and CN. The CO and OH ones are vibration-rotation lines within the same electronic state. It can be seen in their Fig. 4b that the deepest feature in the spectra at 1.62 μm is primarily due to CO(6,3) and the one at 1.59 μm is mostly Si 1.5888 μm; at 1.71 μm it is relatively pure Mg 1.7109 μm. OH at 1.625 and 1.690 μm may be present in the cooler stars.

4 Features in the near-IR as indicators of stellar populations

For stellar population studies both the flux distribution and the intensity of strong absorption features are used. In both cases a combination of information from observed and synthetic spectra can lead to the derivation of a description of stellar populations, their metallicities and abundance ratios in globular clusters and normal galaxies.

For such studies, in the last years we have been applying efforts to build synthetic spectra of the main strong features. The atomic line list used is based on identified lines in the solar spectra, and the atomic constants are obtained through a fitting to both the solar and Arcturus spectra. Concerning infrared atomic lines different complementary lists can be found such as those by Biémont et al. (1985) and Solanki et al. (1990). Some of our results and those from other authors are discussed below.

Most features supposed to be gravity-sensitive, used in the study of elliptical galaxies, are found in the near-IR:

The NaI doublet at λ8183, 8195 Å (e.g. Faber & French 1980), the FeH Wing-Ford band at λ9900 Å (e.g. Whitford 1977), the CaII triplet at λ8489, 8542, 8662 Å (Jones, Alloin & Jones 1984, JAJ84), TiO bands with several strong bandheads, in particular at λ7150 and 7600 Å (e.g. Terndrup et al. 1990), and CO bands at 1.65 and 2.29 μm (e.g. Persson et al. 1980).

In the last years we have been studying different strong features in detail, and we have found that their behaviour as a function of stellar parameters is different from what is stated in the literature. These features are analysed below:

4.1 The CaII infrared triplet (CaT)

Based on a sample of 62 stars observed in the wavelength range $\lambda\lambda$ 7000-9300 Å, JAJ84 concluded that CaII is a very sensitive gravity indicator; however, their sample was constituted of solar abundance stars, and a metallicity dependence could not be seen. The strong metallicity dependence of this feature was demonstrated by Alloin & Bica (1989, AB89) through observations, and by Erderlyi-Mendes & Barbuy (1991, EB91) and Jorgensen, Carlsson & Johnson (1992, JCJ92) through calculations of synthetic spectra: Figs. 4 of EB91 show a mild temperature dependence; Fig. 3 of EB91 and

Fig. 1 of JCJ92 show an also not very pronounced dependence on gravity (varying with metallicity), and Fig. 2 of EB91 and Fig. 3 of JCJ92 show the strong metallicity dependence. Therefore this feature shows a biparametrical behaviour as a function of metallicity and gravity.

4.2 TiO bands

Strong TiO bands of the $A^3\Phi$ - $X^3\Delta$ (γ system) are seen in M giants at λ6230, 6735, 7050, 7600 Å and of the δ and ϕ systems for wavelengths redwards of 9000 Å; although diluted by hot and/or metal-poor stellar populations present in galaxies, these features are seen in the spectra of galaxies (see Figs. 1,2,3 of Terndrup et al. 1990).

Milone & Barbuy (1994) have carried out calculations of synthetic spectra of TiO bands at λ 7105 Å in order to study their behaviour as a function of stellar parameters. The following triparametrical relation was found:

$$\log W(\mathrm{TiO}) = 2.19\,\theta_{\mathrm{eff}} + 0.32[\mathrm{Fe/H}] + 0.002 \log g - 1.24$$

note the very weak dependence on gravity: contrary to a general belief in the literature, TiO bands are insensitive to gravity. Possibly the often mentioned presence of giants connected to the presence of TiO bands means presence of cool giants against hot (turn-off) dwarfs, in which case the inference is correct, but the presence of TiO bands could also be due to cool dwarfs.

In Milone, Barbuy & Bica (1995) we have built composite spectra of single-aged populations, using isochrones for the populations more metal-poor than [Fe/H] $\leq$ -0.5, and using observations of colour-magnitude diagrams of bulge clusters for the more metal-rich ones; we have obtained composite populations for both $[\alpha/\mathrm{Fe}] = 0.0$ and $+0.3$, and a relation W(TiO) = f([Fe/H],[α,Fe]) was obtained:

$$\ln W(\mathrm{TiO}) = 3.39 + 1.21[\mathrm{Fe/H}] + 1.02[\alpha/\mathrm{Fe}]$$

Relations such as this one can be useful to study composite spectra of globular clusters and galaxies, especially if combined to similar expressions for other features, such as those of widespread use like Mg_2, for which calculations given in Barbuy (1994) lead to the relation:

$$\mathrm{Mg}_2 = 0.206\exp(0.749[\mathrm{Fe/H}] + 0.996[\mathrm{Mg/Fe}])$$

4.3 NaI 8190 Å infrared feature

Spinrad & Taylor (1971, hereafter ST) used the NaI doublet at λ8183, λ8195 Å, which is a relatively strong feature in spectra of normal galaxies, having suggested the presence of a strong dwarf component in M31 and M81. Whitford (1977) and Hardy & Couture (1988) used the gravity discriminator FeH Wing-Ford bands to conclude that ST suggestion could not be confirmed. Cohen (1978) observed the infrared NaI doublet, CaII triplet, FeH and TiO bands in M31 and M32, concluding that no enhancement of cool dwarf population is required to explain these features, but that instead, M31 shows a higher metallicity than M32. Faber & French (1980) have used the intensity of this feature as an indicator of dwarf/giant ratio, returning to the same conclusion of ST: an excess of M dwarfs in M31. The question was further studied by Persson et al. (1980) through the observation of CO and H_2O narrow-band indices in M31, having measured the indices from the bulge to the nucleus; the CO data do not support a dwarf-enriched nucleus, but the infrared data and the NaIλ8195 Å are consistent with a metallicity increase of a factor 3 between the bulge and the semistellar nucleus.

Alloin & Bica (1989, hereafter AB89) studied the behaviour of this feature in individual stars and in the integrated light of stellar clusters and galaxies. They concluded that the enhancement of the feature in galaxies is due to another absorber, possibly TiO molecular bands at λ8205 Å. Xu et al. (1989), dealing with the same feature, also claimed that it is actually a blend of TiO, FeI and MgI lines with the NaI doublet, more sensitive to metallicity, which produce the observed absorption and that the presence in galaxies of a significant contribution of late dwarfs is not required.

Boronson & Thompson (1991) found strong evidence that points to a dwarf-dominated light in NGC 4472, suggesting that population gradients may be present, in contradiction with the work of Tendrup et al. (1990).

Delisle & Hardy (1992) observed 10 bulges of spirals and ellipticals in the spectral range 8000-10000 Å, concluding again that the Na feature is probably a result of a blend and that a metallicity effect may be present.

The interpretation of the NaI λ 8190 Å feature in galaxies still is surrounded by a controversy, whether it is predominantly a dwarf/giant ratio indicator or a metallicity sensitive feature.

In Barbuy, Rossi & Piorno Schiavon (1995, in preparation), we have computed synthetic spectra of the NaI doublet and surrounding lines, including CN and TiO molecular lines, for a range of stellar parameters. It appears that, in agreement with Xu et al., atomic lines present from 8200 to 8205 Å get enhanced in cool giants, whereas the NaI lines themselves are strong in dwarfs, but disappear in giants. When measuring the whole feature, the behaviour of NaI is masked by the other atomic lines. The behaviour of the NaI lines and contaminating lines is illustrated in Fig. 1. But the conclusion is that the NaI doublet lines are an excellent dwarf/giant indicator, meaning that if the NaI lines are strong, the integrated line would have a substantial contribution from cool dwarfs. In order to measure the clean NaI lines one has to measure the equivalent width with a limit wavelength in the red side at about 8197 Å, in order to avoid the contaminating atomic lines. It is not clear in which kind of galaxy the NaI doublet would be strong, since normally the integrated light is dominated by giants or turn-off (hot) dwarfs.

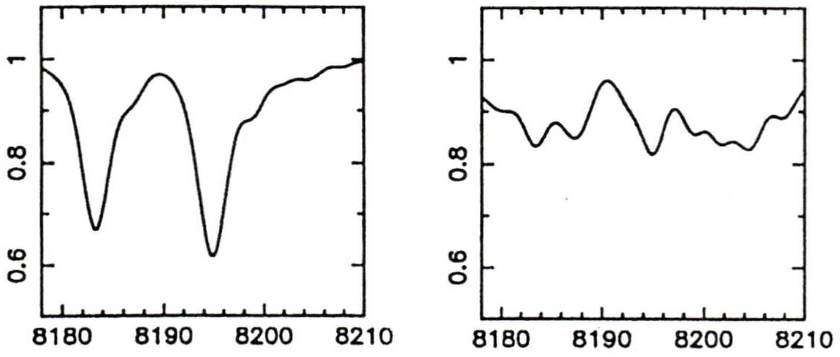

Fig. 1. NaI 8183, 8195 Å doublet computed for (T_{eff}, log g, [Fe/H]) = (4000, 4.5, 0.0) and (4000, 1.0, 0.0).

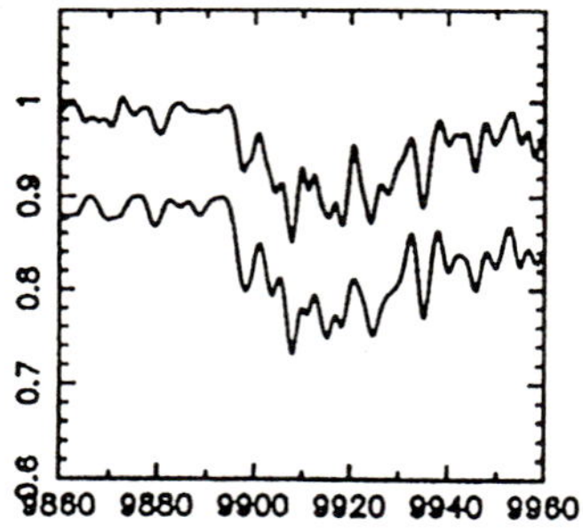

Fig. 2. Comparison between observed and synthetic spectra for the cool M dwarf Gl 273

4.4 FeH Wing-Ford bands

Couture & Hardy (1993) obtained spectra in the range 9800-10252 Å for 28 'groups' of stars, comprising the range dF7 to dM8 and gK2 to gM9, plus C and S stars. They concluded from their observations that in the Wing-Ford feature, FeH does not contribute significantly relative to TiO in giants, but does in dwarfs (theirs Figs. 2,3). They also present the first clear detection of the Wing-Ford band in galaxies (their Fig. 4).

In Piorno Schiavon & Barbuy (1995, in preparation) we find that FeH should be the dominant component of the Wing-Ford band, but this is not yet conclusive. The band is stronger in dwarfs, and could be indeed a reliable dwarf/giant indicator. In Fig. 2 (Piorno Schiavon & Barbuy 1995) we show the comparison between observed and synthetic spectra for a M dwarf, where the feature is entirely due to FeH.

4.5 CO bands

The main CO first overtone vibration-rotation feature with bandhead at 2.29 μm, is shown to be positively correlated with metallicity in cool giants of similar temperatures (similar J-K) for globular cluster and old disk cluster giants, following the relation: CO = 0.16 + 0.074[Fe/H] valid for [Fe/H] $\leq$ 0.0 cf. Frogel et al. (1983, 1990). Frogel et al. (1990) find that extrapolation of this relation for more metal-rich bulge giants is not linear. According to Terndrup et al. (1991) CO metallicity estimates are consistent with those using the NaI doublet at 2.207 μm and the CaI triplet lines at 2.263 μm, where the bandhead at the K band CO_K would be more appropriate than that in the H band CO_H. On the other hand, CO bands show dependence also on temperature and gravity: CO strengthens with decreasing temperature and increasing luminosity (Persson et al. 1980, KH86). LR92 find empirically a relation between the strenght of CO in the region 2.29-2.4 μm and colour temperature: W(CO) = 4.32×10^{-2} - 4.58×10^{-6} T_c.

Lambert & Ries (1981) observed vibration-rotation transitions of the electronic ground states of the CO, OH and NH molecules pointing out that the fundamental (Δv = +1) sequence of CO V-R transitions falls near 5 μm; the 1.6 μm bands are too weak in the late-G stars whereas the 2.3 μm bands are too saturated in the late-K stars.

A study of the behaviour of the different features at the H-band with stellar parameters was also presented by OMO93. They tried to find features around 1.6 μm that could be used to determine the temperature and/or the gravity of cool stars. The H-band is preferable to the K-band in the sense that dust emission is still important at 2.29 μm but it is very small below 1.8 μm. The selected features were Si (1.59 μm), CO(6,3) 1.62 μm and CO(2,0) 2.29 μm. CO bands are sensitive to luminosity being stronger in supergiants, but H_2O would be a better indicator of gravity. On the other hand, Figs. 5a,b of OMO93 show that these features can be used to deduce the average spectral type (temperatures) of a stellar system. This is more clearly seen in Fig. 6 of OMO93. In particular, the 1.62/1.59 ratio is a sensitive temperature indicator (Fig. 5c of OMO93).

Origlia et al. (1994, hereafter OMO94) have applied their method to study globular clusters, being able to derive a rank of metallicities: the 3 features Si 1.59, CO 1.62 and CO 2.29 μm were measured in the spectra of a list of clusters. A 'calibration' of their mean temperatures is given in Fig. 2 of OMO94 through the 1.62/1.59 and 1.62/2.29 ratios of equivalent widths. This diagram shows that NGC 6553 is cooler than, and Terzan 5 is hotter than expected if the metallicity ranking by Zinn (1985) is adopted. The clusters metallicities can therefore be ranked through these infrared indices (see Fig.

3 by OMO94). However, care must be taken for clusters with anomalous Red Giant Branches (RGB); in particular, in Ortolani, Barbuy & Bica (1995) it was found that Terzan 5 is very metal-rich, probably with solar metallicity, and slightly more metal-rich than NGC 6553. Possibly the disagreement relative to OMO94 regarding Terzan 5 is due to a somewhat anomalous RGB, where the giants are faint due to their strong opacities, and in integrated light their contribution is less important. It can also happen, in the case of core-collapse clusters that the RGB is underpopulated.

4.6 H_2O, OH, NH, C_2

The main H_2O feature occurs at 1.9 μm, but H_2O lines occur throughout the JHK bands (Alexander et al. 1989). Strong H_2O bands are centered near 1.4, 1.9 and 2.7 μm (KH86, LR92), but its presence is less obvious than that of CO. They are very deep and broad in the coolest dwarfs and giants (but less in supergiants, cf. LR92). LR92 find a relation between the equivalent width of H_2O, corresponding to the regions 1.25-1.8 and 2.0-2.1 μm and colour temperature: $W(H_2O) = 0.184 - 4.578 \times 10^{-5}\ T_c$ The behaviour of H_2O bands at 1.9 μm as a function of CO and JHK colours was studied by Aaronson et al. (1978 - see their Fig. 2).

The OH (2,0) vibration-rotation spectrum of OH (Maillard et al. 1976) seem to contribute to the spectra of cool giants and supergiants (LR92).

The strongest NH lines belong to the $\Delta v = +1$ sequence between 3 and 4 μm. NH is insensitive to gravity, and moderately sensitive to temperature according to Lambert & Ries (1981).

Finally, C_2 Ballick-Ramsay bands at 1.88 μm characterize Carbon stars.

5 Conclusions

The infrared spectrometers now available allowing to obtain high quality spectra in the near-IR enlarge possibilities in spectroscopic work. A larger number of certain species (e.g. MgI) combined to those in the visible allow a better chemical composition analysis of individual stars. Independent analyses using only infrared spectra are also possible. For stellar population work, several strong molecular bands and atomic features can be used to constrain the solutions regarding the stellar populations in composite systems, in particular for a better understanding of their M giants component. Besides, a coverage from UV to IR in flux distribution also impose stronger constraints.

References

Aaronson, M., Frogel, J. A., Persson, S. E. (1978): ApJ 220, 442–448

Alloin, D., Bica, E. (1989): A&A 217, 57–65

Arnaud, K. A., Gilmore, G., Collier Cameron, A.: (1989): MNRAS 237, 495–511

Barbuy, B. (1994): ApJ 430, 218–221

Bessell, M. S., Brett, J. M., Scholz, M., Wood, P. R. (1989): A&A 213, 209–225

Bica, E., Alloin, D. (1987): A&A 186, 49–63

Biémont, E., Brault, J. W., Delbouille, L., Roland, G. (1985): A&AS 61 107

Couture, J., Hardy, E. (1993): ApJ 406, 142–157

Diaz, A. I., Terlevich, E., Terlevich, R. (1989): MNRAS 239, 325–345

Fluks, M. A., Plez, B., Thé, P.S., De Winter, D., Westerlund, B. E., Steenman, H. C.: 1994, A&AS, 105, 311

Frogel, J. A., Cohen, J. G., Persson, S.E. (1983): ApJ 275, 773–789

Frogel, J. A., Terndrup, D. M., Blanco, V. M., Whitford, A. E. (1990): ApJ 353, 494–523

Gunn, J. E., Stryker, L. L. (1983): ApJS 52, 121–253

Gustafsson, B., Bell, R. A., Eriksson, K., Nordlund, A. (1975): A&A 42, 407–432

Johnson, H. R., Bernat, A. P., Krupp, B. M. (1980): ApJS 42, 501–522

Jones, J., Alloin, D., Jones, B. (1984): ApJ 283, 457–465

Jorgensen, U. G., Carlsson, M., Johnson, H. R. (1992): A&A 254, 258–265

Kleinmann, S. G., Hall, D. N. B. (1986): ApJS 62, 501–517

Kurucz, R. L. (1992): in The Stellar Populations of Galaxies, IAU Symp. 149, Eds. (B. Barbuy & A. Renzini), Kluwer Academic Pub., 225-232

Kurucz, R. L. (1995a): eds. W.L. Wiese & S. Adelman, in press

Kurucz, R.L. (1995b): Laboratory and Astronomical High Resolution Spectra eds. A. Sauval, R. Blomme & N. Grevesse, ASP Conf. Ser. 81, 583-588

Lambert, D. L., Ries, L. M. (1981): ApJ 248, 228–248

Lançon, A., Rocca-Volmerange, B. (1992): A&AS 96, 593–612

Lázaro, C., Hammersley, P. L., Clegg, R. E. S., Lynas-Gray, A. E., Mountain, C. M., Zadrozny, A., Selby, M.J. (1994): MNRAS 269, 365–393

Maillard, J. P., Chauville, J., Mantz, A. W. (1975): J. Mol. Spec. 63, 120–141

Milone, A., Barbuy, B. (1994): A&AS 108, 449–454

Milone, A., Barbuy, B., Bica, E. (1995): A&AS, in press

Ortolani, S., Barbuy, B., Bica, E.: 1995, A&A, in press

Pickles, A. J. (1985): ApJ 296, 340–364

Plez, B. (1992): A&AS 94, 527–552

Plez, B., Brett, J. M., Nordlund, A. (1992): A&A 256, 551–571

Origlia, L., Moorwood, A. F. M., Oliva, E. (1993): A&A 280, 536–550

Origlia, L., Moorwood, A. F. M., Oliva, E. (1994): The Messenger 75, 21–24

Persson, S. E., Cohen, J. G., Sellgren, K., Mould, J., Frogel, J. A. (1980): ApJ 240, 779–784

Piorno Schiavon, R. P., Barbuy, B. (1995): in *New Light on Galaxy Evolution*, IAU Symp. 171, Eds. (R. Bender & R. Davies), Kluwer Academic Pub., in press

Solanki, S. K., Biémont, E., Mürset, U. (1990): A&AS 83, 307–315

Terndrup, D. M., Frogel, J. A., Whitford, A. E. (1990): ApJ 357, 453–476

Terndrup, D. M., Frogel, J. A., Whitford, A. E. (1991): ApJ 378, 742–755

Zinn, R. J. (1985): ApJ 293, 424–444

New spectrophotometric population synthesis models of ellipticals and early type spirals

Alexandre Vazdekis[1], Reynier Peletier[2], Emilio Casuso[1] & John Beckman[1]

[1]Instituto de Astrofísica de Canarias,E-38200 La Laguna, Tenerife, Spain,
[2]Kapteyn Institute, Groningen, the Netherlands

1 Introduction

We have developed a model for calculating colours and absorption line indices for composite stellar systems. The model can synthesize observations for single age, single metallicity stellar populations, but can also incorporate chemical evolution, following the evolution of a galaxy from an initial gas cloud to the present time. We have obtained accurate observations in a number of colours and line indices of three well-studied galaxies: NGC 3379, NGC 4472 and NGC 4594, and have used these to calibrate our models.

2 The model

The model is based on the isochrones of Bertelli *et al.*(1994). We have developed our own method to convert the isochrone theoretical parameters to observable colours. The stellar libraries of Worthey *et al.*(1994) are used to calculate synthetic line strengths. Our chemical evolution is calculated following broadly the formalism of Arimoto & Yoshii (1986) and Casuso (1991). Details can be found in Vazdekis *et al.*(1995).

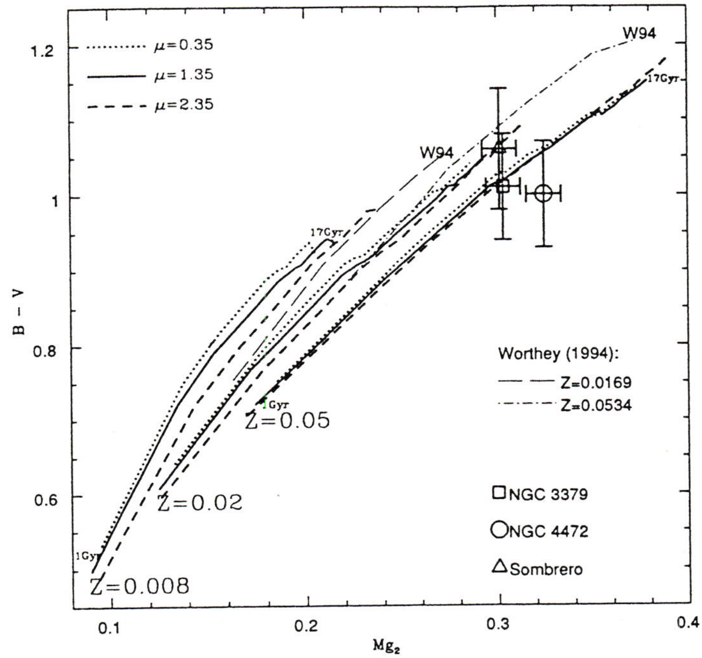

Fig. 1. The synthetic diagram of B-V vs. Mg_2 obtained by calculating single-age stellar populations for different ages (1 to 17 Gyrs. along each curve shown), metallicities and IMF slopes: 0.35, 1.35 (Salpeter case) and 2.35. Also indicated are the values for the three galaxies at 5 arcsec from the centre.

3 Results

In Fig. 1 one sees that reasonable fits are obtained for Mg_2 vs. B-V with metallicities larger than solar but that models with $Z < Z_\odot$ do not fit these 3 galaxies (Casuso *et al.*1995). A set of observed spectral indices is shown in Fig. 2 compared with predictions of single-age models. Fits are good for indices except those of Fe and Ca. Including α-enhancement improves the fit for Fe, but worsens the fit for the NaD index, using the ratios of Weiss *et al.* (1995). For an evolutionary scheme with a single constant IMF slope, in a closed-box approximation, the Mg_2 index from the models always falls short of the observed values (Vazdekis *et al.* (1995). A smilar result was found by Arimoto & Yoshii (1986) who could barely obtain the red $V - K$ colours necessary for ellipticals. We obtain much better fits if we introduce a significant change in the IMF slope, favouring massive stars in the early stages of galactic evolution, and low-mass stars for the remaining time.

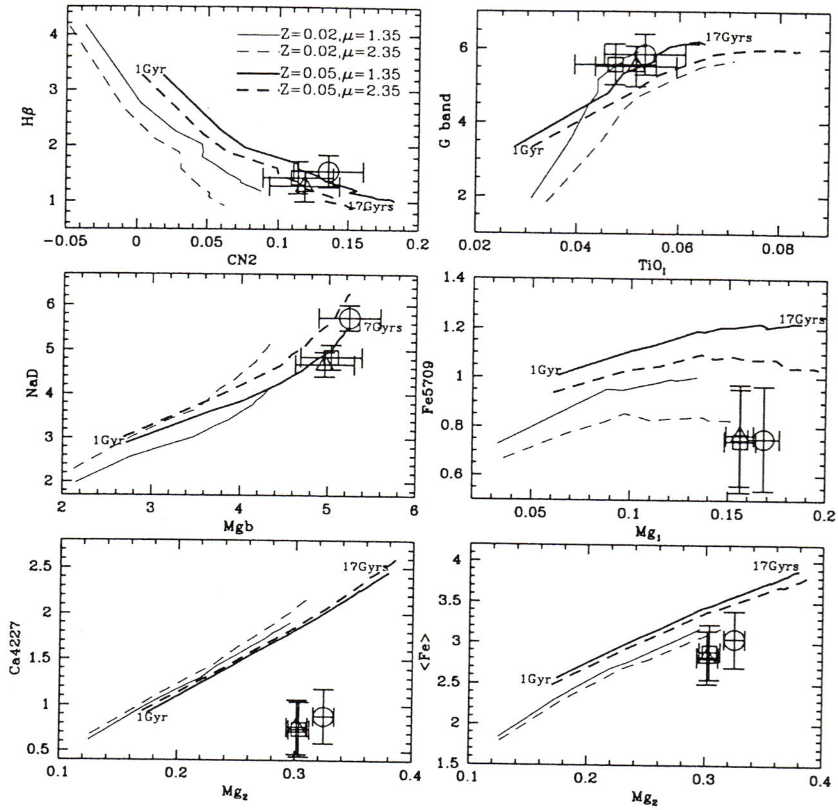

Fig. 2. Plot of selected index-index diagrams, for solar and supersolar metallicities for different IMF slopes. Notice that features due to Fe and Ca do not yield good fits to the points for the three galaxies observed.

References

Arimoto, N., & Yoshii, Y., 1986, A&A, 164, 260

Bertelli, G., Bressan, A., Chiosi, C., Fagotto, F. & Nasi, E., 1994, A&A Suppl. 106, 275

Casuso, E., 1991, Ph. D. Thesis, Univ. of La Laguna

Casuso, E., Vazdekis, A., Peletier, R. & Beckman, J.E., 1995, ApJ, In press.

Vazdekis, A., Casuso, E., Peletier, R. F., & Beckman, J. E., 1995, submitted

Weiss, A., Peletier, R.F. & Matteucci, F., 1995, A&A, 296, 73

Worthey, G., Faber, S., González, J. & Burstein, D.: 1994, Ap.J Supp. 94, 687

Worthey, G., 1994, ApJ Supp. 95, 107

Infrared Colour Gradients in Spiral Galaxies

Donald M. Terndrup[1]

[1] Deptartment of Astronomy, The Ohio State University, 174 W. 18th Ave., Columbus, Ohio 43210 USA

Abstract. Several recent studies have pointed out that there are serious degeneracies in broad-band colours, which make it difficult if not impossible distinguish the effects of metallicity and age in the integrated light. In this paper, I discuss the difficulties involved in breaking these degeneracies, particularly in situations where there is significant reddening from dust. Corrections for dust absorption in spiral galaxies are likely to be highly model dependent, particularly for edge-on systems.

1 Are Colour Gradients Informative?

One of the more common things to do with a data set of multicolour surface photometry of galaxies is to measure colour gradients along various axes. There is a long history of such work in elliptical galaxies, and recently there is increasing evidence that the colour gradients reflect a reduction in mean metallicity with increasing radius (e.g., Davies *et al.* 1993). Thus, as new infrared imaging surveys of spiral galaxies began to come along, there have been attempts to discern population gradients, say along the major axis of the disk or the minor axis of the bulge, or to explore whether the stellar population in the disk differs from that in the bulge.

It has becoming increasingly clear, however, that broad-band colours are often a poor population discriminant because of strong degeneracies in age and metallicity. We have collectively understood for quite some time, of course, that a given population gets redder if the age or average metallicity increase, or if the population is reddened by dust. Even obtaining spectra to measure line-strength gradients may not help break the degenerate nature of the integrated light of other galaxies. For example, Silva & Elston (1994) and Worthey (1994) have quantitatively explored the effects of metallicity and age on the integrated light of model stellar populations; they have conclusively shown that with a very few exceptions age and metallicity effects are indistinguishable.

When we turn to spiral galaxies the situation is even more complicated because of the effects of dust. For example, in our recent study of 43 spiral galaxies in rJK (Terndrup *et al.* 1994), we were forced to make very general conclusions about the sizes of metallicity or age gradients or about how the populations of the disks and bulges differ. We found that on average the disks in our sample were ≈ 0.1 mag bluer in $r - K$ than the bulges. This colour difference could arise if the disks are $2 - 4 \times 10^9$ yr younger *or* have lower metallicities by Δ[Fe/H] ≈ -0.2. The sizes of the colour gradients in the bulges are similarly difficult to interpret, though we found that gradients of Δ[Fe/H] ≈ -0.2 (more metal-poor outward) are common in the bulges in our sample. Metallicity gradients of this size have been detected in the bulge of the Milky Way (e.g., Frogel *et al.* 1990; Terndrup *et al.* 1990; Tiede *et al.* 1995), though at least one recent study (Ibata & Gilmore 1995) has concluded that the metallicity gradient in the galactic bulge may be negligible.

Aside from the problems of degeneracies, there were two reasons why we were not able to make more detailed conclusions about the interpretation of the colour gradients in our earlier study. First, the data in that paper are limited in quality, primarily because they were obtained with the first generation of infrared-sensitive arrays, which have very small formats. The detector we employed was a 58×62 InSb array, which had a field of view of only $82''$. With this small field, there are significant errors in the estimation of the sky value in each filter, which translate into rather large errors in the colour gradients. Second, we did not have any optical photometry which can help distinguish reddening from population effects when used in combination with infrared colours (see the talks by Peletier and by de Jong in this conference).

I'm sorry that I have to report that there are in addition significant scale errors in the photometry of Terndrup *et al.* (1994). After queries from F. Simien and H.-W. Rix, to whom I am very grateful, we discovered a calibration error in the surface brightness profiles $\mu(K)$ and in the calculated $r - K$ colours. The source of the error was an incorrect transformation from mag pixel^{-1} to mag arcsec^{-1}. That transformation was applied twice, which made the reported $\mu(K)$ profiles too faint, and the $r - K$ colours too blue, by ≈ 0.65 mag. The $J - K$ colours are unaffected by this error. Also, we assumed that the transformation between $r - K$ and (Cousins) $R_C - K$ was small; in fact the transformation is $R_C = r - 0.34$ (Bell & VandenBerg 1987). These errors do not change our conclusions of the colour gradients, which are independent of errors in the photometric zero points. Nevertheless, our paper should be read with caution. A formal erratum is in press.

2 New Surveys

Several groups are beginning extensive new combined optical-infrared surveys of spiral galaxies, which may provide more leverage in sorting out dust from population effects in other galaxies. In this part of my talk, I will be describing our efforts at Ohio State University. We shall no doubt hear about parallel work from other investigators elsewhere in this conference.

The goal of our survey is to construct a digital atlas of high-quality optical and near-IR images of $\sim$ 300 spiral galaxies. The galaxies are distributed in both hemispheres, and sample the Hubble sequence with minimal bias over galactic type, inclination angle, and luminosity. The images from the survey will be photometrically calibrated, and will be digitally available to the community shortly after the survey is completed. The survey is currently about half complete.

We are conducting the survey with wide-field infrared arrays, which gives us a much better estimate of the sky on each frame. Most of the infrared images are being obtained with the Ohio State Infrared Imager and Spectrograph (OSIRIS), which is described by DePoy *et al.* (1993).. In Figure 1, I show surface brightness contours in J for one of the survey galaxies, NGC 4818. The contours are at intervals of 0.5 mag, and the lowest contour is at $\mu(J) \approx 22$ mag arcsec^{-2}. The two large boxes around the galaxy show the area of the OSIRIS field at the 1.5 m telescope at Cerro Tololo and at the 1.8 m Perkins Telescope at Lowell Observatory in Flagstaff, Arizona. The small box centred on the nucleus of NGC 4818 is the area of the tiny InSb detectors used in our and other previous surveys. Clearly the new data will give a much better view of nearby galaxies than was previously possible, allowing better investigations of the substructure within galaxies.

3 The Effects of Dust

Once we have obtained new, high-quality broad-band images in several different wavebands, we will certainly wish to correct them for the effects of dust. Here I wish to discuss some recent work by Leslie Kuchinski at Ohio State, which may indicate that such a correction may be difficult and dependent on assumptions about the geometry and/or the wavelength dependence of dust absorption and scattering. Adolf Witt and others at this meeting will also have something so say on this subject.

For her dissertation research, Kuchinski is investigating the photometric properties of a large sample of (nearly) edge-on spirals, some of which have round bulges and some of which have boxy- or peanut-shaped bulges. She also has detailed longslit high-resolution spectra of many of these galaxies, from which the projected kinematics can be measured. Recently, Kuijken & Merrifield (1994) have used similar data to conclude that boxy-shaped bulges probably are in systems with central bars. The galaxies were selected to have

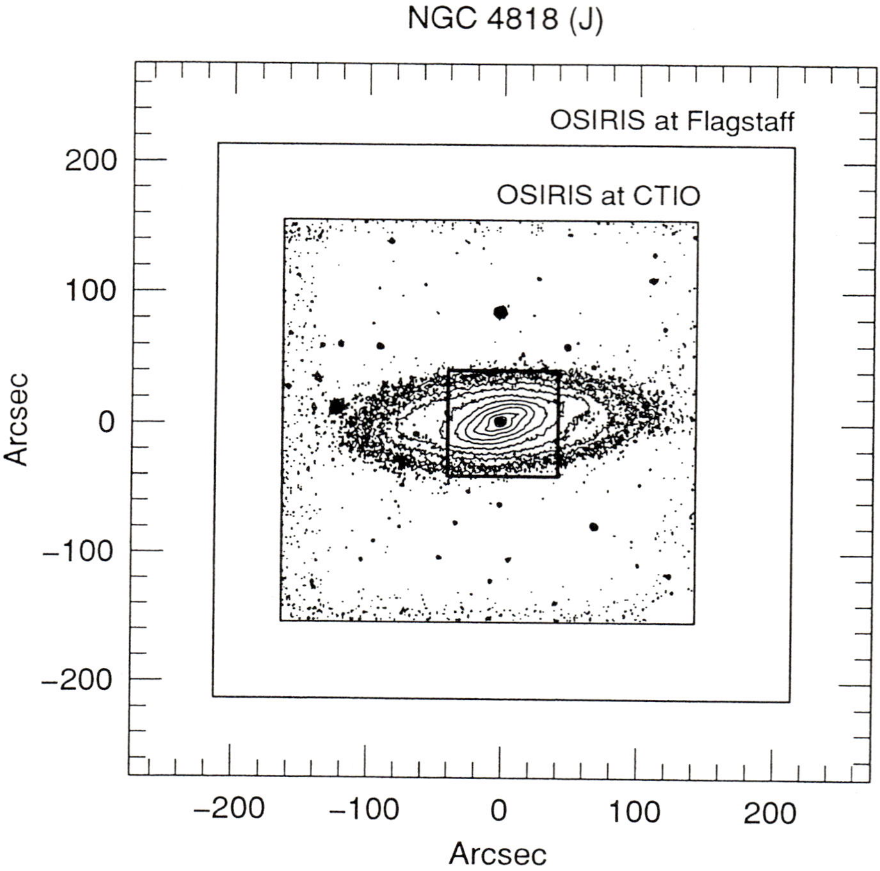

Fig. 1. The sizes of galaxy images in the new Ohio State survey compared to that from previous work using small-format infrared detectors. The contours show the J surface brightness of NGC 4818.

about the same field of view and angular scale (parsec pixel^{-1}) as the COBE map of the Milky Way's (boxy) bulge and inner disk, so ultimately we will have photometry and kinematics for a fair number of galaxies for comparison to our own.

Because these systems are edge-on (galaxies with boxy-shaped bulges are naturally only seen at high inclination), the central parts of the bulges are strongly affected by the dust in the disk. In an initial study of the effects of the dust (Kuchinski & Terndrup 1995), she has explored several simple analytic and numerical models of radiative transfer from the study of Witt *et al.* (1992). Each of these models predicts colour changes as a function of the optical depth at a particular wavelength, say $\tau(V)$. The models are compared to the colour changes along the minor axis in each galaxy as the dust lane bisects the bulge.

An example of Kuchinski's initial results is shown in Figure 2. The points with error bars are the $J - H$, $H - K$ colours at various distances along the minor axis; the blue points are for the outer bulge, while the red points are for the bulge near the disk (data for both sides of the bulge are plotted). The error bars include the effects of photon noise in the sky and the galaxy, but do not include the errors in the photometric calibration, which are shown in the lower right of Figure 2.

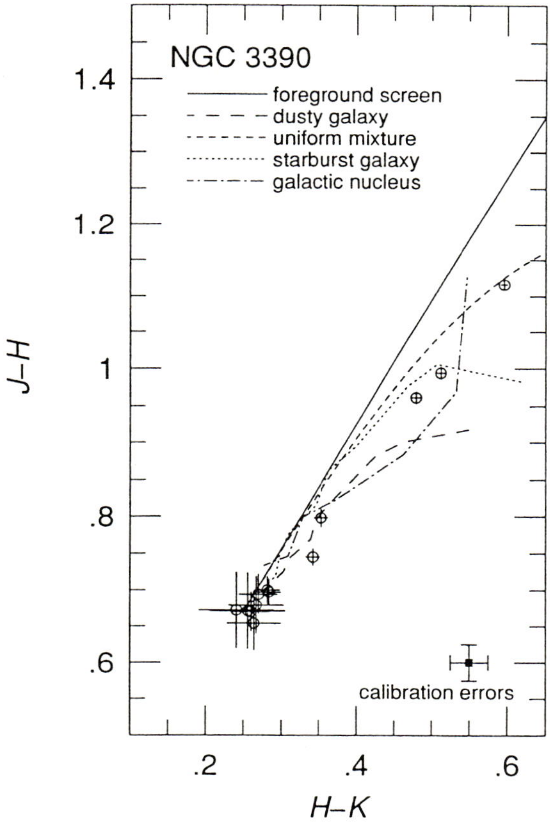

Fig. 2. Minor-axis colours (open symbols) in the bulge of NGC 3390, compared to various dust-absorption models from Witt *et al.* (1992).

The several lines in Figure 2 show reddening trajectories for various models of radiative transfer. They are all pinned to the colours in the outer part of the bulge of NGC 3390, and represent different predictions for what colour changes would be produced by dust in the disk. The foreground screen model, which is traditional, assumes that all the outgoing light from the galaxy is attenuated; the relative extinction in the different bands is in this case given by the Rieke & Lebofsky (1985) extinction law. The line labeled 'uniform mixture' shows the reddening trajectory for a uniform slab of stars and dust; it has a different trajectory from the foreground screen law because sources near the outer edge suffer less attenuation that those deep within. The 'dusty galaxy' is a sphere of uniformly mixed stars and dust that differs from the simple uniform mixture model due to its spherical geometry and the inclusion of scattering [Witt *et al.* (1992) tabulate values for fraction of light observed directly and the fraction scattered into the line of sight for several wavelengths ranging from the ultraviolet to near-infrared]. The combination of light scattering into the line of sight and sources near the surface suffering little extinction results in much less reddening than is produced by the simple foreground screen. The 'starburst galaxy' has a centrally concentrated spherical distribution of stars with a uniform sphere of dust embedded in it. Most of the light in this model comes from the central dusty regions, so reddening effects are more pronounced than for the dusty galaxy model. The 'galaxy nucleus' model contains a sphere of stars surrounded by a spherical shell of dust, with no mixing of dust and light sources in either region. This model differs from the simple foreground screen mainly due to the inclusion of scattering, which compensate for some of the absorption to produce slightly less reddening than the screen model.

Note that none of these models have the right geometry for describing the effects of dust in a spiral galaxy; instead one should use a model with a disk and bulge and a plane layer of extinction (e.g., Walterbos & Kennicut 1988; Evans 1994). They do, however, differ from one another primarily in geometry (i.e., the relative spatial distributions of the dust and the starlight) and *not* in the properties of the dust (absorption and scattering). Despite these caveats, I believe we can draw some interesting conclusions from Figure 2. For one thing, the simple foreground screen is probably not an adequate model for the attenuation by dust in these edge-on systems, since the minor-axis colours depart significantly from the predictions of the screen model near the plane of this galaxy. More importantly, each of these models have a substantially different relationship between $\Delta(J-H)$ and $\Delta(H-K)$ and $\tau(V)$. The optical depths required to produce the observed reddening varies from $\tau(V) = 4.4$ for the screen model, to $\tau(V) \sim 15$ in the uniform mixture model. In this galaxy, more than half a magnitude of the K light from the bulge is extinguished by the dust lane. That the optical depths are so model dependent strongly suggests that even in the infrared, one must make appropriate radiative transfer models if one is attempting to turn the surface brightness profiles into mass-to-light ratios. Such model dependence may have implications for the interpretation of the dynamics of the central regions of edge-on galaxies.

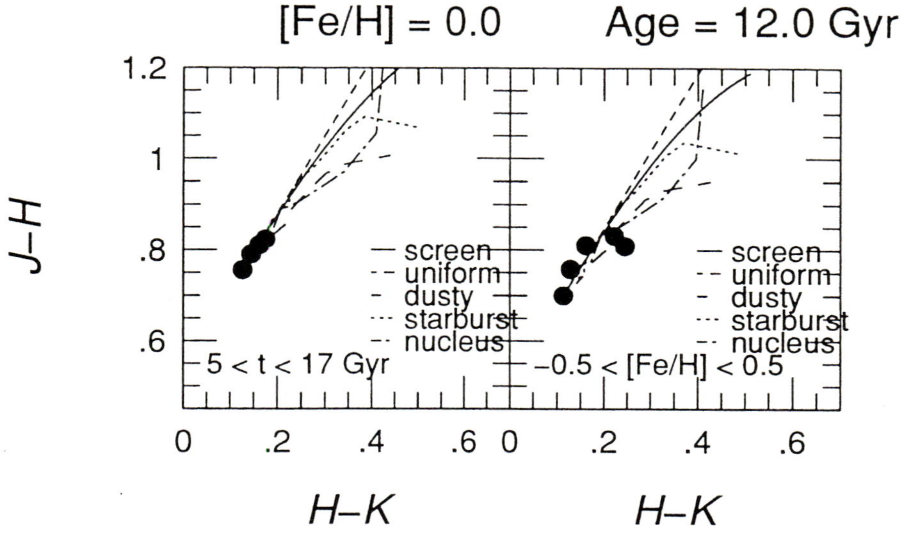

Fig. 3. Comparison of the colour changes from the Witt *et al.* (1992) dust-absorption models to the colours of the single-burst models of Worthey (1994 and references therein). The left panel shows the effects of age at constant metallicity, while the right side shows the effects of metallicity at constant age.

The colour changes produced by the dust in these edge-on galaxies are often much larger than those which would be produced by metallicity or age gradients within these systems. I illustrate this in Figure 3, which shows the model trajectories in the two-colour plane along with the integrated colours from Worthey's (1994) single-burst population models. The left panel shows the effects of age changes at constant metallicity, displaying ages from 5 to 17×10^9 yr at [Fe/H] = 0.0. The right panel shows metallicity effects at an age of 12×10^9 yr, where [Fe/H] runs from −0.5 to +0.5. In both cases, the full range of age and/or metallicity changes, which are much larger than anything which has been proposed in our own bulge, are nevertheless much smaller than the colour changes which can be produced by dust.

4 What Now?

My conclusion is very brief: The effects of age and metallicity are very hard to distinguish with broad-band colours. In addition, we are now learning that we have significant uncertainties in the geometry and scattering properties of dust in spiral galaxies: models with entirely different geometries can lead to the same colour changes and so even the optical depth in dust can be very uncertain. Undoubtedly this means we will probably have to model the radiative transfer in any system in which we wish to learn about the mass-to-light ratio, and will probably have to combine longslit spectra with multicolour photometry if we are to learn about population gradients.

References

Bell, R.A., & van den Bergh, S. 1987, ApJS, 63, 335
Davies, R.L., Sadler, E.M., & Peletier, R.F. 1993, MNRAS, 262, 650
DePoy, D.L., Atwood, B., Byard, P., Frogel, J., & O'Brien, T. 1993, SPIE, 1946, 667
Evans, R. 1994, MNRAS, 266, 511
Frogel, J.A., Terndrup, D.M., Blanco, V.M., & Whitford, A.E. 1990, ApJ, 353, 494
Ibata, R.A. & Gilmore, G. 1995, MNRAS, in press
Kuijken, K., & Merrifield, M.R. 1995, ApJL, 443, L13
Kuchinski, L.E., & Terndrup, D.M. 1995, AJ, submitted
Rieke, G.J. & Lebofsky, M.J. 1985, ApJ, 288, 618
Silva, D.R., & Elston, R. 1994, ApJ 428, 511
Terndrup, D.M., Davies, R.L., Frogel, J.A., DePoy, D. L. & L. A. Wells 1994, ApJ 432, 518
Terndrup, D.M., Frogel, J.A., & Whitford, A.E., 1990, ApJ, 357, 453
Tiede, G.P., Frogel, J.A., & Terndrup, D.M. 1995, AJ, in press
Walterbos, R.A.M. & Kennicutt, R.C. 1988, A&A, 198, 61
Witt, A.N., Thronson, H.A., & Capuano, J.M. 1992, ApJ, 393, 611
Worthey, G. 1994, ApJS, 95, 107

Colour Gradients in the Optical and Near-IR

Roelof S. de Jong[1,2]

[1]Univ. of Durham, Dept. of Physics, South Road, Durham DH1 3LE, UK
[2]Kapteyn Institute, P.O.box 800, 9700 AV Groningen, The Netherlands

Abstract. For many years broadband colours have been used to obtain insight into the contents of galaxies, in particular to estimate stellar and dust content. Broadband colours are easy to obtain for large samples of objects, making them ideal for statistical studies. In this paper I use the radial distribution of the colours in galaxies, which gives more insight into the local processes driving the global colour differences than integrated colours. Almost all galaxies in my sample of 86 face-on galaxies become systematically bluer with increasing radius. The radial photometry is compared to new dust extinction models and stellar population synthesis models. This comparison shows that the colour gradients in face-on galaxies are best explained by age and metallicity gradients in the stellar populations and that dust reddening plays a minor role. The colour gradients imply M/L gradients, making the 'missing light' problem as derived from rotation curve fitting even worse.

1 The colour gradients

A sample of 86 spiral galaxies was imaged in the B, V, R, I, H and K passbands to study light and colour distributions as a function of radius. Full details of sample selection and data reduction are described in de Jong & van der Kruit (1994). The galaxies were selected to be face-on and to have a diameter of at least $2'$. The sample is statistically complete and can be corrected for selection effects. It can therefore be used to analyze the nature of the Freeman law (Freeman 1970) and this analysis has been reported elsewhere (de Jong 1995a, 1995b).

The luminosity profiles were determined in the usual way by measuring the average surface brightness on annuli of increasing radius. Radial colour profiles were created by combining profiles in different passbands. The run of colour as function of radius is put on a common scale for all galaxies in Fig. 1, where the average B–K colour at each radius is plotted as function of the average R surface brightness at this radius.

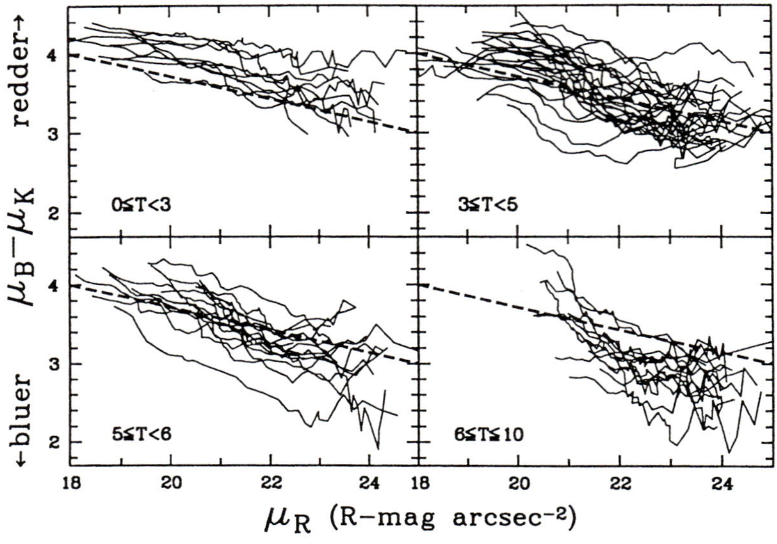

Fig. 1. The average B–K colour at each radius as function of the average R surface brightness at this radius. The galaxies are divided into 4 morphological RC3 T-type bins. The dashed lines are provided to have a common reference among the bins.

Two observations can readily be made from this diagram. Firstly, all galaxies become bluer going radially outward, correlating strongly with the average surface brightness at each radius. Secondly, even at the same surface brightness, late type spiral galaxies are bluer than earlier types (use the dashed lines to guide the eye). Furthermore it should be noted that there is a smooth transition in colour from the bulge to the disk region. The colours of bulges are nearly identical to the colours of inner disk regions (see also Peletier, these proceedings).

2 The dust and stellar population synthesis models

A possible explanation for the colour gradients is reddening due to dust extinction. As galaxies are intricate mixtures of stars and dust, we cannot describe the reddening by a simple extinction law, but have to calculate the separate contributions of absorption and scattering. To predict reddening profiles due to absorption and scattering full 3D Monte Carlo simulations were made (de Jong 1995a) of galaxies with smooth exponential dust and stellar distributions in both radial and vertical directions. The main free parameters are the dust to stellar scaleheight (z_d/z_s) and scalelength (h_d/h_s) ratios, the central optical depth ($\tau_{0,V}$) and the properties of the dust particles. The (poorly determined) Galactic dust properties were assumed, since extragalactic albedo and phase scattering functions have never been measured.

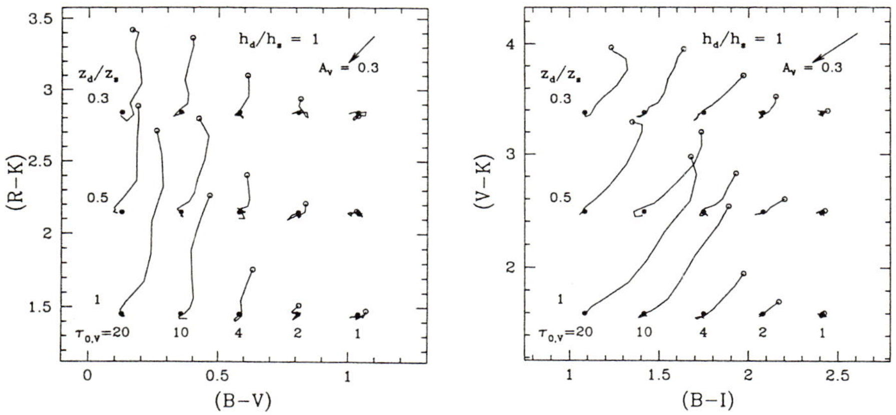

Fig. 2. The radial colour-colour reddening profiles resulting from the Monte Carlo simulations of stars and dust in exponential disks for different z_d/z_s and $\tau_{0,V}$ values. The arbitrary unreddened colours are indicated with filled circles for each model, the galaxy centres indicated by open circles.

Figure 2 shows a number of model colour-colour reddening profiles for different z_d/z_s and $\tau_{0,V}$ values. The positions of the models in the diagram are arbitrary (depending on the colours of the underlying population), but the shape of the reddening profiles are determined by the distribution and the properties of the dust. Note that all reddening profiles point in about the same direction, independent of the dust configuration, and that this direction is different from the standard screen model extinction vector (arrow).

I have used two sets of stellar population synthesis models in the comparison with the data. The Solar metallicity models of Bruzual & Charlot (1993) are used to study the colour changes of populations due to different star formation histories. Two extreme cases are considered, a single star burst model and a constant star formation model. The colours of these populations are inspected after 8 and 17 Gyr. The models of Worthey (1994) are used to study the effects that age and metallicity have on the colours of a population.

3 The comparison between models and data

Figure 3 shows the colour-colour diagrams of the models and the galaxies, again divided into four T-type bins. Note that the *same* model should fit the data in *all* colour combinations, thus in both Figs. 3a & 3b. The thin lines represent the galaxy data; the central galaxy colours are indicated by the open circles, the lines show the run of colours as function of radius. All galaxies with type T<6 are confined to a small region in these diagrams, only the later types show a considerable larger spread.

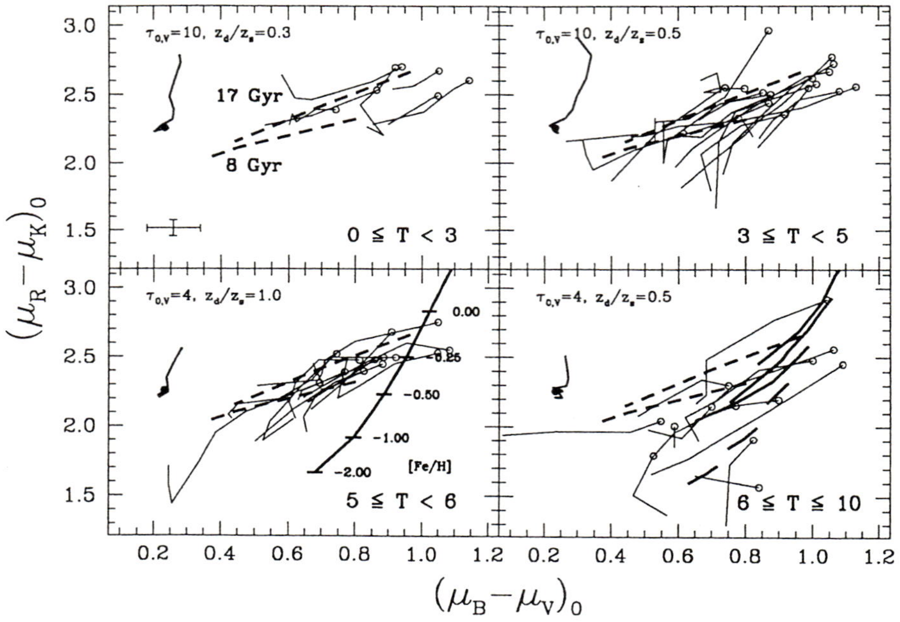

Fig. 3a. The run of B–V versus R–K colour as function of radius for the galaxies (thin lines, centre indicated by open circle). Dust models in the top left corner, Bruzual & Charlot (1993) models after 8 and 17 Gyr thick dashed lines, Worthey (1994) models for indicated metallicities thick solid lines in the two bottom panels.

Some dust model profiles are indicated in the top-left corner of the panels. As mentioned in Sect. 2, these profiles can be placed anywhere in the diagrams and their direction depends mainly on the dust properties, not on the relative distribution of dust and stars. Clearly, the colour gradients cannot be caused by reddening alone, assuming that the dust properties used are correct.

In the $5 \leq T < 6$ panels the colours predicted by Worthey's models after 12 Gyr are shown for a range of indicated metallicities. The metallicity-colour trend runs in the same direction for other ages and apparently, a metallicity gradient alone cannot explain the observed colour gradients.

The effects of different star formation histories are indicated by the two dashed lines in the centre of the panels. The red ends of these lines indicate the colours of a single burst population, the blue end of a population of constant star formation rate. Both the position and the direction of these solar metallicity models seem to agree reasonably well with the data, and the most simple explanation for the colour gradients would be age gradients across the disks of spiral galaxies. Still this cannot be the whole story, as galaxies still have star formation in their central regions, which means that a single burst is a bad approximation. Furthermore it is known from measurement of HII regions that spiral galaxies have metallicity gradients in their gas content,

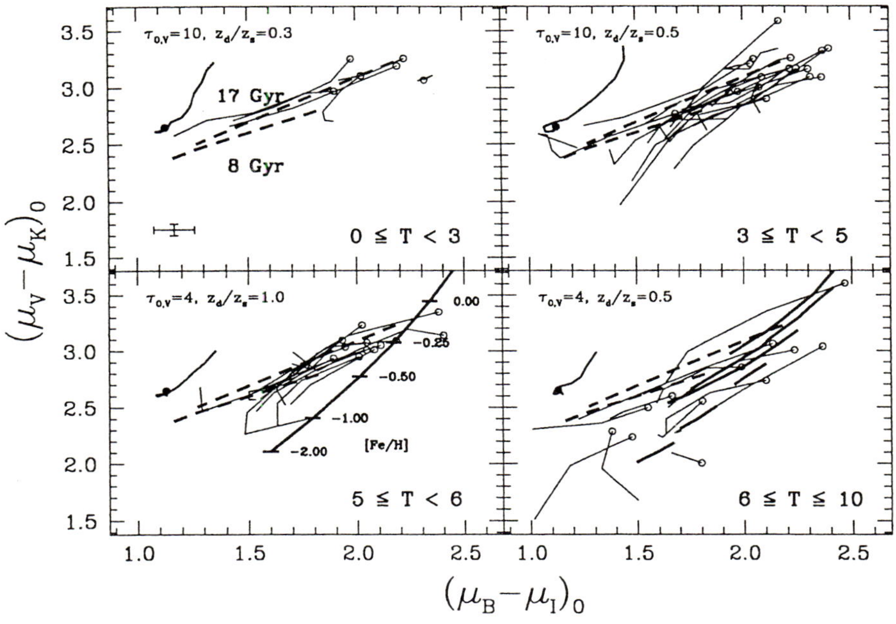

Fig. 3b. As Fig. 3a, but for B–I versus V–K.

which most likely is partly reflected in the stellar component. So the most consistent picture is one where colour gradients are caused by both age and metallicity gradients, with the central regions of galaxies being on average quite old and having a range of metallicities, whereas the outer parts are young and have low metallicities.

The large spread in colours of the late type galaxies can be explained by stellar population changes as well. The single-burst age evolution is indicated by the thick, solid lines in the 6≤T≤10 panels, for the metallicities indicated in the 5≤T<6 panels. The colours indicate that the stellar population in some of these galaxies are on average very young and have a low metallicity.

The colour gradients imply large M/L gradients for the optical passbands, making the 'missing light' problem as derived from rotation curve fitting even worse, irrespective whether they are caused by dust or population changes.

References

Bruzual G.A., Charlot S. 1993, ApJ 405, 538
de Jong R.S. 1995a, Ph.D. Thesis, Univ. of Groningen, The Netherlands
de Jong R.S. 1995b, A&A, submitted
de Jong R.S., van der Kruit P.C. 1994, A&AS 106, 451
Freeman K.C. 1970, ApJ 160, 811
Worthey G. 1994, ApJS 95, 107

Stellar Populations of Bulges and Disks: New Insights from Near-IR Colours

Reynier Peletier[1,2] & Marc Balcells[1]

[1]Kapteyn Institute, Groningen, the Netherlands
[2]Instituto de Astrofísica de Canarias,E-38200 La Laguna, Tenerife, Spain

Abstract. Colours of spiral galaxies generally are affected by the underlying old stellar population, younger stars, and extinction by dust. Old and young populations generally can be disentangled using a combination of blue and near-infrared colours. Extinction effects are very hard to take into account, except for galaxies with special orientations. In this paper we give some results of one of the first studies of galactic bulges and disks in various optical and near-infrared bands, and its implications for the stellar populations of spirals.

1 Introduction

When one compares spiral galaxies with ellipticals, one can notice that much less is known about the stellar populations of the former than of the latter, When one then looks at a grand-design spiral, one immediately understands the reason for this. Spirals not only seem to have an old, underlying population, but many times they also contain young stars, as well as dust between the stars. For ellipticals one can in general get a good idea of the metallicity, just from an optical colour like $B-V$. For spirals with a combination of young and old stellar populations one colour is not sufficient to be able to separate them. Even a combination of several optical colours (e.g. $U-V$ and $B-V$) is generally not sufficient, since one cannot distinguish between a lower metallicity of the old population, and a larger fraction of young stars. To overcome this problem one has to go to the near-infrared. Tully *et al.*(1982) and especially Frogel (1985) has shown that the $U-V$ vs. $V-K$ diagram is a very useful tool for detecting young stellar populations in spirals. Since $V-K$ is a very red colour, it happens that young stars do not contribute as much in

$V-K$ as in $U-V$, and that for a given $V-K$ the $U-V$ of an old population is very different from the $U-V$ of a combination of old and young stars. On the other hand Frogel noticed that there is a considerable scatter in the $U-V$ vs. $V-K$ diagram for spirals, much larger than for ellipticals, and that many spirals were redder than the reddest ellipticals.

These two effects very likely are due to extinction by dust. Images of edge-on galaxies, like NGC 891, show many magnitudes of extinction by dust in the central regions of the disk. Because of this, the integrated colours of this galaxy are very red (Wirth & Shaw 1983). In Balcells & Peletier (1994,hereinafter called Paper I) we have shown that the colours of the stellar populations themselves in these objects are not red, in general even bluer than those of ellipticals of the same size.

Since that paper only discusses data in the optical, the information presented about stellar populations in bulges was limited. For that reason we have obtained images in J and K for the sample, with a new two-dimensional detector with a resolution that is sufficient to spatially select regions that are not or almost not affected by extinction. The colours (optical and optical-infrared) of these regions have been analyzed in the way described by Frogel (1985), and in this paper we discuss the implications for stellar populations in spiral galaxies.

2 Sample and observations

We have investigated a sample of early-type spiral galaxies, ranging from S0's to Sbc. These galaxies generally have bulges that can be separated rather easily from their surrounding disk. Furthermore, except for the latest type, the amount of extinction by dust in these galaxies is limited, so that the colours will still be able to contain information about the stellar populations. The galaxies have inclinations larger than 50°, and are the brightest galaxies in a certain part of the sky. These two properties enable us to also on the basis of their morphology, and not just on their surface brightness profiles, separate disk and bulge.

The sample that we observed are galaxies of Table 1 of Paper I, except for 2, which had declinations larger than 60°, and so could not be observed at UKIRT. Optical data (presented in that paper) were obtained for all galaxies. All galaxies of the *dustfree* subsample were observed both in J and K. The other galaxies were all observed in K, but not necessarily in J.

The optical data consist of U, B, R and I surface photometry, obtained in June, 1990, on the 2.5m INT telescope at La Palma. The data have been described in Paper I. The pixelsize of the data was 0.549", and the effective seeing on the images lies between 1.2" and 1.6". The images were taken under photometric conditions.

The near-infrared data were observed in June, 1994, at the 4m United Kingdom Infrared Telescope at Hawaii, using IRCAM3 (Puxley & Aspin

1994), an infrared camera equipped with a 256 × 256 InSb array. The observations will be described in detail in a subsequent paper (Peletier & Balcells, in preparation). The array has a pixelsize of 0.291", as measured on the images, so that the field is about 75" on the side. Cosmetically, the array is very clean, with less than 1% bad pixels. For every object we took images in 10 positions, which each consisted of several readouts, making up a total integration time of 100s per position. The object was moved around on the chip on 6 of these exposures, while 4 consisted of blank sky, about 10' from the galaxy. The data were flatfielded using median sky flat fields, and a final mosaic was made aligning the individual frames. The effective seeing on the final frames was between 0.8" and 1.0". Here also the frames were taken under photometric conditions, with maximum zero point errors of 0.1 mag in J and K.

In Fig. 1 we show for a typical galaxy a composite $U - R - K$ map, and two colour maps. These show the bulge and disks, and the regions with the largest extinction and star formation. In Fig. 2 we show a cut along the minor axis in the $U - R$ vs. $R - K$ diagram. One side of the galaxy is well behaved, and $U - R$ is increasing for increasing $R - K$, but on the other side the combination of extinction, scattering and star formation makes this diagram very hard to interpret.

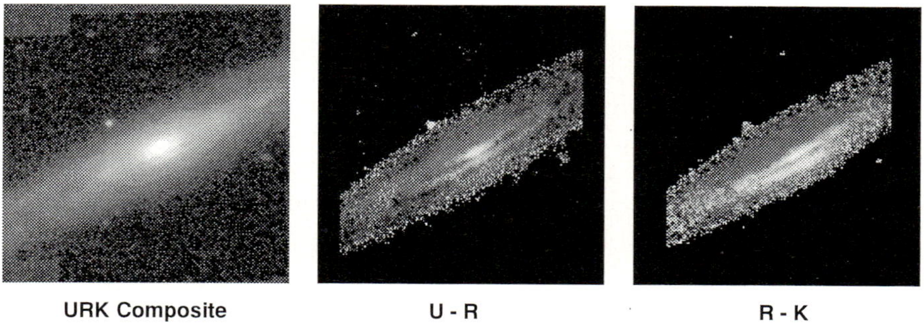

Fig. 1. Greyscale image and two colour maps of IC 1029, showing the geometry and position of the dustlanes in a typical galaxy (a) and the lack of colour differences between bulges and disks (b and c).

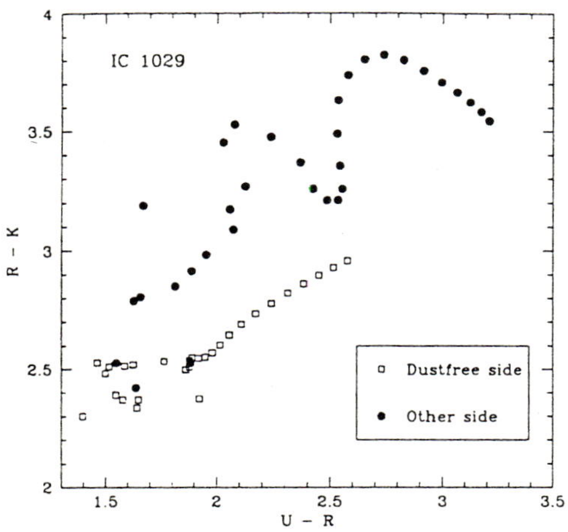

Fig. 2. Colour-colour diagram of a minor axis cut of IC 1029. Both sides are indicated by different symbols.

3 Dust in spiral galaxies

Colours contain useful information about stellar populations only when they have been corrected for extinction. Since the effects of extinction can be severe, the errors introduced by correcting for it are often so large that it is better to choose regions of the galaxy that are not, or very little, affected by extinction. In paper I we describe the procedure we employ to find extinction-free regions of bulges. For disks the situation is more complicated. Since dust and stars here are mixed almost everywhere, one cannot find dustfree regions. From a colour map (e.g. Fig. 1) one can see that the extinction in these early-type spirals is almost all concentrated on one side of the galaxy. This is the case for all galaxies that are not seen close to edge-on. One minimizes the extinction if one measures the disk colours on the other side. Since we are interested in radial and not vertical disk profiles we have measured the colours in wedges, at 15° from the major axis, with a width of 10°.

For this procedure to work the disk itself should contain relatively little extinction, since otherwise both sides are severely affected. A way to measure the extinction in the disk is to measure its radial colour gradient between a band that is affected by extinction and one that is not, e.g. B and K (Peletier *et al.*1995). If the density ratio of dust and stars is more or less constant the central extinction will be much larger than the extinction in the outer parts, purely because of the difference in stellar density. In dusty galaxies this will cause the scale lengths in B to be much larger than those in K. In Fig. 3 we plot scale length ratios between B and K as a function of galaxy type. The open squares are points from Peletier *et al.*(1994). In that paper also details are given about how the radial scale lengths have been determined. Fig. 3 is very instructive. One can see that the scale length ratios for galaxy types up to 2 (Sab) are almost never larger than 1.4. Scale length ratios for galaxies of type 6 and larger also are small. Only galaxies of type 3-5 are sometimes very dusty. This is in good agreement with the surface-brightness vs. inclination test (Valentijn 1990). An independent estimate for the scale length ratios due to stellar population gradients is 1.1 – 1.2 (Peletier *et al.*1995). This means that disks of galaxies of type 2 or smaller are likely not to be very dusty, and so the effects of dust in this paper will in general not be very large. We may be reasonably confident to use the colours for stellar population measurements.

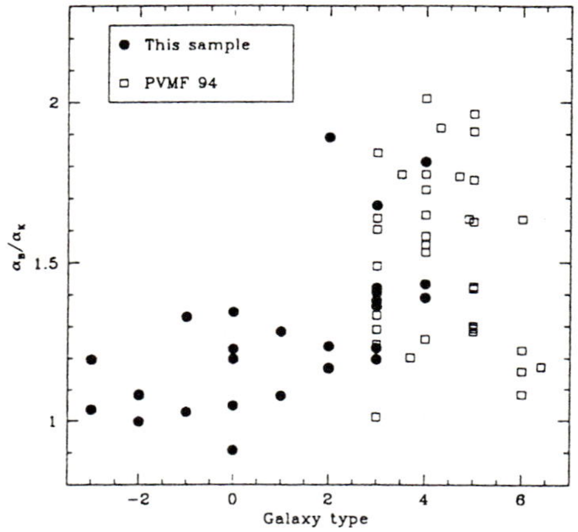

Fig. 3. Scale length ratios between B and K of galaxy disks as a function of galaxy type.

4 Stellar populations from dustfree colours

In Paper I a discussion is given of optical ($U-R$, $B-R$ and $R-I$) colours and colour gradients as a function of other parameters like bulge luminosity. It was found that about half the bulges have the same colour as elliptical galaxies of the same luminosity, and that the other half is bluer in $U-R$ and $B-R$. This was interpreted as a sign of the presence of younger stars, under the assumption that all ellipticals are old (15 – 17 Gyr). Now, with the combination of optical and near-infrared colours, we don't have to make that assumption any more and can directly determine the average age of the stars.

First we present, in Fig. 4, some colour-colour diagrams for the bulges. 30 galaxies are plotted, at $r_{\rm eff}/2$, or 5", if $r_{\rm eff} < 10$", but only 20 in the diagram in which J is included, since 10 galaxies were not observed in J. Also included are lines of single burst models of constant age, with metallicity varying.

These models are from Vazdekis *et al.*(1995) described elsewhere in these proceedings. The models are made by converting theoretical parameters like effective temperature and gravity to colours, choosing an initial mass function and integrating along a theoretical HR diagram, similar to Peletier (1989) and Worthey (1994). In this case however, we have taken a lot of care with the temperature–metallicity–colour calibration that the integrated models fit the data for giant elliptical galaxies. For example, for a given $B-V$ the $V-K$ given by Worthey (1994) is much too red to fit the giant elliptical galaxies. In Fig. 4 models are plotted for ages of 17, 12, 8, 4 and 1 Gyr. Along the lines metallicity ranges from Z=0.05 to Z=0.0004. One can see that only the diagrams that contain optical and optical-infrared colours can separate age and metallicity. $R-K$ and $J-K$ are both sensitive to more or less the same kind of stars, so in this diagram metallicity and age cannot be separated. The same holds for the $U-R$ vs. $B-R$ diagram. The systematic error in the model colours, due to uncertainties in the stellar evolution theory, or lack of template stars, seems to be on the order of 0.1 or 0.2 mag, although this value can be smaller, since the reddest bulges not necessarily have to be 17 Gyrs old. Despite these errors some galaxies in the $U-R$ vs. $R-K$ diagram cannot be fitted by a model of 17 Gyr, and for some the best fitting model only has an age of 1 Gyr. This does not mean that there is no underlying old stellar population present, but just that a much younger stellar population is dominant.

We see that most of the reddest, and also largest (Paper I) bulges are old, and that some have to be younger. It is also noteworthy that all bulges here are redder than $R-K=2.4$. This means that the metallicity of all these bulges is larger than about 0.3 $Z_\odot$. This is probably due to the way the sample was chosen - bright nearby galaxies, automatically excluding all dwarfs, and galaxies with low metallicities.

The colour - colour diagrams for disks are very similar. The differences in colour between the bulge at 0.5 r_{eff} and the disk at two scale lengths is much smaller than the range in colour indicated in Fig. 4. In Peletier & Balcells (1995) it is explained that this means that disks are at most 2-4 Gyr younger than bulges. The colours don't exclude an equal age, which would imply that the colour differences are due to metalicity gradients.

In Fig. 5 a plot is given of $R-K$ colour gradients in the bulges as a function of total K-band magnitude. Since $R-K$ is an indicator of old stellar populations measuring the bulge gradients in this colour is a good way to measure the metallicity gradients in bulges. Two different symbols are used for our galaxies. The filled circles indicate the galaxies of the dustfree sample of Paper I. The filed squares are those that may contain some dust. There are no significant differences to be seen between the $R-K$ gradients of both groups, showing that dust extinction is not a major factor influencing the gradients. The gradients are somewhat larger than the visual-infrared gradients of elliptical galaxies (Peletier *et al.*1990, Silva & Elston 1994). Our average $R-K$ gradient is $\Delta(R-K)/\Delta(\log r)$ = -0.232, and for $U-R$ -0.399.

For ellipticals average values are -0.14 for $R-K$ and -0.20 for $U-R$, although not much data are available for $R-K$. In Paper I we concluded from the similarity between gradients in the optical between bulges and disks that this is a strong argument to show that bulges and ellipticals formed in the same way, and that the subsequent disk-formation did not substantially affect the bulge. We can say the same now using the $R-K$ colour. A recent review by Minniti (1995) shows that possibly also the bulge of our own galaxy has a similarly small colour/metallicity gradient, even though earlier measurements by Terndrup (1988) were indicating large values. Fisher *et al.*(1995) find that for a sample of S0 galaxies the vertical gradients in the bulge in $B-R$ and in absorption line strength are much larger than the radial gradients. Although we don't have many galaxies in our sample that are really edge-on we also see hints for the same effect, and this might be the reason of Terndrup (1988)'s large metallicity gradient in our Bulge. Using colour - metallicity conversions by Worthey (1994) we have converted our average $R-K$ and $U-R$ gradients to metallicity gradients, assuming constant age. From $R-K$ we find that $\Delta(\log Z) = -0.123$ per dex in radius, and from $U-R$ we find a value of -0.243. The fact that the $R-K$ gradients give on the average a smaller metallicity gradient than $U-R$ shows that some young stars must be present in the outer parts of the bulges, possibly due to interactions with the disk.

Also plotted in Fig. 5 is the sample of Terndrup *et al.* (1994) of similar kinds of galaxies. Even though almost all $R-K$ gradients of our sample are negative, those of Terndrup *et al.* can be both positive and negative. We presume that the errors in that sample are larger than indicated by the errorbars, something which is possible, since they used a small detector covering a small field.

As a final point we would like to talk a bit about colours of disks as a function of galaxy type. Here the colours are the ones that have been determined in such a way as to minimize the effects of extinction (see above). 4 different colours are plotted in Fig. 6. The $J-K$ and $R-K$ colours here are indicators of the old underlying stellar population that constitutes most of the mass. $B-R$ and $U-R$ also depend on additional young stars. Noteworthy is the fact that there is not much change in the colour, when varying galaxy type. One can say that ellipticals, S0's and spirals up to type 2 (Sab's) have more or less the same colours, indicating few young stars and similar metallicities. Some galaxies of type 3 obviously suffer from extinction effects, while in all colours, but especially in $U-R$ galaxies of type 4 (Sbc's) are bluer than the others. This figure shows again that a considerable fraction of the stars in Sbc's and probably also later type spirals is young.

5 Conclusions

The most important results that have been discussed in this paper are:

- By investigating colour gradients in disks between B and K for a sample of 70 galaxies we find that galaxies of type 3-5 are much more affected by extinction than are others. In the galaxies of other types, it is possible to find regions that are only slightly affected by extinction, and for which the colours can be used to study stellar populations.
- Using colour-colour diagrams of optical and optical-infrared colours, and comparing them with single age, single metallicity stellar population models, we find that a considerable fraction of the bulges in the sample has to be younger than elliptical galaxies, or contain a large fraction of stars that are younger than those in elliptical galaxies. The differences in colour from one galaxy to another can be much larger than the colour difference between the bulge and disk of the same galaxy. For that reason the inner disks of these galaxies cannot be more than 2-4 Gyrs younger than their bulges.
- Radial $R-K$ colour gradients of bulges are negative and small. Bulges generally are redder in the center than in their outer parts, and the average colour gradient is about 0.2 mag per dex. These values are due to metallicity, or possibly age gradients, but not due to extinction by dust. Metallicity gradients inferred from $R-K$ gradients are smaller than those from $U-R$, indicating that there are probably also age gradients playing a role here.
- Dustfree colours in disks are remarkably constant. Only for galaxies of type 3 and later disks are significantly bluer than those of S0's.

References

Balcells, M. & Peletier, R.F., 1994, AJ 107, 135 (**Paper I**)
Fisher, D., Franx, M. & Illingworth, G., 1995, ApJ, submitted
Frogel, J.A., 1985, ApJ 298, 528
Minitti, D., 1995, ESO preprint
Peletier, R.F., 1989, Ph.D. Thesis, Univ. of Groningen
Peletier, R.F., Valentijn, E.A. & Jameson, R.F, 1990, A&A 233, 62
Peletier, R.F., Valentijn, E.A., Moorwood, A.F.M. & Freudling, W., 1994, A&A Suppl. 108, 621.
Peletier, R.F. & Balcells, M., 1995, ApJ, submitted
Peletier, R.F., Valentijn, E.A., Moorwood, A. & Freudling, W., Knapen, J.H. & Beckman, J.E., 1995, A&A 300, L1
Puxley, P. & Aspin, C., 1994, Spectrum 2
Silva, D,R. & Elston, R., 1994, APJ 428, 511
Terndrup, D.M., 1988, AJ 96, 886
Terndrup, D.M., Davies, R.L., Frogel, J.A., Depoy, D.L. & Wells, L.A., 1994, ApJ 432, 518

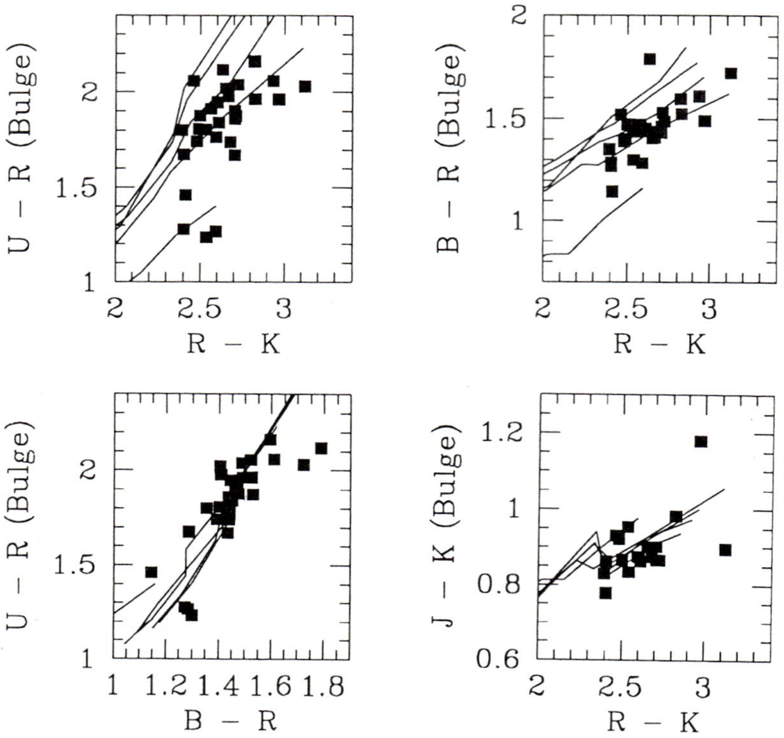

Fig. 4. Colour-colour relations for the bulges. Also drawn are lines of constant age by Vazdekis *et al.*(1995). Plotted are ages 17, 12, 8, 4 and 1 Gyr.

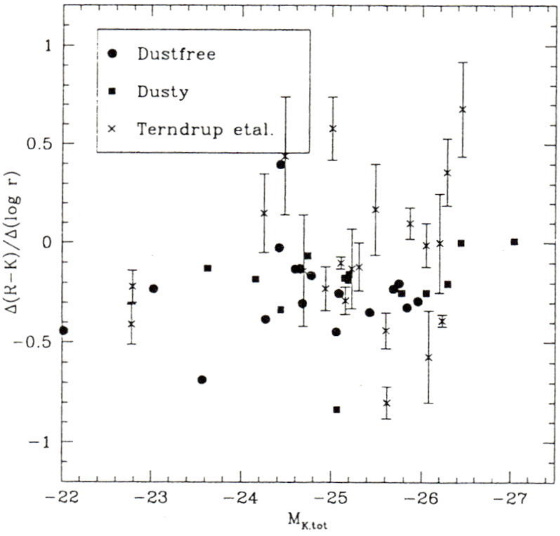

Fig. 5. $R - K$ gradients for the bulges. Also given are the data of Terndrup *et al.*(1994).

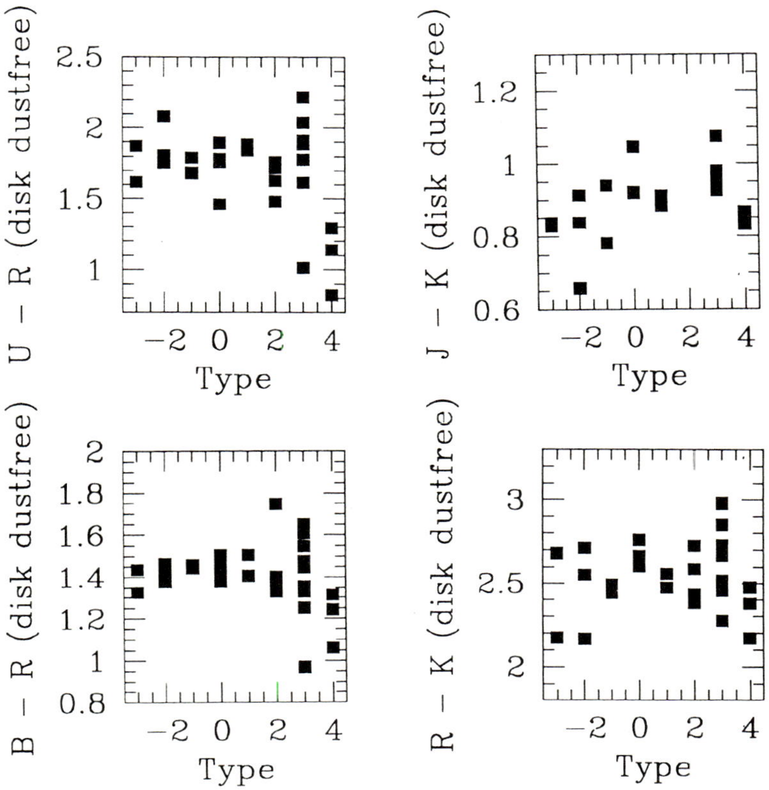

Fig. 6. Disk colours as a function of galaxy type.

Tully, R.B., Mould, J.R. & Aaronson, M., 1982, ApJ 257, 527.

Valentijn, E.A., 1990, Nature 346, 153

Vazdekis, A., Casuso, E., Peletier, R.F. & Beckman, J.E., 1995, in preparation

Wirth, A. & Shaw, R., 1983, AJ 88, 171

Worthey, G., 1994, ApJ Suppl 95, 107

Determining the Contribution of Young Stars to Near Infrared Light

James E. Rhoads

Princeton University Observatory, Princeton, NJ 08544, USA

Abstract. As part of an effort to determine the stellar origins of near infrared (NIR) light, we have measured the NIR photometric CO absorption index in different regions of the galaxy NGC 1309. This index measures the strength of the gravity-sensitive $2.3\mu m$ stellar absorption band and is a good indicator of young, cool supergiant stars. Our data suggest that such young supergiants dominate the $2\mu m$ light from active star forming regions in NGC 1309. The galaxy's quiescent regions, in contrast, do not show evidence of young supergiants. It follows that the $2\mu m$ light comes from different stellar populations in different places, and large changes in the $2\mu m$ surface brightness need not imply correspondingly large features in the galaxy's mass distribution.

Introduction: Because near infrared (NIR) light is relatively insensitive to both dust absorption and emission and to the presence of hot young stars, there has been recent interest in using the NIR bandpasses as photometric tracers of mass. Observations of disk galaxies in increasingly red broadband filters (Schweizer 1976 at $0.64\mu m$; Elmegreen & Elmegreen 1984 at $0.83\mu m$; Rix & Rieke 1993, Rix & Zaritsky 1995 at $2.2\mu m$) have shown that spiral patterns can have large amplitudes in surface brightness at these wavelengths. This has been interpreted as evidence that the spiral pattern involves old stars (e.g., Rix 1993). Because old stars comprise most of the mass in galactic disks, this would mean spirals are large amplitude, nonlinear features.

We have sought to test this interpretation by looking for spatial variations in the carbon monoxide (CO) index. Rotational-vibrational transitions of CO in stellar atmospheres result in an absorption band at wavelengths $\geq 2.3\mu m$, whose strength can be measured by the photometric index $CO = 2.5 \log_{10}(F_{2.2\mu m}/F_{2.36\mu m}) + c$. (Here F_λ is the flux in a bandpass with central wavelength λ, and the constant c is determined by comparison with photometric standard stars). For fixed effective temperature T_{eff} and metallicity z, the strength of this band depends strongly on stellar surface gravity

g. Consequently, it has been used extensively as a luminosity class indicator. The band strength also depends on temperature and metallicity (for tables, see Bell & Briley, 1991), so its interpretation is not entirely straightforward. Nonetheless, if the CO index of two stellar populations differs significantly, the populations themselves must also differ, and it is likely that their mass to $2.2\mu m$ light ratios differ.

The galaxy: We chose the galaxy NGC 1309 for study based on its morphology: Prominent bright patches are seen along the northern spiral arm in K' band (Rix & Zaritsky [1995]), and these patches are bright blue in the 3-color optical UBV image by Wray (1988). Thus, they are presumably regions of active star formation, and NGC 1309 is a good place to look for the signature of young stars at $2\mu m$.

NGC 1309 is a nearly face-on Sbc galaxy with a redshift $cz = 2135\,\mathrm{km/s}$ measured from HI observations (de Vaucouleurs et al 1991). A weak bar may be present in my data. Total or aperture flux measurements have been published at wavelengths of 21 cm (Bottinelli, Gougenheim, & Paturel 1982); 20 cm (Condon et al 1990); 100, 60, 25, and $12\mu m$ (Soifer et al 1989); K and H bands (Devereux 1989); and V, B, and U bands (Véron-Cetty 1984). Rix & Zaritsky (1995) obtained I and K' band images, and measure the K' band exponential scale length ($11.1''$) and principal Fourier moments.

Expectations: If K band light traces mass, it is [probably] dominated by the same stellar population everywhere, and spatial variations in the CO index will be absent or small. However, if K band light from active star forming regions includes large contributions from young supergiants, we would expect large spatial variations in the CO index, with a larger index in star forming regions than in the background disk.

Observations: We observed NGC 1309 on 18 December 1994 using the GRIM II camera at the Apache Point Observatory 3.5 meter telescope in remote observing mode. Characteristics of the observations are summarized in table 1.

Table 1. Characteristics of the observations. λ_c and $\Delta\lambda$ are the central wavelength and bandpass (FWHM) of the filter. t_i is the exposure time for individual frames, and Σt is the total on-source exposure time in that band.

Filter	λ_c (μm)	$\Delta\lambda$ (μm)	t_i (s)	Σt (s)	s/n (peak)
K'	2.124	0.337	225 s	15 s	75
CO cont	2.22	0.089	1080 s	40, 60 s	75
CO band	2.36	0.094	880 s	40 s	28

The GRIM II camera was used in f/5 mode, giving pixel size $0.482''$ and field of view $123''$.

Exposures of fields 4–5 arcminutes from the galaxy were obtained along with the on-source exposures to allow sky subtraction. Conditions were photometric, and I observed four NIR photometric standard stars (HD 225023, G158-27, HD 40335, & G77-31) from the list of Elias et al (1982).

Data reduction followed mostly standard procedures and was done using the IRAF software package.

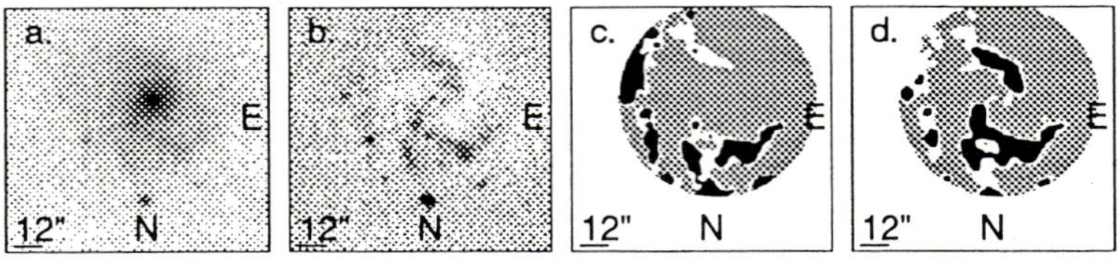

Fig.1: a: The K' band image of NGC1309, displayed with a logarithmic intensity transformation. b: The nonaxisymmetric K' band light, displayed with a linear intensity transformation. c, d: The masks used to determine the CO indices of various regions. c: mask 1 (black) corresponds to star forming regions. d:mask 2 (black) corresponds to the spiral arms. The grey regions are control masks (c: mask 3; d: mask 4).

Analysis: To analyze our results, we first use the nonaxisymmetric part of the broadband (K') light to define pixel masks, corresponding to physically distinct regions of the galaxy. We then add together all CO band filter and all CO continuum filter flux in each mask, and from these total fluxes define the total CO index for the masked regions.

This *"masked aperture photometry"* combines the physical insights of a spatially resolved map with the small statistical errors of aperture photometry.

In this work we have used two general types of mask – relative (mask 1) and absolute (mask 2) thresholds in the nonaxisymmetric light. We have also varied the outer radii of the masks, so that for each threshold criterion we can make a cumulative plot of the CO index as a function of radius.

In detail, call the K' band surface brightness distribution $S_{K'}(r,\phi)$ and the axisymmetrized K' band surface brightness $\overline{S}_{K'}(r)$. Then mask 1 is the set $\mathcal{M}_1 := \{(r,\phi) : S_{K'}(r,\phi) > 1.4 \times \overline{S}_{K'}(r)\}$, and mask 2 is $\mathcal{M}_2 := \{(r,\phi) : S_{K'}(r,\phi) - \overline{S}_{K'}(r) > 27 ADU/pixel \equiv 20.7 K'\ mag/\text{arcsec}^2\}$. (The thresholds are chosen so that masks 1 and 2 have equal areas.) Two control regions were also made, namely $\mathcal{M}_3 := \{(r,\phi) : S_{K'}(r,\phi) < 1.2 \times \overline{S}_{K'}(r)\}$ and $\mathcal{M}_4 := \{(r,\phi) : S_{K'}(r,\phi) - \overline{S}_{K'}(r) < 10.5 ADU/pixel \equiv 21.7 K'\ mag/\text{arcsec}^2\}$.

Results: CO indices in different regions appear to be different. $CO(\mathcal{M}_1) = 0.195 \pm 0.018$ corresponds roughly to the CO index of the star forming regions, $CO(\mathcal{M}_2) = 0.125 \pm 0.009$ to the spiral arms, and $CO(\mathcal{M}_3) = 0.064 \pm 0.006 \approx CO(\mathcal{M}_4) = 0.059 \pm 0.008$ to the background disk.

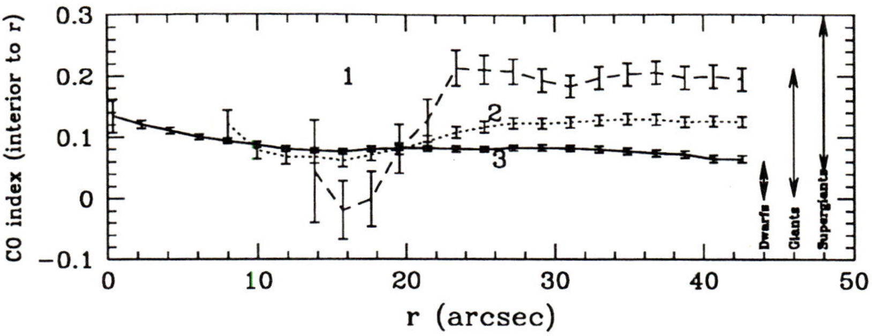

Fig.2: Cumulative CO index profiles for masks 1–3 in figure 1; the curve for mask 4 resembled that for mask 3.

These results are subject to the following caveats: First, sky and bias level subtraction are imperfect, and may introduce large systematic errors. Second, color terms may be present in the transformation from instrumental to standard magnitudes, as our photometric standards do not extend to large CO indices. The zero point of the CO index is well constrained, but there may be a scaling error $\lesssim$ 10%. Third, dust can both absorb and emit $2\mu m$ light. We have insufficient wavelength coverage to determine how important these effects are in NGC 1309.

Interpretation: One approach to interpreting these results is population synthesis, which allows detailed modelling. There are substantial difficulties due to physical uncertainties (primarily convective overshooting and mass loss) that make the late stages of a massive star's life difficult to predict.

Another approach: examine the age-CO index relation for star clusters. We use Large Magellanic Cloud star clusters with published photometry from Persson et al (1983) and age estimates from Elson & Fall (1985, 1988). The result is a correlation that suggests stellar populations with $CO \sim 0.2$ (like the bright patches in NGC 1309) have age $\tau \sim 10^7$ yr, while those with $CO \sim 0.06$ (like the interarm regions of NGC 1309) have $\tau \sim 10^9$ yr. These should be compared with the typical rotation period of a few 10^8 yr for stars 2–3 scale lengths from the center of a spiral galaxy.

These age estimates for the stellar populations of NGC 1309 are approximate for several reasons, among them the scatter in figure 3 and probable differences between the LMC and NGC 1309. Still, none of the uncertainties seem large enough to alter the conclusion that the $2\mu m$ light from bright patches is dominated by a stellar population younger than the galactic rotation period, while $2\mu m$ light from the background disk is dominated by a population older than the rotation period.

Acknowledgements: I would like to thank Jim Gunn, Michael Strauss, Ed Fitzpatrick, Bohdan Paczyński, Stephane Charlot, David Spergel, and Neil Tyson for discussions and other help during the course of this work.

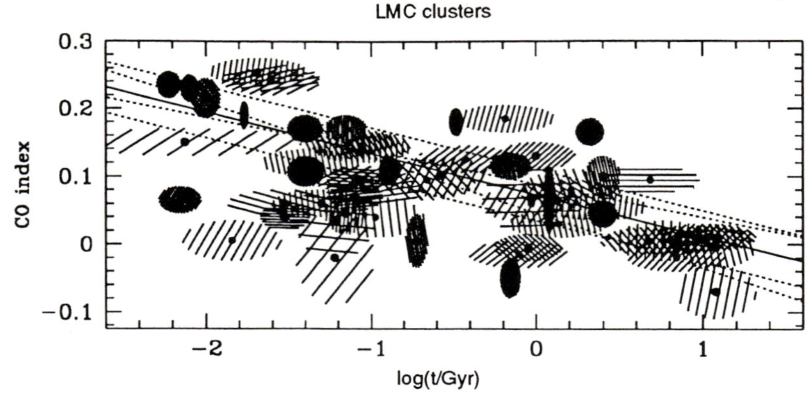

Fig.3: The Age - CO index relation for star clusters in the LMC. Each point is a cluster; the 1σ error ellipses are shaded.

This research has been supported by NSF grant AST 91-17388 and NASA grant ADP NAG5-269. This research has made use of the Simbad database, operated at CDS, Strasbourg, France.

References

Bell, R.A. & Briley, M.M. 1991, AJ 102, 763

Bottinelli, L., Gougenheim, L., & Paturel, G. 1982, A&AS 47, 171

Condon, J.J., Helou, G., Sanders, D.B., & Soifer, B.T. 1990, ApJS 73, 359

de Vaucouleurs, G., de Vaucouleurs, A., Corwin, H. G., Buta, R.J., Paturel, G., & Fouqué, P. 1991, *Third Ref Catalog of Bright Galaxies*, New York: Springer-Verlag

Devereux, N.A. 1989, ApJ 346, 126

Elias, J.H., Frogel, J.A., Matthews, K., and Neugebauer, G. 1982 AJ 87, 1029

Elmegreen & Elmegreen 1984, ApJS 54, 127

Elson, R.A.W. & Fall, S.M., 1985, ApJ 299, 211

Elson, R.A.W. & Fall, S.M., 1988, AJ 96, 1383

Persson, S. E., Cohen, J. G., Matthews, K., Frogel, J. A., Aaronson, M. 1983 ApJ 266, 105

Rix, H.-W. 1993, PASP 105, 999

Rix & Rieke 1993, ApJ 418, 123

Rix, H.-W., & Zaritsky, D. 1995 submitted

Schweizer 1976, ApJS 31, 313

Soifer, B.T., Boehmer, L., Neugebauer, G., & Sanders, D.B. 1989, AJ 98, 766

Véron-Cetty, M.P. 1984, A&AS 58, 665

Wray, J. D. 1988, *Color Atlas of Galaxies*, Cambridge: Cambridge University Press

How Much Do Young Stars Contribute to the K-band Light?

Cesare Chiosi [1] and Antonella Vallenari [2]

[1] Department of Astronomy, Vicolo Osservatorio 5, Padova, Italy
[2] Astronomical Observatory, Vicolo Osservatorio 5, Padova, Italy

Abstract. Near infrared, integrated magnitudes and colours derived from population synthesis are presented and applied to star clusters to check their ability to reproduce the observational data. In addition to this, the luminosity and colour evolution of very simple galactic models for spirals and elliptical galaxies are shown to illustrate the role of different star formation histories. Finally, the contribution of young stars to the infrared light is discussed for very simple models of galactic evolution.

1 Introduction

In this paper we describe magnitudes and colours obtained from population synthesis paying particular attention to the near infrared pass-bands. The material is taken from the libraries of single stellar populations (SSP), i.e. coeval, chemically homogenous assemblies of stars with age t and chemical composition Y (helium) and Z (metallicity), recently calculated by Bertelli et al. (1994), Bressan et al. (1994, 1995), and Tantalo et al. (1995). In order to show how closely the theoretical results reproduce the observational data for stellar aggregates of various complexity, we briefly examine a few test examples, the detailed discussion of any of these being beyond the present aims. As the fundamentals of Populations Synthesis are known from long time (e.g. Renzini 1986), we limit ourselves to a few key points that are at the base of the present results. Equally for the properties of the stellar models used to derive isochrones and integrated magnitudes and colours of the SSPs. Suffice it to mention that stellar models are still affected by several points of uncertainty (for instance the real extension of convective regions, mixing mechanism, loss by stellar winds in various phases, etc...). A recent review of the subject is by Chiosi et al. (1992) to whom we refer.

2 Basic Population Synthesis

The integrated monochromatic flux generated by the stellar content of a galaxy of age T can be expressed as

$$F_{\lambda}(T) = \int_{0}^{T} \Psi(t, Z)\ F_{sp,\lambda}(\tau, Z)\ dt \tag{1}$$

where $\Psi(t, Z)$ is the star formation rate (SFR) and $F_{sp,\lambda}(\tau, Z)$ is the contribution from a SSP of age τ and metallicity Z and

$$F_{sp,\lambda}(\tau, Z) = \int_{M_L}^{M_U} \phi(M)\ f_{\lambda}(M, \tau, Z)\ dM \tag{2}$$

where $f_{\lambda}(M, \tau, Z)$ is the monochromatic flux of a star of mass M, metallicity $Z(t)$, and age $\tau = T - t$, and $\phi(M)$ is the initial mass function (IMF) usually expressed by the Salpeter law

$$\phi(M) = A \times M^{-x} \tag{3}$$

with $x = 2.35$. The IMF is normalized in the following way; we assume $A = 1$ and fix the minimum (0.15 $M_{\odot}$) and maximum (120 $M_{\odot}$) mass of the integration interval. Therefore a SSP has an intrinsic total mass for which magnitudes are derived. To be applied to real clusters they must be re-scaled to the current mass of the clusters under examination. Finally, the SFR of equation (1) can be expressed as the total mass converted into stars per unit of time, i.e. in $M_{\odot} yr^{-1}$.

3 Libraries of Stellar Models and Isochrones

In order to calculate the flux $F_{sp,\lambda}(\tau, Z)$ emitted by a SSP, we must construct isochrones in the CMD. The more accurate this calculation, the more precise are the fluxes for the whole galaxy. The precise shape of the isochrone and relative number of stars along these depend on the properties of the underlying evolutionary tracks (kind of input physics) and the IMF. The isochrones in usage are taken from the library of Bertelli et al. (1994). These are calculated using the stellar models of the Padua group (Alongi et al. 1993; Bressan et al. 1993; Fagotto et al. 1994a,b,c), in which a moderate amount of overshoot from the convective regions is allowed. The isochrones span the age range from a few 10^6 yr to beyond 18 Gyr, and cover all evolutionary phases from the zero age main sequence up to either the stage of PN formation or central carbon ignition depending upon the mass of the most evolved star in the isochrone. The chemical compositions are in the range [Y=0.23, Z=0.0004] to [Y=0.475, Z=0.1] and obey the enrichment law $\Delta Y/\Delta Z = 2.5$. Among the many physical factors that bear very much on the isochrone calculations, we recall:

i) The enrichment law $\Delta Y/\Delta Z$. This important parameter is known with a large uncertainty (cf. Pagel et al. 1992). Our choice correspond to the estimate inferred from the solar abundances and constitutes a sort of lower limit.
ii) The efficiency of mass loss during the RGB and AGB phases, in particular whether or not it depends on the metallicity. See the analysis of the clump stars in M67 by Carraro et al. (1995). The isochrones in usage here are for standard assumptions of the mass loss efficiency, as described in the studies of population synthesis by Bressan et al. (1994) and Tantalo et al. (1995).

4 Libraries of Stellar Spectra

Like in Bressan et al. (1994), the main body of the spectral library is from Kurucz (1992), however extended in the high and low temperatures ranges. For $T_{eff} > 50,000$ K pure black-body spectra is assigned For $T_{eff} < 3500$ K, the new catalog of stellar fluxes by Fluks et al. (1994) is adopted. It contains observed spectra for all M-spectral subtypes in the wavelength range $3800 \leq \AA \leq 9000$, and photospheric synthetic spectra in the range $9900 \leq \AA \leq 12500$. The scale of T_{eff} in Fluks et al. (1994) is similar to that of Rigdway et al. (1980) for spectral types earlier than M4 but deviates from it for later spectral types. Since Ridgway' et al. (1980) scale does not extend beyond the spectral type M6, no comparison for more advanced spectral types is possible. The problem is further complicated by the effects of metallicity. The Ridgway T_{eff} scale is based on stars with solar metallicity ($Z \sim 0.02$) and empirical calibrations of the T_{eff} - scale for $Z \neq 0.02$ are not available. In contrast, the libraries of SSPs span the range of metallicity $0.0004 \leq Z \leq 0.1$. To cope with this difficulty, Tantalo et al. (1995) have introduced the metallicity - T_{eff} relation of Bessell et al. (1989,1991) using the (V–K) color as a temperature indicator. Finally, integrated magnitudes and colours of the Johnson system are calculated following the method described by Bressan et al. (1994).

5 Single Stellar Populations: the Building Blocks

Much can be learned from the study of SSPs. In particular, we may address the following questions:
i) How much do the single evolutionary phases contribute to the total flux in various pass-bands and hence colours ?
ii) Is there any feature marking the age of a SSP ?
iii) What is the fading rate of a SSP ?
The discussion is limited to the template SSP with [Y=0.28, Z=0.020].

Single Phases. The contributions of the main sequence (MS), red giant branch (RGB), and asymptotic giant branch (AGB) phases to the bolometric luminosity and integrated (V–K) colour as function of the age are shown in Figs. 1 and 2, respectively. All the trends are as expected from simple

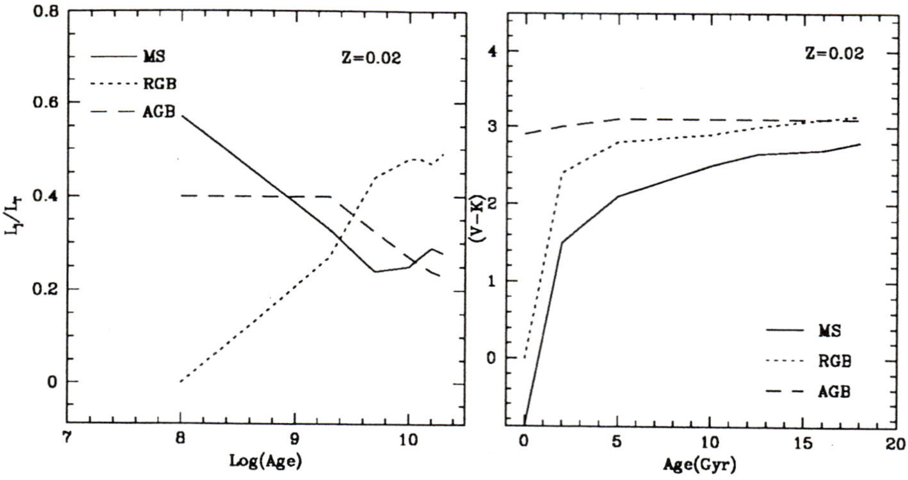

Fig. 1-2. **Fig. 1** (left): Fractionary contribution to the total luminosity of a SSP with [Y=0.28, Z=0.02] from stars in different evolutionary stages (MS main: sequence; RGB: red giant branch; AGB: asymptotic giant branch). **Fig.2** (right): The contribution of the same stars as in Fig. 1 to the colour (V–K)

considerations of stellar evolution theory (cf. Renzini & Buzzoni 1986 for a discussion of idealized SSPs). Remarkably, the colour (V–K) is built up by MS and RGB stars and to much less extent by AGB stars. The contribution from these latter gets indeed smaller and smaller at increasing age.

Phase Transitions. The most interesting feature of broad band magnitudes and colours to look at is the variation expected to occur at the appearance of the first AGB and RGB stars in a SSP. According to elementary theory of stellar evolution, the onset of the AGB and RGB phases occur at two precise values of the SSP age, $t(M_{UP})$ and $t(M_{HeF})$ respectively. They correspond to the initial mass of the star able to develop past core He-burning a highly degenerated carbon-oxygen core (AGB phase) or a highly degenerated helium core prior to core He-flash (RGB phase). Since $t(M_{UP})$ is much smaller than $t(M_{HeF})$, in a SSP AGB stars appear earlier than RGB stars, and for ages older than $t(M_{HeF})$ both are present. Finally, their appearance is expected to cause *rapid* variations in the near infrared colours of the SSP, that Renzini & Buzzoni (1986) have named *Phase Transitions*. Looking at the template SSP with Z=0.02 shown in Fig.3 (identical considerations apply to other compositions), we notice the following behaviour of the colours as a function of the age. At very young ages the colours are dominated by the bright massive stars, which evolve as blue objects because of the very efficient mass loss by stellar wind. At slightly older ages (a few 10^7 yr), the detailed shape of the colour-age relation depends on the path of massive stars in the CMD, i.e. whether or not extended loops may occur during the core He-burning phase whose characteristics are always hard to predict. At ages older than a few 10^8 yr, the colours are dominated by the properties of AGB stars, in particular by those during the thermally pulsing regime, i.e. the evolutionary rate, lifetime, and maximum luminosity. At ages of about 1.1 Gyr the phase transition of RGB stars should occur in the infrared colours, whereas there is no sign of it. The explanation of this resides in the Fuel Consumption Theorem of Renzini & Buzzoni (1986). Indeed the relative bolometric luminosity of the AGB stars tends to decrease in coincidence with the increasing contribution from the RGB stars. The opposite trends tend to balance each other so that the appearance of RGB is masked by the already existing AGB stars. Finally, at very old ages only the effect of RGB stars is visible.

Fading Rate. The rate of magnitude change (shown in Fig.4) in SSPs older than a few 10^8 yr is fairly regular and can be approximated by a linear relationship

$$M_\lambda = 2logt + C_\lambda$$

with C_λ=-18.2 for the K-band, -18.6 for the M-band, etc.... for Z=0.02. Similar relations hold for the other metallicities.

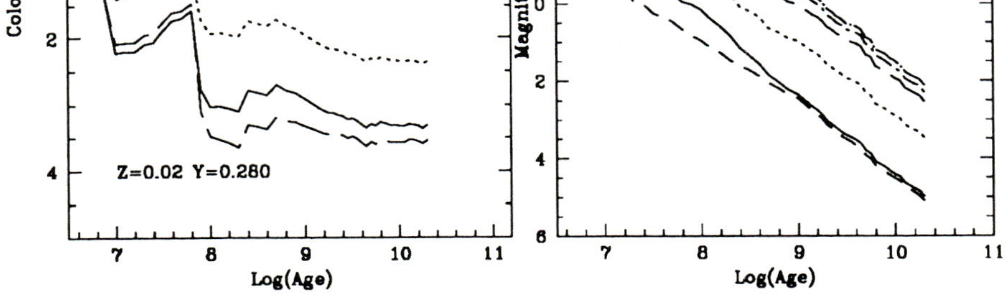

Fig. 3-4. Fig.3 (left): Colours as function of the age for the SSP with [Y=0.28, Z=0.02]. **Fig. 4** (right): Magnitudes as function of the age for the same SSP

6 Integrated Magnitudes and Colours of Star Clusters

In this section we briefly compare the integrated magnitudes and colours of star clusters of the LMC (large age range), Milky Way (old globulars), M31 (old globulars) with their theoretical counterparts.

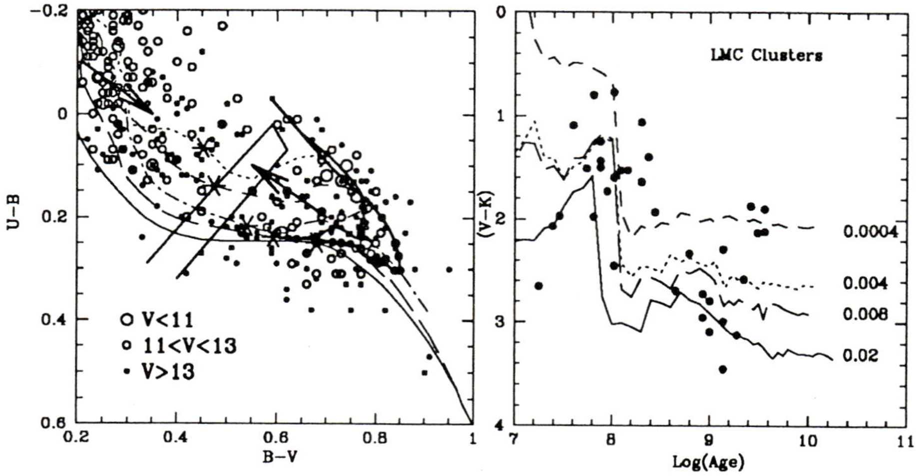

Fig. 5-6. Fig.5 (left): The plane of integrated (U–B) versus (B–V) colours for LMC clusters. The path of SSPs with different compositions, the dispersion vectors, and the Bica et al. (1995) gap are shown. **Fig.6** (right): The integrated (V–K) colours versus age relation for LMC clusters compared with SSPs of different metallicity

Integrated Blue Colours of LMC Clusters. The integrated (U–B) and (B–V) colours for the family of LMC clusters sampled by Bica et al. (1995) have recently been studied by Girardi et al. (1995) re-confirming the results of the previous analysis of Chiosi et al. (1988). Main results of the Girardi et al. (1995) study are:

i) The large dispersion in the observed colours can be accounted for by stochastic effects on the IMF (the effect is dominant for the blue clusters), and an additional dispersion's of 0.2 dex in metallicity and of 0.05 mag in colour excess for the red clusters.

ii) The bimodal distribution in the (B–V) colour is reproduced by a sequence of clusters almost evenly distributed in the log of the age, whose metallicity is governed by a normal age-metallicity relation.

iii) The gap noticed by Bica et al. (1995) in the (U–B) versus (B–V) plane can be explained by the particular direction along which colours are dispersed in that region of the plane. The gap is shown in Fig. 5 together with SSPs and the dispersion vectors.

iv) The age distribution function shows periods of enhanced star cluster formation at about 100 Myr and 1-2 Gyr, and a gap in the age range 3 to 12-15 Gyr.

Integrated Red Colours of LMC Clusters. Fig. 6 shows the comparison of the infrared colours for LMC clusters of Persson et al. (1983) with the theoretical data. The ages assigned to the clusters are derived from the so-called S parameter according to the revision made by Girardi et al. (1995). The calibration in use is $log(t) = (0.0733 \pm 0.0032) \times S + (6.277 \pm 0.096)$ Considering the scatter induced by stochastic effects on the IMF (cf. Chiosi et al. 1998, Girardi et al. 1995) and the range of possible metallicities, the agreement is remarkably good.

Integrated Colour-Magnitude Diagram. Star clusters can be studied in the colour magnitude diagram of their integrated properties. To this purpose see the comparison made by Vallenari et al. (1995, this volume) for the Galactic and M31 globular clusters.

7 Models of Spiral and Elliptical Galaxies

The Models. Chemo-spectro-photometric models for elliptical galaxies have recently been presented by Bressan et al. (1994, 1995) and Tantalo et al. (1995). The models allow for infall of primordial gas at a suitable rate in order to simulate the process of galaxy formation. The infall scheme in usage here follows the prescription proposed long ago by Chiosi (1981) and since then adopted in many studies of chemical evolution (cf. Matteucci 1991 and references). The SFR $\Psi(t)$ is assumed to be proportional to the current value of gas mass $M_g(t)$, i.e. $\Psi(t) = \nu Mg(t)\ M_{\odot}\ \mathrm{yr}^{-1}$, where the ν represents the inverse star formation time scale. Under the effect of infall, the SFR as a function of time starts small, grows to a maximum, and then declines with a time scale that is comparable with the infall time scale.

Although the Bressan et al. (1994) and Tantalo et al. (1995) models have been particularly designed for elliptical galaxies, for the purposes of this paper they can be also applied to spiral galaxies by suitably choosing the infall time scale.

In the case of spiral galaxies these are simplified to a sort of solar vicinity pool and the models are imposed to obey the constraints holding for the solar neighborhood (metallicity, gas fraction, G-dwarf problem, etc...). The total mass is about $10^{10} M_{\odot}$, the present day gas fraction is 0.05, the metallicity is about $2 \times Z_{\odot}$, the age of the disk is about 10 Gyr. These constraints are matched assuming the time scale of infall to be 4 Gyr. This means that star formation is an ever continuing process, whose SFR declines by roughly a factor of 10 during the life time of the disk.

In the case of elliptical galaxies the models are constrained to reproduce colour-magnitude relation of Bower et al. (1992) and the UV properties (Burstein et al. 1988, Ferguson et al. 1991). The goal is met using short time

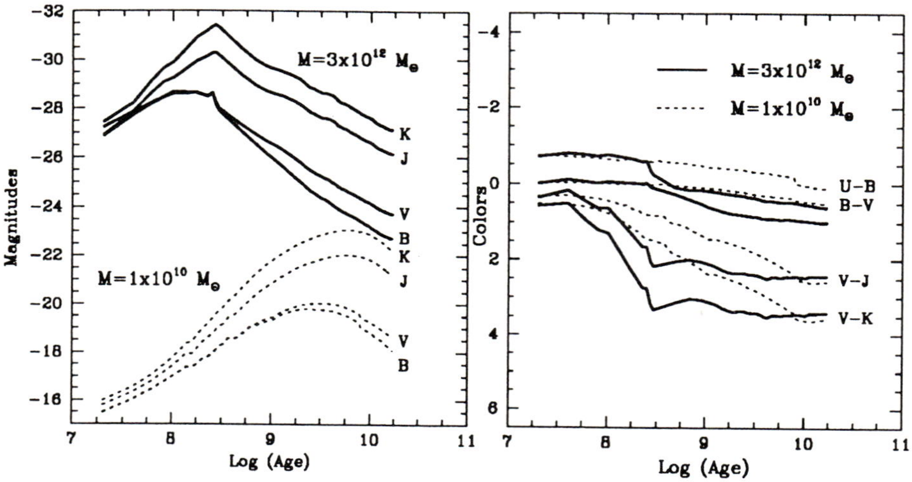

Fig. 7-8. Fig.7 (left): Luminosity evolution of models for spirals (dashed lines) and ellipticals (solid lines). **Fig. 8** (right): Colour evolution of models for spirals (dashed lines) and ellipticals (solid lines)

scales of mass accretion (about 0.1 Gyr) – this is approximately the age interval during which star formation occurs (an initial spike) – and allowing for the occurrence of galactic winds triggered by the energy deposit into the interstellar gas due to supernova (I and II) explosions of and stellar winds from massive stars. When the energy storage overwhelms the gravitational binding energy of gas this is supposed to be expelled from the galaxy thus halting further star formation.

Spirals and Ellipticals in the Infrared. Figs. 7 and 8 show the time dependence of the integrated magnitudes B, V, J, and K, and integrated colours (U–B), (B–V), (V–J), and (V–K), respectively.

Spiral galaxies reach their peak luminosity at about 5 Gyr. The range of magnitude variations during the last 10 Gyr is about 2 mag in all the pass-bands. The colours (U–B), (B–V), (V–J), and (V–K) monotonically increase toward the red at increasing galaxy age.

Elliptical galaxies evolve passively. They reach the peak luminosity at about 0.2-0.4 Gyr and since then steadily fade. The range of magnitude variation is about 65÷6 mag in the blue bands B, V and about 4÷ mag in the infrared pass-bands J and K. The colours (U–B), (B–V), monotonically increase toward the red at increasing galaxy age, whereas the infrared colours (V–J) and (V–K) flash to red at the onset of the AGB phase of the stars formed in the rather short initial period of activity and then remain flat for the remaining galaxy life (fully mimicking the behaviour of a SSP with suitable metallicity, i.e. the mean value of the stellar mix).

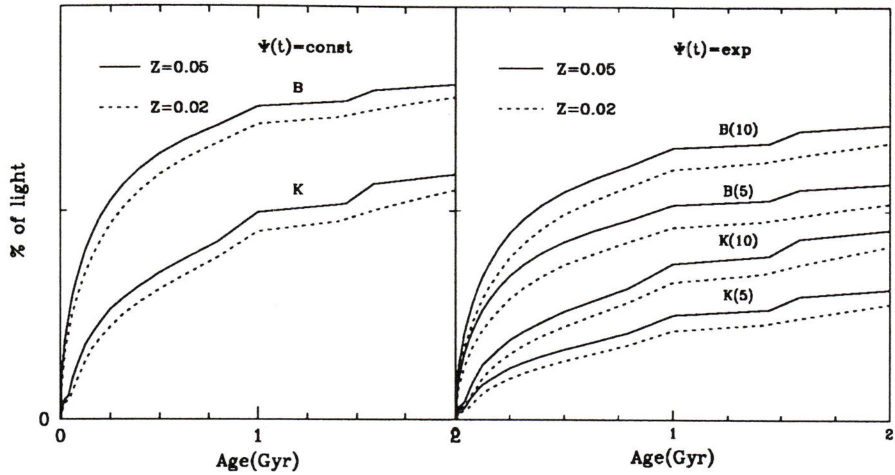

Fig. 9-10. Fig.9 (left): Fractionary contribution to the B and K light from stars younger than a certain value of the age in a galactic model with constant metallicity as indicated and constant $\Psi(t)$. Fig.10 (right): The same as in Fig.9 but for model galaxies with exponentially decreasing $\Psi(t)$ of the e-folding time scale as indicated in brackets

8 How much do young stars contribute to the light in different pass-bands ?

The easiest way of looking at this problem is by means of the SSPs. Following Ritz (1995), for any value of the age t we define the contribution from stars younger that t with respect to the total in any given pass band λ as

$$\frac{F_\lambda(<t)}{F_\lambda(T)} = \frac{\int_0^t \Psi(\tau)F_{sp,\lambda}(\tau)d\tau}{\int_0^T \Psi(\tau)F_{sp,\lambda}(\tau)d\tau}$$

where $F_{sp,\lambda}$ is the current flux of a SSP, T is the age of the galaxy, and $\Psi(t)$ is the star formation rate in suitable units. Using the above formalism, we implicitly reduce the stellar mix of a real galaxy to a unique SSP of fixed composition. The extension to an admixture of many SSPs with different chemical compositions is trivial and not of interest here. The results for $\Psi(t)$=constant and $\Psi(t) = exp[-(T-t)/\tau]$ with $\tau = 5$, 10 Gyr are shown in Figs. 9 and 10, respectively, limited to the B and K pass-bands. In our examples with constant or mildly declining star formation rate ($\tau = 10$ Gyr), the contribution from stars younger than about 0.5 Gyr can be as high as 50-60 % (B band) and 20-25 % (K band) for solar composition. These percentages vary with the metallicity.
Detailed tabulations of magnitudes and colours for SSPs with different composition and age are available from C. Chiosi upon request.

This study has been financially supported by the Italian Ministry of University, Scientific Research and Technology (MURST) and the Italian Space Agency (ASI).

References

Alongi, M., Bertelli, G., Bressan, A., Chiosi, C., Fagotto, F., Greggio, L., Nasi, E. (1993): A&AS 97, 851

Bertelli, G., Bressan, A., Chiosi, C., Fagotto, F., Nasi, E. (1994): A&AS 106, 275

Bessell, M. S., Brett, J. M., Scholz, M., Wood, P. R. (1989): A&AS 77, 1

Bessell, M. S., Brett, J. M., Scholz, M., Wood, P. R. (1991): A&AS 89, 335

Bica, E., Claria, J. J., Dottori, H., Santos, Jr. J. F. C., Piatti, A. (1995): A&A, in press

Bower, R. G., Lucey, J. R., Ellis, R. S. (1992): MNRAS 254, 601

Bressan, A., Chiosi, C., Fagotto, F. (1994): ApJS 94, 63

Bressan, A., Chiosi, C., Tantalo, R. (1995): A&A, submitted

Bressan, A., Fagotto, F., Bertelli, G., Chiosi, C. (1993): A&AS 100, 647

Burstein, D., Bertola, F., Buson, L. M., Faber, S. M., Lauer, T.R. (1988): ApJ 328, 440

Carraro, G., Girardi, L., Bressan, A., Chiosi, C. (1995): A&A, in press

Chiosi, C. (1981): A&A 83, 206

Chiosi, C., Bertelli, G., Bressan, A. (1988): A&A 196, 84

Chiosi, C., Bertelli, G., Bressan, A. (1992): ARA&A 30, 305

Fagotto, F., Bressan, A., Bertelli, G., Chiosi, C. (1994a): A&AS 100, 647

Fagotto, F., Bressan, A., Bertelli, G., Chiosi, C. (1994b): A&AS 104, 365

Fagotto, F., Bressan, A., Bertelli, G., Chiosi, C. (1994c): A&AS 105, 39

Ferguson, H. C., Davidsen, A. F., Kriss, G. A., et al. (1991): ApJ 382, L69

Fluks, M. A., Ples, B., Thé, P. S., de Winter, D., Westerlund, B. E., Steenman, H. C. (1994): A&AS, 105, 311

Girardi, L., Chiosi, C., Bertelli, G., Bressan, A. (1995): A&A 298, 87

Ritz, H.-W. (1995): ApJ, in press

Kurucz, R. (1992): private communication

Matteucci, F. (1991): in Frontiers of Stellar Evolution, D. L. Lambert (ed.), ASP Conference Ser. vol. 20, p. 439

Pagel, B. E. J., Simonson, E. A., Terlevich, R. J., Edmunds, M. G. (1992): MNRAS 255, 325

Persson, S. E., Aaronson, M., Cohen, J. G., Frogel, J. A., Matthews, K. (1983): ApJ 266, 105

Renzini, A. (1986): in Stellar Populations, C. A. Norman, A. Renzini, M. Tosi (eds.), Cambridge: Cambridge University Press, p. 213

Renzini, A., Buzzoni, A. (1986): in Spectral Evolution of Galaxies, C. Chiosi & A. Renzini (eds.), Dordrecht: Reidel, p. 195

Tantalo, R., Chiosi, C., Bressan, A., Fagotto, F. (1995): A&A, submitted

II

IR OBSERVATIONS OF LOCAL GROUP GALAXIES

The Brightest Stars in the Galactic Bulge

Patricia Whitelock

South African Astronomical Observatory, P O Box 9, Observatory, 7935, South Africa. e-mail: paw@saao.ac.za

Abstract. The brightest stars in the Galactic Bulge are Mira variables and OH/IR stars. It is now potentially possible to use these stars as probes of low-mass populations in other spiral galaxies. The properties of these large-amplitude long-period variables are reviewed with particular emphasis on the period-luminosity and period-colour relationships. Figures are given for the maximum bolometric and K luminosity of Miras in the Bulge. The period-colour relations differ in different metallicity environments. Although the same may be true of period-luminosity relations, current evidence is inconclusive on this point.

1 Introduction

In the context of this meeting we are interested in the galactic Bulge from the point of view of how it compares with, and how it helps us to understand, the bulges of other spiral galaxies. Of particular interested in recent years has been the possibility of resolving and studying individual stars in the bulges of other galaxies. So we naturally ask the questions: what are the brightest stars in the Galactic Bulge, what is their parent population, what is their evolutionary condition and how can we use them as probes of the structure and kinematics of the Bulge? These are the issues I attempt to address in this review.

The most luminous stars in the Bulge, both bolometrically and in the infrared, are the Mira variables and OH/IR stars which I will refer to collectively as Miras. I want to concentrate on Miras in the outer Bulge but I will say a few words about similar stars found near the Galactic Centre. The review does not cover the supergiants which are found only very close to the Galactic Centre (see Genzel's contribution to these proceedings). I want to emphasise the fact that the stars discussed here are low-mass asymptotic giant branch (AGB) stars with absolute bolometric magnitudes, $M_{bol} > -6$.

They should not be confused with the high-mass supergiant variables with $M_{bol} < -7$. Miras and supergiants are different beasts and there is no such thing as a "supergiant Mira"—a confusing name which has just made its appearance in the literature.

Low and intermediate mass stars reach their maximum luminosity at the tip of the AGB just prior to their evolution across the HR diagram to become planetary nebulae and white dwarfs. At this stage, at least within metal-rich populations, an instability in the stellar atmosphere results in large-amplitude long-period variations which make the star easily identifiable as a Mira variable. In fact Miras have by far the largest visual and near-IR amplitudes of any type of pulsating star. The K (2.2μm) light curves of Bulge Miras are illustrated in Fig. 4 of Whitelock et al. (1991). The K amplitudes range from about 0.4 mag to over 2 mag. In visual light the amplitudes are even larger, ranging from 2.5 to over 10 mag. So, very accurate photometry is unnecessary in deciding if such stars are variable. However, several observations spaced over a year or two are essential. If accurate near-infrared photometry is available then Miras can be separated from others stars with a high degree of success on the basis of their colours alone.

2 Miras in the Bulge

It is generally assumed that the Bulge contains no carbon-rich Miras, although the oxygen dominance has been confirmed for only a small fraction of the known variables. The following discussion refers to oxygen-rich stars except where carbon-Miras are explicitly mentioned.

The near-infrared colours of the Miras are distinctive as can be seen from the $J-H/H-K$ two-colour diagram for Miras in the galactic Bulge (Fig. 1). The unusual colours are the result of strong water-vapour absorption in the IR spectrum and, in the most extreme cases, reddening due to circumstellar shells. The shells are the result of heavy mass loss and there is a strong correlation between mass-loss rate and $K-[12]$ colour, as illustrated in Fig. 2, which enables a good estimate of the rate to be obtained from a measurement of the combined near-infrared and IRAS colour.

One of the most useful properties we can measure for a Mira is its pulsation period. There are period-luminosity, -colour and -metallicity relationships —all potentially rather powerful. Within the globular clusters, Miras are found only in the more metal-rich environments ($\log z \gtrsim -1.0$) and the period of the Mira seems to be, at least approximately, a function of the metallicity of the parent cluster (Feast 1981). NGC 5927 is the cluster containing the longest period Mira known (viz. V3 with a period of 310 day although its membership of the cluster is not beyond dispute); it has a metallicity slightly less than solar. Our knowledge of the period distribution of Miras in the Bulge is rather limited as there are only a few regions for which there are complete samples; these were discussed by Whitelock (1993). The period

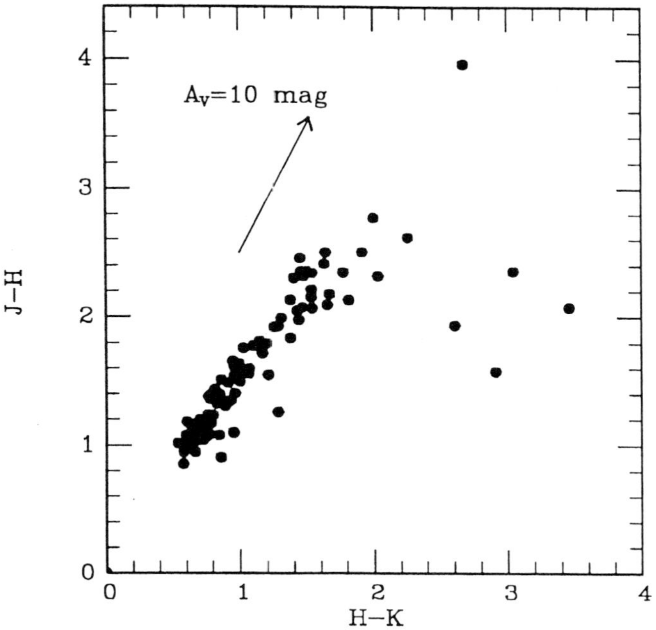

Fig. 1. A near-infrared two-colour diagram showing the IRAS Miras from Whitelock et al. (1991)

distribution of Miras in the Bulge covers the same range as does that of the clusters (150 to 310 day) but extends to much longer periods (over 500 day). Slightly larger numbers of long-period pulsators are found at lower latitudes, with periods up to 700 day, although the majority of the stars have periods less than 500 day. A very similar distribution is found for Miras in the solar neighbourhood (e.g., Whitelock et al. 1994). It is only in the galactic plane (and Centre) that stars with periods in excess of 700 day are found and even there they are rare. Notice that the period distribution in the Bulge is unlike that of the clusters; the maximum period found among the clusters is close to the median of the Bulge distribution. Another point to note is that the longest period stars, which are also the most luminous will have had more massive progenitors than their lower luminosity counterparts, and at least some of them must be the result of binary mergers rather than single star evolution (Renzini & Greggio 1990).

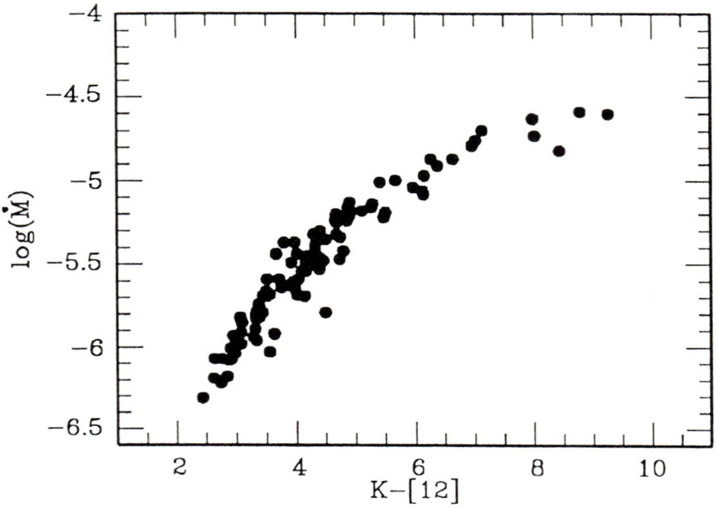

Fig. 2. The mass-loss rate as a function of the infrared colour for the IRAS Miras from Whitelock et al. (1991)

3 The Period-Luminosity Relation

The period luminosity (PL) relation for Mira variables is critical to their use as distance indicators and probes of the structure of the Bulge. Although the PL relation is the most important method for determining the distances of Miras, it is still a source of considerable confusion. This arises because of suggestions that there are theoretical reasons for expecting metal-rich Miras to be fainter than their less metal-rich counterparts and that we might therefore anticipate that Miras from the LMC will be brighter than Miras in the solar neighbourhood (e.g., Wood 1990; Jura & Kleinmann 1992). The problem is complicated by the fact that the only Miras for which the metallicities are known are those in globular clusters with known abundances; there is as yet no direct way of measuring the metallicities of the Miras themselves. In fact although we know a considerable amount about the metallicity differences between young objects in the LMC and the solar neighbourhood, we know rather little about the metallicity of the older stars and of the Miras in particular. Whitelock et al. (1994) and Whitelock (1995) discussed the PL relation in some detail and concluded that there is as yet no justification for applying corrections to the observed PL relation in order to use it in different environments.

Wood (1995) has shown that the PL relation derived for LMC Miras also fits stars from the SMC (his Fig. 5). The large line-of-sight depth to the SMC is responsible for the scatter and makes this a somewhat less sensitive test than it might otherwise be. Notice that Wood's diagram includes both carbon- and oxygen-rich Miras In fact carbon- and oxygen-rich stars fit the same PL relation at 2.2 μm. Wood has suggested that this agreement of the PL relations among stars from different environments is contrary to theoretical expectations. Let's look a bit more closely at this idea and in particular at the expression derived by Feast (1995) from Wood's (1990) AGB relation and the pulsation equation ($P\sqrt{\rho} = \text{const}$):

$$M_{bol} = 2.88 - 2.04 \log P + 0.73 \log z - 2.05 \log M + 2.04 \log Q,$$

where P, z and M are respectively the period, metallicity and mass of the star and Q is the pulsation constant for first overtone pulsation. In fact this equation suggests no clear-cut prediction for what will happen at different z values. In particular it is not reasonable to assume that, if we increase z while leaving M constant, P will remain the same and that a fainter bolometric magnitude will be the inevitable consequence. Stars with the same initial mass and different z will evolve differently; they will, e.g., have different mass-loss histories. There is therefore no a priori reason to assume that they will have the same mass at the point at which they become Miras. It would seem that we need a more holistic understanding of these late stages of stellar evolution to make any sensible predictions of the effects of metallicity on the PL relation.

4 The Period-Luminosity Relation for Bulge Miras

We would like to make a direct comparison of Miras in the Bulge with those from the LMC to test the PL relation. The large line-of-site depth of the Bulge makes it misleading to simply plot the luminosities of individual stars on a PL relation. We must compare the distribution of magnitudes with models, as discussed below. The only place where this is unnecessary is very close to the Galactic Centre where we might reasonably assume the stars are at the distance of the Centre. As this potentially simplifies the problem it is worth considering these Miras first.

4.1 The Miras Near the Galactic Centre

Lindqvist et al. (1992) discovered 134 OH-sources within 100pc of the Centre and van Langeveld (1992), and later Jones et al. (1994), measured luminosities and periods for a few of them. They seem to be rather different from the Miras further out in the Bulge in that they include stars with periods up to 1000 day. This is presumably indicative of rather more massive stars than are found in the outer Bulge, which could be the consequence of a younger population or the result of numerous binary mergers in the dense environment around the Centre. The luminosities measured by van Langeveld suggest either that the distance to the Galactic Centre exceeds 10 kpc, or that the Miras near the Galactic Centre are distinctly fainter than an extrapolation of the LMC PL relation would imply. In contrast, the luminosities measured by Jones et al. are considerably brighter. Because there still seems to be some uncertainty in the luminosities and periods of these stars it is premature to conclude that they are very much fainter than an extrapolation of the LMC PL relation. To be fair to everyone involved I should make it clear that these are very difficult measurements to make. The crowding in the Galactic Centre is horrendous and the corrections for interstellar extinction, which typically amounts to tens of visual magnitudes, are extremely uncertain. Nevertheless it is well worth attempting a more detailed study of these sources despite the difficulties.

4.2 Miras in the Bulge

Glass et al. (1995) fitted models to the distribution of distance moduli of Miras in the Sgr I window of the Bulge. Using the LMC PL relations they deduced distances to the centre of the stellar distribution of 8.7 kpc and 9.0 kpc from the K and M_{bol} relations, respectively, with an uncertainty, which is difficult to tie down because of systematic effects, of the order of 0.7 kpc. Thus, *if the distance the Galactic Centre is 8.0 kpc*, it appears that the Bulge Miras are fainter than their Magellanic Cloud counterparts; by about 0.2 to 0.3 mag. However, the question of the distance to the centre is still sufficiently uncertain to preclude the possibility of any definitive statement on the subject.

5 Period-Colour Relationships

One of the interesting things to come out of the recent Bulge study is a difference in the period-colour relations between Miras in the LMC and in Sgr I (Glass et al. 1995, Feast 1995). This is illustrated in Fig. 3 for $J - H$ and $H - K$. The straight lines are fits to the LMC data from Feast et al. (1989). In the Bulge there are stars with large $J - H$ that fall well above the line. These are stars with moderately thick dust shells for which the colours are modified by significant circumstellar reddening. The bulk of the rest of the Bulge Miras fall below the LMC line in the $J - H$ diagram and above it in the $H - K$ one. Thus the effect cannot be rectified simply by changing the magnitude of the (uncertain) reddening correction for the Bulge stars, as that will affect both colours in the same sense.

Feast (1995) has recently derived a new effective temperature-colour calibration for the Miras based on measured angular diameters. Using this together with an expression derived by combining Wood's AGB models with the pulsation equation Feast suggests that the different period-colour relations can be understood if LMC Miras with periods in the 200 to 300 day range have a mean metallicity, $\log z \sim -0.6$, while those in the Bulge, in the same period interval, have $\log z \sim -0.2$.

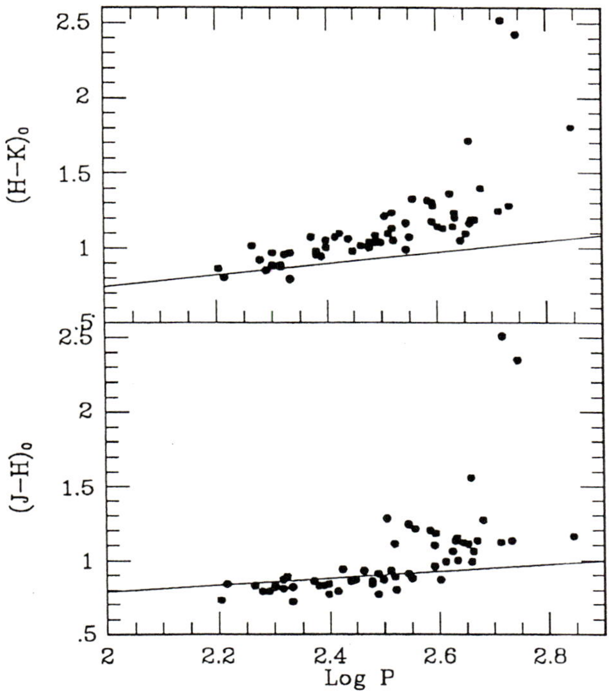

Fig. 3. The colours of Bulge Miras (Glass et al. 1995). The straight lines are least squares fits to the colours of the LMC Miras (Feast et al. 1989). This explanation of these colour differences are discussed by Feast (1995)

A detailed comparison of these results with the colours of Miras in the solar neighbourhood, which is complicated by uncertainties in the interstellar reddening, has not yet been completed. But Fig. 4 shows the colours of IRAS-selected Miras in the South Galactic Cap, which we can assume are unaffected by interstellar reddening. These are very similar to the colours of the Bulge Miras and different from those of the LMC and globular cluster stars. There are of course rather few short-period Miras in this IRAS-selected sample and it is possible that short-period stars will have colours more like those of the cluster Miras; this point is under investigation. One note of caution must be registered with reference to what is meant by metallicity. The Bessell et al. (1989) models with which Feast made his colour comparisons and deduced metallicity differences between the various populations use a simple scaling of the solar abundances. The colours of the Miras are likely to be strongly influenced by the oxygen abundance via the steam absorption features in the near-infrared.

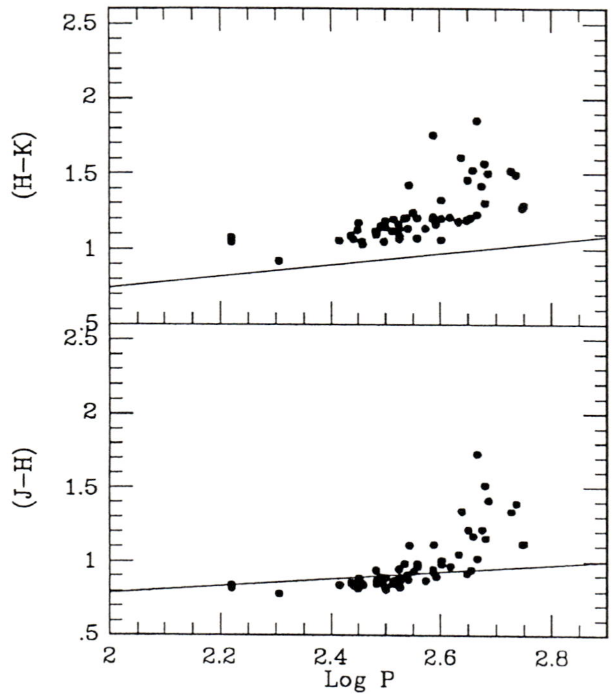

Fig. 4. The colours of IRAS selected Miras from the South Galactic Cap (Whitelock et al. 1994) The straight lines are least squares fits to the colours of the LMC Miras (Feast et al. 1989).

6 The Structure of the Bulge

Miras have played a key role in the recent discovery of the barred structure associated with the Bulge. Nakada et al. (1991) examined the distribution of colour-selected IRAS sources (most of which are probably Miras) and found a longitudinal asymmetry which they interpreted as a tilt in the Bulge. Whitelock & Catchpole (1992) looked at IRAS Miras, for which Whitelock et al. (1991) had determined periods and luminosities. They proved, from the number distribution as a function of distance modulus, that the half of the Bulge which is at positive galactic longitude is closer to us than the other half. This has been confirmed from recent analysis of the data from the DIRBE experiment on the COBE satellite (Dwek et al. 1995).

Menzies (1990) obtained optical spectra of the brightest of these IRAS Miras and demonstrated that at latitudes of $7^\circ < |b| < 8^\circ$ the Bulge was rotating at 70 $\mathrm{kms^{-1}kpc^{-1}}$. A large number of IRAS sources in the Bulge have been detected as SiO Masers by Izumiura et al. (1995 and references therein).

They determined a rotation velocity for the bar of 75 $kms^{-1}kpc^{-1}$, with no evidence for a departure from cylindrical rotation, although the velocity dispersion is high (76 kms^{-1}) and the number of sources in a given latitude strip small. Interestingly this rotational velocity is close to the pattern speed, 63 $kms^{-1}kpc^{-1}$, of the Bulge mass distribution as modelled by Gerhard & Binney (1993) to explain the kinematics of the gas near the Galactic Centre. The OH/IR sources near the Galactic centre are of course rotating much more rapidly at 1200 $kms^{-1}kpc^{-1}$ (Lindqvist et al. 1992), which is one reason why suggestions have been made that they might be more closely related to the disk than to the Bulge.

7 Absolute Magnitude of the Brightest Stars

Although there are Miras in the Bulge with periods up to 700 day, periods above 500 day are rare and probably the result of binary mergers and therefore unlikely to be common in the outer parts of extragalactic bulges which is where we will be looking for individual stars in the first instance. So taking a period of 500 day we expect $M_{bol} \sim -5.3$ mag, and $M_K \gtrsim -8.4$ mag. The K magnitude will be fainter if the star is surrounded by a thick circumstellar shell. These numbers assume the Miras in the Bulge have the same luminosity as those in the LMC. If they are actually fainter then the absolute magnitudes change to $M_{bol} \sim -4.9$ and $M_K \gtrsim -8.1$, or thereabouts. A number of people including Wendy Freedman, Michael Rich, Dante Minniti and their colleagues, have made detections of luminous stars in the bulges of M31 and M33. I am looking forward to seeing measurements of the pulsation periods of these stars, which will confirm that they are luminous single stars and not multiple images and enable a real comparison to be made with the Milky Way Bulge.

Acknowledgements I am grateful to Michael Feast and John Menzies for their helpful discussion and comments on this manuscript.

References

Bessell, M.S., Brett, J.M., Scholz, M., Wood, P.R. (1989): *A&AS* **77**, 1–30

Dwek, E., Arendt, R.G., Hauser, M.G., Kelsall, T., Lisse, C.M., Moseley, S.H., Silverberg, R.F., Sodroski, T.J., Weiland, J.L. (1995): *ApJ* **445** 716–730

Feast, M.W. (1981): *Physical Processes in Red Giants* Eds. Iben, I., Renzini, A., (Reidel, Dordrecht) 193–204

Feast, M.W. (1995): *MNRAS* in press

Feast, M.W., Glass, I.S., Whitelock, P.A., Catchpole, R.M. (1989): *MNRAS* **241**, 375–392

Gerhard, O., Binney, J. (1993): *Galactic Bulges* IAU Symp. 153, Eds. Dejonghe, H., Habing, H.J., (Kluwer, Dordrecht) 275–282

Glass, I.S., Whitelock, P.A., Catchpole, R.M., Feast, M.W. (1995): *MNRAS* **273**, 383–400

Izumiura, H., Deguchi, S., Hashimoto, O., Nakada, Y., Onaka, T., Ono, T., Ukita, N., Yamamura, I. (1995): *ApJ* in press

Jones, T.J., McGregor, P.J., Gehrz, R.D., Lawrence, G.F. (1994): *AJ* **107**, 1111–1119

Jura, M., Kleinmann, S.G. (1992), *ApJS* **79**, 105–121

Lindqvist, M., Winnberg, A., Habing, H., Matthews, H.E. (1992): *A&AS* **92**, 43–62

Menzies, J.W., (1990): *Bulges of Galaxies* Eds. Jarvis, B.J., Terndrup, D.M. ESO-CTIO Workshop No. 35 (ESO) 115–117

Nakada, Y., Deguchi, S., Hashimoto, O., Izumiura, H., Onaka, T., Sekiguchi, K., Yamamura, I. (1991): *Nature* **353**, 140–141

Renzini, A., Greggio, L. (1990): *Bulges of Galaxies* Eds. Jarvis, B.J., Terndrup, D.M. ESO-CTIO Workshop No. 35 (ESO) 47–63

van Langevelde, H.J. (1992): thesis (Leiden)

Whitelock, P.A. (1993): *Galactic Bulges* IAU Symp. 153, Eds. Dejonghe, H., Habing, H.J., (Kluwer, Dordrecht) 39–54

Whitelock, P.A. (1995): *Astrophysical Applications of Stellar Pulsation* Eds. Stobie, R.S., Whitelock, P.A. ASP Conf. Ser. **83**, (ASP Provo) 165–175

Whitelock, P.A., Catchpole, R.M., (1992): *The Center, Bulge, and Disk of the Milky Way* Ed. Blitz, L. (Kluwer,Dordrecht) 103–110

Whitelock, P.A., Feast, M.W., Catchpole, R.M. (1991): MNRAS **248**, 276–312

Whitelock, P.A., Menzies, J.W., Feast, M.W., Marang, F., Carter, B.S., Roberts, G., Catchpole, R.M., Chapman, J. (1994): MNRAS **267**, 711–742

Wood, P.R. (1990): *From Miras to Planetary Nebulae: Which path for Stellar Evolution?* Eds. Mennessier, M.O., Omont, A. (Editions Frontiéres, Gif-sur-Yvette) 67–84

Wood, P.R. (1995): *Astrophysical Applications of Stellar Pulsation* Eds. Stobie, R.S., Whitelock, P.A. ASP Conf. Ser. **83**, (ASP Provo) 127–138

Imaging of Local Group Bulge Populations in the Near Infrared

R. Michael Rich[1,2]

[1]Department of Astronomy, Columbia University, 538 W. 120th St., NY NY 10027, USA [2]Alfred P. Sloan Fellow

Abstract. Infrared imaging from the ground and using the Hubble Space Telescope continues to find evidence for evolved luminous stars that cannot plausibly arise from a globular cluster-aged stellar population. Giants with $M_{bol} < -5$ are found in M31, M32, and M33, and the HST imaging strongly supports the idea that ground-based observers are not being confused by blended images of stars. On the other hand, a direct measurement of the age of the Galactic bulge (in Baade's Window) finds evidence that our bulge is as old as the halo globular clusters. Yet surveys of the Galactic bulge continue to find large numbers of long period Miras, which should in principle be younger objects. There is likely to be a wide range in the ages of bulge populations; we point out the case of M33 where there is almost certainly a large age range in the populations of the central 100 pc.

1 Introduction

In the original stellar populations paradigm, bulges of galaxies are old (even metal poor) and considered to be candidates for the sight of the first burst of star formation. From an intuitive standpoint, this makes logical sense. The center of a proto galaxy should have the highest density and the greatest turbulence, conditions that should favor a burst of star formation. The collapse time is $\sim 1/\sqrt{\rho}$; star formation should begin in the central regions first. In recent years, Renzini (1992, 1993, 1994) has championed this traditional viewpoint, which appears to be confirmed by the most sophisticated cosmologial simulations (e.g. Steinmetz & Müller 1994). The age differences that people argue about are important in formation history: Is the bulge 2-5 Gyr younger than the halo, therefore formed much later? Unfortunately, the difference between a 10 and 14 Gyr old metal rich population is extremely difficult to detect in the integrated spectrum, or even in the color-magnitude diagrams of resolved populations. As a result of this situation, the present lively state of lively debate has evolved.

Many properties of bulges distinguish themselves from the globular clusters that are well studied and are accepted to have the oldest datable stars. It has been known since the days of early satellite flights (Code & Welch 1979) and ground-based photometry (Aaronson et al. 1978) that elliptical galaxies have a hot ultraviolet rising flux that is generally not accounted for in the globular clusters. Further, Johnson (1966) was the first of many to point out that ellipticals have a cool, luminous stellar component not found in the clusters. A wider abundance range, and particularly the presence of old, metal rich stars can explain both features (cf. Renzini 1992), yet the differences should give us pause in considering these issues.

Of greater concern for the age issue are the long period Mira variables and the population of luminous AGB stars in the Milky Way spread over more than 100pc (Catchpole *et al.* 1990, Glass *et al.* 1995), an issue I will take up at length later (and which forms much of the basis for Rich's (1992) arguments in *favor* of a bulge that is younger than the extreme halo population.

Another observation that favors a bulge younger than the halo is the Solar abundance of s-process elements (made in earlier generations of AGB stars) over the entire bulge abundance range (McWilliam & Rich, 1994). If the bulge formed first in an early rapid starburst, some generations of stars would have formed in the first 10^8 yr before the first white dwarfs came into being; such hypothetical stars would entirely lack s-process elements, yet they are not found.

In addition, "Jeans-like" instabilities have recently been discovered that will cause a cold bar to vertically thicken, asuming the appearance of a peanut-shaped bulge (cf. Pfenniger & Norman, 1990; Raha *et al.* 1991; Merritt & Sellwood, 1994). This at once explains the prevalence of triaxial bulges, and opens the possibility for a number of mechanisms that could make a bulge from a starburst.

Can bulge formation scenarios be constrained by imaging of high-redshift galaxies? The restored HST produces clear images of spirals at $z = 0.5$, half the Hubble time; bulges can still be discerned at z=1. Lilly et al. (1996) have taken the critical step of obtaining HST images of a large sample of field galaxies with redshifts. At $z \approx 0.5$, half of the "disk" galaxies have red bulges; some well-formed spirals were in place 5-8 Gyr ago (Schade *et al.* 1995). However, many luminous disk galaxies are seen with strong blue central condensations, and it is interesting to speculate that these might be proto-bulges. The problem with lookback studies is that we cannot trace the evolution of any individual galaxy in time, a major problem if bulge formation is extended over a full Gyr. So the lookback studies will have to be complemented by inferences gained from the study of the Milky Way and its near neighbors. As we extend our gaze beyond our own galaxy, the infrared is critical in the study of stellar populations.

1.1 Why the Problem Is Difficult

For populations older than 10 Gyr, the main sequence turnoff isochrones crowd together, so that relative age differences of < 2 Gyr are (in practice) indistinguishable. Even with HST, at M31 the turnoff point is at $V = 29$; turnoff age dating for Local Group bulge populations is unlikely to ever occur. At the bright end, the bolometric magnitudes and pulsation periods of luminous asymptotic giant branch (AGB) stars offer a more measurable indicator of intermediate-aged populations. In a thorough discussion of this issue, Renzini (1992) raises the point that in metal rich populations, extended giant branches may be the result of extreme He and metal abundances, or may be the evolved progeny of blue stragglers (merged binaries that evolve as more massive stars). However, AGB stars are luminous in the IR, and they give us our best hope for finding signs of intermediate-age stars in bulges. Therefore, we will proceed to discuss what has been learned from ground and space-based studies of these populations.

2 Ground-Based Near-IR Imagery

2.1 The Milky Way Bulge

Frogel & Whitford (1987) obtained IR photometry of Blanco, Blanco, & McCarthy's (1984) optically-selected sample of M giants in Baade's Window. This study has dominated thinking about the bulge population; few bulge giants appeared with $M_{bol} < -4$, the nominal cutoff for the oldest stellar populations. However, this study did not include the reddest stars in the Milky Way bulge, such as the Miras observed by Whitelock *et al.* (1991).

Large scale surveys of OH/IR stars by Habing and his students (e.g. Blommaert 1992) find large numbers of OH/IR stars toward the bulge; some are bolometrically bright and no counterparts are found in Galactic globular clusters. Surveys of the Miras (Whitelock *et al.* 1991; Glass *et al.* 1995) find that about half the Miras have periods longer than 300 days. Since the work of Feast (1963) it has been known that in the Solar neighborhood, such long period stars rotate with the disk population; only short period Miras have halo kinematics. Following McWilliam & Rich (1994), we know that the mean abundance of the bulge is not vastly different from the Solar neighborhood, thus removing the most attractive means of explaining away Miras that run up to 700 days in period. Not all can be evolved blue stragglers; some of the longest period stars may be, but if all metal rich stars evolve to become visible as Miras, then 1% or less of all Miras should be the progeny of blue stragglers.

The most striking observation of the bulge, however, is the Catchpole *et al.* (1990) infrared survey of the Galactic Center region. Luminous AGB stars ($M_{bol} < -5$) are markedly concentrated toward the Galactic Center, over scales of $\approx$ 100pc. While the distribution of these stars is flattened, it is hard to attribute this population to young stars in the plane or associated with the Galactic Center. Taken with the evidence from the Miras, one may conclude that there must be an intermediate-age population in the bulge.

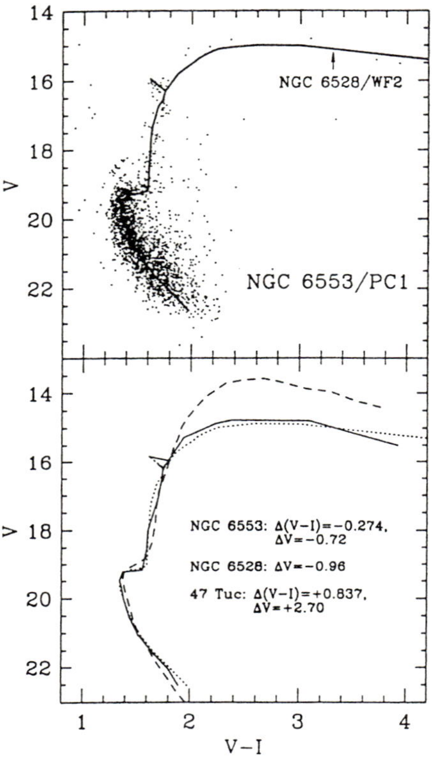

Fig. 1. The metal rich globular clusters in the bulge are as old as 47 Tuc. *Upper panel:* NGC 6553 main sequence photometry observed with WFPC2 (Ortolani *et al.* 1995). *Lower pane:* V, I loci for the metal rich clusters NGC 6528 and 6553 align nearly perfectly with 47 Tuc. $\Delta V_{TO}^{HB} = 3.6 \pm 0.1$mag, a value typical of the oldest Galactic globular clusters.

While the HST main sequence photometry in Baade's Window also seems to show an intermediate age population (Holtzmann *et al.* 1993), Ortolani *et al.* (1995) have found striking new evidence that the bulge (at least in the Baade's Window field) is as old as the Galactic Center globular clusters. Figure 1 illustrates the most important conclusion, that the turnoff to horizontal branch separation (ΔV_{TO}^{HB} is the same in the Solar metallicity clusters NGC 6553 and 6528 as it is in an old halo cluster such as 47 Tuc. The luminosity function of the bulge field is then compared with the clusters by tying field and cluster at the HB, so as to eliminate reddening and distance uncertainties. Allowing for the spatial distribution of the bulge along the line of sight, one still finds that the age of the bulge is identical to that of the clusters and that its age dispersion must be $< 10\%$. If the bulge is indeed as old as the halo, in addition to the implcations for galaxy formation one must ask why the AGB indicators seem to be telling a different story. Whitelock's Miras, for example, are found 7-8° from the nucleus. If the evolved stars fail us in the Milky Way, we can hardly have any faith in their reliability in more distant populations.

2.2 Local Group Bulges in the Near-IR

Ground-based near-infrared imaging finds consistent evidence for luminous AGB stars in Local Group bulges. The first imaging of the M31 bulge by Rich & Mould (1991) was followed by discovery of luminous giants in M32 by Freedman (1992) and confirmed by Elston & Silva (1992). Minniti *et al.* (1993) found a bulge population of luminous AGB stars in M33, confirmed by McLean & Liu (1994). Responding to suggestions that their infrared-bright stars belong to the disk (Davies *et al.* 1991), Rich *et al.* (1993) imaged 5 fields with varying bulge/disk population in M31, once again finding strong evidence for stars with $M_{bol} < -5$ associated with the M31 bulge. Assertions that these luminous stars are actually unresolved blends have appeared in the literature (Depoy *et al.* 1993) and will probably only be definitively settled when the ongoing *HST* projects described below in §3 are completed. However, it is instructive to look at Minniti & Bedding's (1994) figure, which shows that the J, K color-magnitude diagrams of the 3 Local Group populations look remarkably similar.

While one cannot ignore the possibility that these stars are blended images, we note that it requires blends of more than two equally bright stars to get boosts of 1 mag. Further, Rich (1993) did extensive artificial star and artificial stellar population simulations and measurements, and confirmed the presence of an extended giant branch in his field. Ground-based efforts are currently underway to use adaptive optics in the K band to resolve this problem; this will work only if the correction is nearly perfect, without large wings to the stellar profiles (the analogy of pre-repair *HST* point spread functions). The best images by Rich, Pahre, & Mould (1995) were taken in 0.4 arcsec seeing and despite the improved seeing, the bright stars persist. While the ultimate test will come when *HST* has K-band imaging, exciting preliminary results are coming from HST.

3 *HST* Imaging of Bulge Populations

Becuase of the blending issue, there is still widespread skepticism about the reality of the "bright" AGB stars in the M31 bulge. In a fundamental conceptual advance, Bica *et al.* (1991) point out that because metal rich giants are so faint, we should see no old, metal rich stars in M31 with $I < 22$. Any resolved stars brighter than $I = 20.7$ (globular cluster giant branch tip) would be candidates for intermediate age stars. Examining Figure 1, it is clear that the NGC 6553 locus falls well below that of 47 Tuc (Ortolani *et al.* 1990); this is due to blanketing of the V band (and ultimately) I by TiO and neutral metals. Using WFPC1, Rich & Mighell (1995) find ≈ 500 stars with $I < 20$ within 150 pc of the nucleus, continuing Mould's (1986) noted tendancy for bright stars to concentrate toward the nucleus. Of course, the difficulties of dealing with WFPC1 data are well known, and the I-band WFPC2 data are under analysis at this time.

In M33, HST studies confirm Minniti *et al.* 's (1993) bulge and strongly suggest that stars with $I < 21$ (approximate globular clusters giant branch tip cutoff) are more spatially concentrated than their fainter counterparts corresponding to a globular cluster-aged population (The KS test gives > 99.5% confidence). In the case of M33, at least, there is very strong evidence for a late central burst of star formation that took place many Gyr after the formation of the oldest stars. It is only reasonable to assume that similar bursts could have occurred elsewhere in the local group.

3.1 *HST* Imaging at 1 Micron

Until the 1997 refurbishment mission, our best hope to use *HST* to image bulges in the infrared is the F1042M filter on WFPC2. Although suffering from the lowest throughput of the non-UV filters, transmission peaks longward of 1 μm so that the coolest giants can be isolated. In collaboration with Ken Mighell and Wendy Freedman, I have imaged the bulges of M31, M32, and M33 thorugh this filter. Since calibration is understood to be difficult, we have used the distant but populous M31 globular cluster G1 as a template old, metal rich stellar population at the distance of M31. WFPC2 imaging of G1 (Rich *et al.* 1995) reveals an old metal rich globular cluster with a red horizontal branch, slightly more metal rich than 47 Tuc (Figure 2). Because we find no extended giant branch, we conclude that this cluster is not intermediate age.

Because we understand G1 reasonably well, we are able to interpret the F1042M = 1 μm luminosity functions of the bulges of M31 and M32, compared to G1. The preliminary analysis illustrated in Figure 3 confirms what has been seen in the I and K bands: a giant branch extended by > 1 mag. Much work remains to double check this result, but crowding alone cannot explain it (G1, at the distance of M31, is also very crowded). While spectral synthesis of the integrated light of M32 has long shown evidence for an

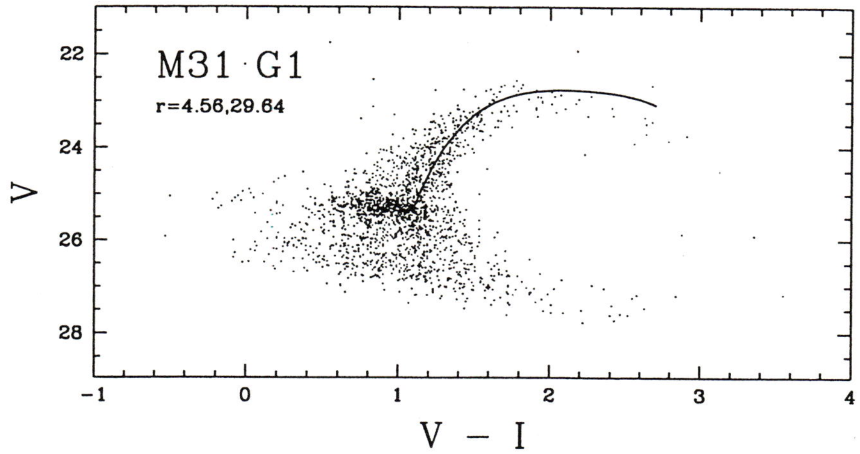

Fig. 2. WFPC2 photometry of the brightest M31 globular cluster, G1 (Rich *et al.* 1995) transformed to Cousins V, I. Solid curve is the locus of the 47 Tuc giant branch from Da Costa & Armandroff (1990). Notice that the close resemblance to 47 Tuc, including the red HB. The luminosity function of this cluster is identical to that of 47 Tuc, to 1 mag below the HB.

age spread, the M31 bulge does not show any obvious sign in its spectrum or integrated colors for even a *trace* population of intermediate age (a point emphasized by Mighell & Rich, 1995). The brightest infrared-bright stars in these two systems are also identical. One emphasizes that even with the HST data, one must still do artificial star tests and perhaps population models before results can be considered secure. We plan to do a similar analysis using the *I* band data, as soon as they are released. Most importantly, we conclude that G1 is useful as a comparison template population in assessing the luminosity functions of the Local Group bulges. The evidence at this time appears to confirm the reality of these extended giant branches, while their cause remains to be understood.

4 Conclusions

Infrared imaging of bulges in the local group has become capable of resolving individual stars near the nuclei of these galaxies. Despite recent evidence that the Milky Way bulge might be as old as 47 Tuc, the extended giant branches have not gone away (even when imaged with *HST*). At the present time, we cannot make inferences concerning relative or absolute ages based on the luminosities of these bright stars. However, there is no doubt that the stellar populations of these spheroids are worthy of careful study. If massive bulges like that of M31 are old, then we will have to find some other explanation for the extended giant branch.

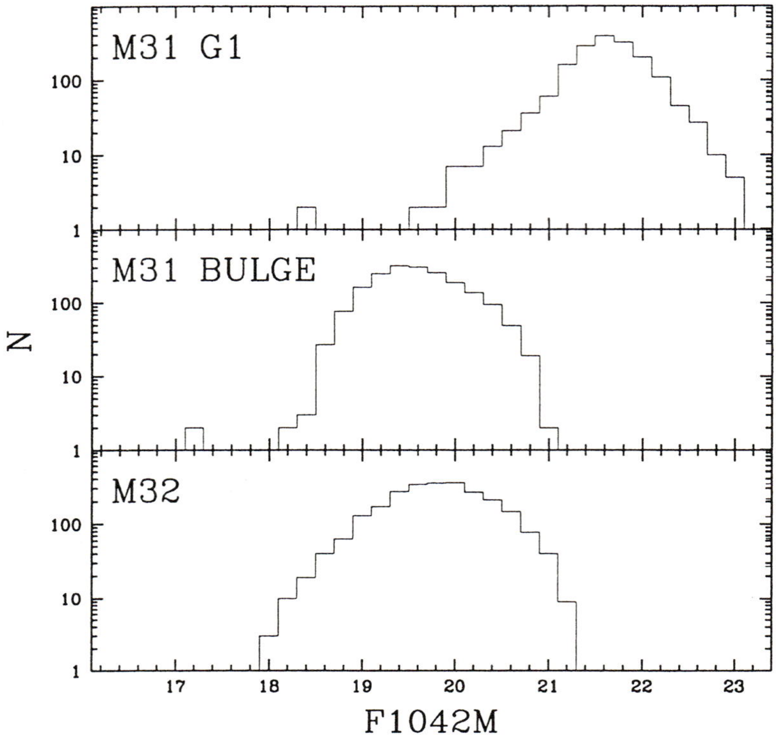

Fig. 3. WFPC2 luminosity functions in the 1 μm filter F1042M. Using M31 G1 as a template old stellar population, we can assess the reality of extended giant branches in the Local Group Populations. For M31 and M32, data are from the PC frame centered on the nucleus; all stars are within 100 pc of the nucleus. The > 1 mag extension of the bright end of the giant branch cannot be due to crowding effects. The bright tail of M32 might be due to an extended period of star formation, but we caution that this photometry is preliminary (Rich *et al.* 1995, in preparation).

In M33, there is now very strong evidence that the centrally concentrated population is many Gyr younger than the oldest halo stars. It is likely that there is a range in bulge ages, just as there seems to be for the few galaxies observed at intermediate redshift. I close this review with a WFPC2 image of the M31 double nucleus, taken with the F1042M filter. The structure found by Lauer *et al.* (1993) is seen at this wavelength, which confirms their conclusion that normal dust cannot be causing the double structure of the nucleus. This striking image gives a taste of what will come when ground and space-based imaging provides us with diffraction limited infrared images of the cores of galaxies.

RMR acknowledges support from NASA grant NAGW-2479 and STScI grant GO 5464.

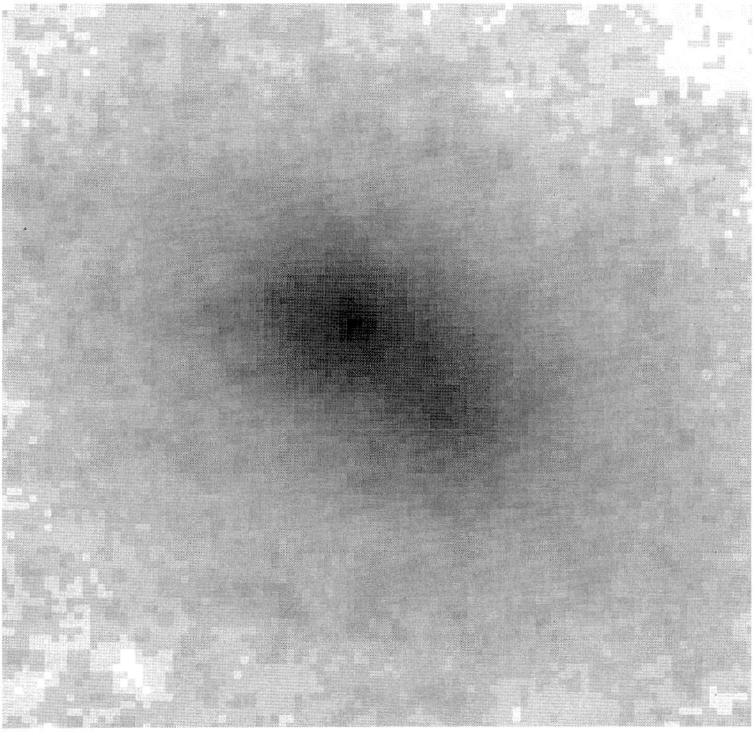

Fig. 4. WFPC2 image of the M31 double nucleus in the 1 μm filter F1042M (Rich *et al.* 1994). The double nucleus structure first noted by Lauer *et al.* (1993) is present at 1 micron, which confirms that normal dust absorption does not cause the double nucleus. A color map finds no evidence for stellar population gradients. If the brighter peak (P1) is the nucleus of a dwarf galaxy or a globular cluster, it is most peculiar that no color differences can be seen between F555W and this image.

References

Aaronson, M., Cohen, J.G., Mould, J., & Malkan, M. (1978): ApJ **223** 824–834

Bica, E., Barbuy, B., & Ortolani, S. (1991): ApJ **382**, L15–L17

Blanco, V.M., Blanco, B.M., & McCarthy, M.F. (1984): AJ, **89**, 636–647

Blommaert, J. (1991): Ph.D. Thesis, Leiden University

Catchpole, R.M., Whitelock, P.A., & Glass, I.S. (1990): MNRAS **247**, 479–497

Code, A.D., & Welch, G.A. (1979): ApJ **228**, 95–104

Da Costa, G.S., & Armandroff, T.E. (1990): AJ **100**, 162–181

Davies, R.L., Frogel, J.A., & Terndrup, D.M. (1991): AJ **102**, 1729–1733

Depoy, D.L., Terndrup, D.M., Frogel, J.A., Atwood, B., & Blum, R. (1993): AJ **105**, 2121–2126

Elston, R., & Silva, D.R. (1992): AJ **104**, 1360–1364

Feast, M.W. (1963): MRAS **125**, 367–390

Freedman, W.L. (1992): AJ **104**, 1349–1359
Frogel, J.A., & Whitford, A.E. (1987): ApJ **320**, 199–237
Holtzmann, J.A., Light, R.M., & WFPC-IDT (1993): AJ **106**, 1826–1838
Lee, Y.-W. (1992): AJ **104**, 1780–1789
Lilly, S.J., LeFevre, O. *et al.* (1996): ApJ (in press)
McWilliam, A., & Rich, R.M. (1994): ApJS **91**, 749–791
Minniti, D., Olszewski, E.W., & Rieke, M. (1993): ApJ **410** L79–L82
Minniti, D., & Bedding, T. (1995): *Science with the VLT* (eds.) J. Walsh & J. Danziger (Springer-Verlag, Berlin Heidelberg) 236–240
Mclean, I., & Liu, T. (1995): Preprint
Merritt, D., & Sellwood, J.A. (1994): ApJ **425** 551–567
Mould, J.R. (1986): *Stellar Populations*, IAU Symp. 149 (eds.) A. Renzini & M. Tosi (Cambridge University Press, Cambridge) 9–27
Johnson, H.L. (1966): ARA&A **4**, 193–215
Lauer, T., & WFPC-IDT (1993): AJ **106**, 1436-1447
Lee, Y.-W. (1992): AJ **104**, 1780–1789
Ortolani, S., Barbuy, B., & Bica, E. (1990): A&A 236, 362–370
Ortolani, S., Renzini, A., Gilmozzi, R., Marconi, G., Barbuy, B., Bica, E., & Rich, R.M. (1995): Nat (in press)
Pfenniger, D., & Norman, C. (1990): ApJ **363**, 391–410
Raha, N., Sellwood, J.A., James, R.A., & Kahn, F.D. Nat **352**, 411–412
Renzini, A. (1992): *Stellar Populations* (eds.) B. Barbuy & A. Renzini (Kluwer, Dordrecht), 325–336
Renzini, A. (1993): *Galactic Bulges*, IAU Symp. 153 (eds.) H.J. Habing & H. Dejonghe (Kluwer, Dordrecht), 151–168
Renzini, A. (1994): A&A **285**, L5–L8
Rich, R.M., & Mould, J.R. (1991): AJ **101**. 1286–1292
Rich, R.M. (1992): *The Center, Bulge, and Disk of the Milky Way* (ed.) L. Blitz (Kluwer, Dordrecht) 47–76
Rich, R.M., Mould, J.R., & Graham, J.R. (1993): AJ **106**, 2252–2280
Rich, R.M., & Mighell, K.J. (1995): ApJ **439**, l45–154
Rich, R.M., Pahre, M.A., & Mould, J.R. (1995): (in preparation)
Rich, R.M., Mighell, K.J., Freedman, W., & Neill, D. (1994): BAAS, (1994): **105** 76.08
Rich, R.M., Mighell, K.J., Freedman, W., & Neill, D. (1995): AJ submitted
Schade, D., Lilly, S.J., Crampton, D., Hammer, F., LeFevre, O., & Tresse, L. (1995): ApJ **451**, L1–L4
Steinmetz, M., & Müller, E. (1994): A&A **281**, L97–L100
Whitelock, P., Feast, M. & Catchpole, R. (1991): MNRAS **248**, 276–312

Origin of Bulges

Alvio Renzini[1,2]

[1]Dipartimento di Astronomia, Università di Bologna, Italy,
[2]European Southern Observatory, Garching b. München, Germany

Abstract. Two main scenarios have been proposed so far for the origin of galactic bulges. The more traditional view sees bulges forming at very early times as part of the spheroid formation process, that includes halo and bulge. In a more recently proposed scenario bulges form late, as a result of the dynamical evolution of a disk that is prone to develop a bar instability. Recent *HST* and *NTT* results are reported that may help chosing among these two alternatives. They show that two bulge metal rich globular clusters and the bulk of stars in a Baade's Window (BW) field have a uniformly old age, undistinguishable from that of metal poor clusters in the galactic halo. This evidence is interpreted as favoring a fast formation of the galactic bulge, at an early cosmological time. A series of other evidences are listed that indicate a seemingly early formation epoch for the bulk of stars in elliptical galaxies, and the necessity to understand how galactic *spheroids* formed so early in the evolution of the universe is emphasized.

1 Introduction

Spiral galaxies consist of a dynamically cold, rotationally supported *disk*, and of a dynamically hot, mostly pressure supported *spheroid*. The spheroid of spirals is further distinguished in its outer part – the halo – and its inner part – the bulge. Sitting at the center of both the spheroid and the disk, the bulge formation can either be related to the formation of the halo or to that of the disk and its evolution. Moreover, the well known similarity of bulges and elliptical galaxies (e.g., Jablonka, these proceedings) makes the case of even wider interest, as understanding the origin of spiral bulges may give us useful clues for elliptical galaxies as well.

There are basically two broad, competing scenarios for the origin of bulges. In one view spheroids come first, including bulges, and disks grow only later. The idea goes back to the classical paper by Eggen, Lynden-Bell & Sandage (1962), and comes in two opposite varieties. In one version star formation starts out in the halo, and progressively metal-enriched gas drips towards the center so that the innermost part of the spheroid – the bulge – forms last as the culmination of the condensation process (Larson 1974). In the other version, star formation first erupts in the central, high density regions of the protogalaxy. There it proceeds at a very high rate, rapidly enriching the gas to solar abundance and beyond, until local supernova heating quenches further star formation (Renzini & Greggio 1990; Larson 1990; Lee 1992; Renzini 1993; Stenmetz & Müller 1994). While all this would last some $\sim 10^8$ yr in the bulge, star formation in the halo may instead continue for a somewhat longer time ($\sim$ 1 Gyr), given the halo lower density and longer dynamical time. Thus, though both versions predict the bulge and halo stars to be nearly coeval, in the *inside-out* version the bulge is (on average) slightly older than the halo, while it is slightly younger in the *outside-in* picture.

In the opposite scenario, disks come first. Then, at some stage in their evolution they may develop a bar instability, the bar causes dynamical heating of the inner disk and a (peanut-shaped) bulge grows out of it (Combes *et al.* 1990; Raha *et al.* 1991; Pfenninger & Friedly 1991; Hasan *et al.* 1993; Sellwood, these proceedings). In addition, bars can funnel gas from the disk all the way to the center of galaxies, possibly leading to starburst activity resulting in a further growth of bulges (Gerhard & Binney 1993).

Thus, the basic alternative can be formulated as follows. Is the bulge of a spiral the core of the old halo, and is its formation, by and large, coeval to it? Or, is the bulge the corrupted inner part of the disk and the bulk of its its formation a rather late event in the evolution of spiral galaxies? One way of distinguishing between the two options is offered by the very different predictions concerning the distribution of stellar ages. In the *spheroid comes first* scenario bulges and halos are about coeval and very old, the same age of halo globular star clusters ($\sim$ 15 Gyr). A wide distribution of stellar ages, from $\sim$ 15 Gyr to just a few Gyr or less is instead expected if bulges are late comers, a result of the dynamical evolution of disks. Therefore, a decisive step towards distinguishing among these competing scenarios must come from age dating stars in galactic bulges.

This can be attempted from the integrated light of bulges with population synthesis methods, which however are prone to the well known age/metallicity degeneracy problem. For the bulge of the nearby spiral M31, insight into the distribution of stellar ages can be gathered from the luminosity function of bright AGB stars (Rich *et al.* 1993; Rich & Mighell 1995), though crowding problems may introduce amiguities that are difficult to remove (Renzini 1993; DePoy *et al.* 1993). Clearly, the best chance is offered by the bulge of our own Galaxy, where the main sequence turnoff (TO) is at about $V \simeq 20$ as opposed to $V \simeq 28$ in M31.

Recent stellar population studies of galactic bulge fields all agree that the bulk of stars in the bulge are very old, older than at least 5-8 Gyr, but diverge on whether the data require the presence of a population substantially younger than 15 Gyr (e.g. Rich 1985; Terndrup 1988; Frogel 1988; Whitelock *et al.* 1991; Renzini 1993, 1994; Holzman *et al.* 1993; Paczyński *et al.* 1994). Precise age dating of bulge stars is indeed complicated by several factors, including crowding, depth effects, variable reddening, metallicity dispersion, and contamination by foreground disk stars. Some of these limitations can be significantly reduced by studying globular star clusters that are physically located in the bulge, and whose high metallicity ensures that they are not halo interlopers. Indeed, cluster stars have all the same distance, age and metallicity, and their high concentration ensures a reduced contamination by disk foreground stars.

2 Dating Galactic Bulge Globular Clusters with HST

In this perspective two very metal rich bulge globular clusters NGC 6528 and 6553 have been selected for HST imaging, thus obtaining the most accurate color-magnitude diagrams (CMD) and age determinations as currently possible for specific components of the galactic bulge (Ortolani *et al.* 1995a). The two clusters are respectively located at $\sim 4^\circ$ and 6° from the galactic center, and their metallicity is about solar, i.e., [Fe/H] $\sim 0.0\pm0.3$ (Zinn 1985; Barbuy *et al.* 1992; Cohen & Sleeper 1994), i.e., close to the average for bulge stars in Baade's Window (McWilliam & Rich 1994). Their radial velocities (+160 km s^{-1} and –24 km s^{-1}, respectively for NGC 6528 and NGC 6553, Hesser *et al.* 1986) are also consistent with their bulge membership, with bulge K giants exhibiting a (radial) velocity dispersion of 105±11 km s^{-1} (Rich 1990; see also Minniti 1995).

¿From HST-WFPC2 observations Ortolani *et al.* have obtained accurate CMD for each cluster and for each of the four CCD chips of the camera. Fig. 1 (top panel) shows the CMD of NGC 6553 as sampled by the Planetary Camera chip (PC1), that combines good statistics, relatively low differential reddening, and and better photometric accuracy. Notice the stubby red horizontal branch (HB) at $V \simeq 16.2$. The modest bright extension of the main sequence above the turnoff is due to contamination of disk stars in the spiral arm at 2 kpc (Paczyński *et al.* 1994), blending of stellar images due to crowding, and (possibly) blue stragglers. In spite of larger differential reddening and somewhat worse photometry caused by the larger pixels of the WFC, very accurate mean cluster loci can be constructed for each of the three WFC chips: they are all identical to that constructed from the PC1 data, apart from the shift due to the different mean reddening in each chip.

In Fig. 1 (lower panel) we compare the mean cluster locus of NGC 6528 (WF2) to that of NGC 6553 (PC1). The two loci are remarkably identical, from the lower main sequence all the way to the tip of the RGB. In particular, the HB to main sequence TO luminosity difference $\Delta V_{\rm TO}^{\rm HB}$ is the same to within ~ 0.1 mag, which ensures that the two clusters have the same age to within $\sim 10\%$, since $\delta\ln(\rm age) \simeq \delta\Delta V_{\rm TO}^{\rm HB}$ (Renzini 1991). Ortolani *et al.* also estimate the metallicity (Z) of the two clusters to be the same within $\sim 10\%$.

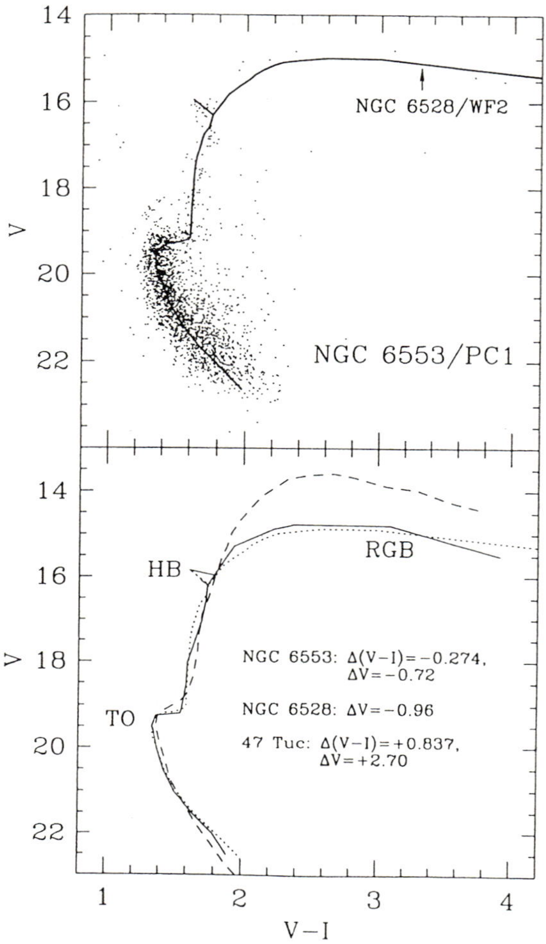

Fig. 1. *Upper panel:* The color-magnitude diagram of the cluster NGC 6553 for the WFPC2-PC1 data. The filters F555W and F814W have been used for V and I. Superimposed is the cluster mean locus obtained from the WFC2 data of NGC 6528. Notice how well the mean locus of NGC 6528 fits the NGC 6553 data. *Lower panel:* The mean cluster loci for NGC 6528 (same as in the upper panel) and NGC 6553 (from PC1), dotted and full line, respectively. The locus of NGC 6553 has been dereddened as indicated [assuming $A_V = 2.63\,E(V-I)$], so to make its TO color equal to that of NGC 6528. The locus of NGC 6528 has been shifted by $\Delta V = -0.96$ to bring the two loci to full coincidence. In this way the two loci have been reduced to the same, lowest reddening and distance (respectively that of NGC 6528 and NGC 6553). Also shown is the locus of the inner halo cluster 47 Tuc (dashed line). It has been shifted to bring into coincidence the blue end of its HB with that of the bulge clusters for the sole purpose of emphasizing the very close similarity in HB to TO luminosity difference (the age indicator), and the brighter upper red giant branch in 47 Tuc, indicating that this cluster is much less metal rich than the two bulge clusters. (Adapted from Ortolani *et al.* 1995a.)

Moreover, the true distance modulus of NGC 6553 is ~ 1 mag smaller than that of NGC 6528. Hence, if NGC 6528 is close to the galactic centre (i.e., at ~ 8 kpc from us), NGC 6553 is at a distance of 5 kpc from us, ~ 3 kpc from the galactic centre, i.e., at the outer fringe of the bulge. Considering the radial velocities, it is likely that NGC 6528 is now close to its pericenter, and NGC 6553 to its apocenter, with the orbits of the two clusters having quite similar energies.

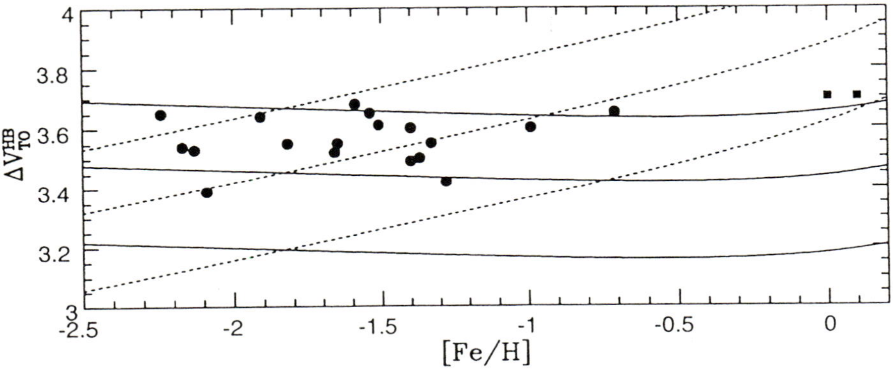

Fig. 2. The luminosity difference between the HB and the turnoff $\Delta V_{\rm TO}^{\rm HB}$ is plotted for a representative set of halo globular clusters (Renzini 1991, filled circles) and for NGC 6528 and NGC 6553 (filled squares). For this display [Fe/H]=0.0 and 0.1 has been adopted for the metallicity of the two clusters. The lines correspond to ages of 18, 15, and 12 Gyr (upper, middle, and lower lines, respectively), with the following assumptions: 1) turnoff luminosity-age relation from Renzini (1991); 2) helium abundance Y=0.23+3.0 Z; 3) no enhancement of α-elements (i.e., solar proportions); 4) HB luminosity-metallicity relations $M_{\rm V}^{\rm HB}$= 1.17+0.39[Fe/H] (Sandage & Cacciari 1990, solid lines), and $M_{\rm V}^{\rm HB}$ = 0.73+0.15[Fe/H] (Walker 1992, dotted line).

Having established their identity, what is the age of the two clone clusters? This is less easy to answer, as subtle ambiguities remain on both the observational and the theoretical side. Cluster ages are best inferred from a semi-empirical calibration of the age-$\Delta V_{\rm TO}^{\rm HB}$ relation (e.g. Renzini 1991). For halo (metal poor) globular clusters the HB luminosity is measured at the TO color, but here the HB does not extend that far to the blue, while its luminosity increases towards the blue, an effect due to a combination of differential reddening and higher TiO blanketing towards the red. While keeping this complication in mind, Ortolani *et al.* estimate $\Delta V_{\rm TO}^{\rm HB} = 3.7 \pm 0.15$ for both clusters. For comparison, Fig. 1 (bottom panel) also shows the mean locus of 47 Tuc (Ortolani *et al.* 1995b), one of the most metal rich clusters in the inner halo, with [Fe/H]$\simeq -0.7$. As it is apparent, the 47 Tuc $\Delta V_{\rm TO}^{\rm HB}$ is $\lesssim$ to that of the two bulge clusters, which ensures that the age of the three clusters must be nearly the same, thus establishing that the two bulge clusters are nearly coeval to the halo clusters.

Next comes the problem of the calibration. For illustration two *extreme* calibrations of the $M_{\rm V}^{\rm HB}$-[Fe/H] relation are displayed in Fig. 2. Not surprisingly, using the steep $M_{\rm V}^{\rm HB}$-[Fe/H] relation the bulge clusters appear slightly older than the halo, metal poor clusters. Using instead the shallow $M_{\rm V}^{\rm HB}$-[Fe/H] relation they appear some Gyr younger than the metal poor clusters, thus continuing an apparent age-metallicity trend already implied for halo clusters (e.g. Walker 1992; Carney 1992). Compelling evidence in favor of one calibration or the other is still lacking, which does not allow yet to distinguish between the inside-out and the outside-in formation options. However, HST observations of globular clusters in the Andromeda galaxy may soon decide the issue. Meanwhile, the continuity of the $\Delta V_{\rm TO}^{\rm HB}$-[Fe/H] relation from the halo all the way to the bulge implies age continuity between halo and bulge globulars, and argues for the whole spheroid being originated in a unique formation process, rather than the bulge being *added* later by a different process. Also worth noting is that the high metallicity of the two clusters was reached in a rather short time ($\lesssim$ few Gyr) over the whole bulge, up to ~ 3 kpc from the galactic center. This evidence appears to favor a fast, dissipative formation of the bulge.

3 The Age of Stars in Baade's Window

The question remains of how representative of the general population of the galactic bulge are the two metal rich clusters. The answer comes from a comparison of their CMDs with that of a field in Baade's Window (BW, projected distance from the Galactic center ~ 500 pc) obtained from ESO/NTT data (Ortolani *et al.* 1995b). With a Montecarlo simulation, the cluster luminosity function has been broadened to mimic the depth effect that is present in the BW field. This comparison reveals that the age and average metallicity of the bulk population in the BW must be very similar to that of the two clusters. This is illustrated in Fig. 3, where the luminosity functions of NGC 6528 (WFC2) and BW are compared to each other. When the HB peaks in the luminosity function are made to coincide, the main sequence luminosity functions coincide with extremely high precision in the brighter, age-dependent part that is less affected by incompleteness and field contamination (i.e., for $19.5 \lesssim V \lesssim 20.5$).

Given that the relative positioning of the two luminosity functions is accurate to within ~ 0.1 mag, Ortolani *et al.* infer that the age of the bulk population in BW is the same as that of the two globulars NGC 6528 and NGC 6553, to within 10%. The identity of the main sequence luminosity functions also ensures that the age dispersion of BW stars must also be very small, probably less than $\sim 10\%$. A larger age dispersion would have resulted

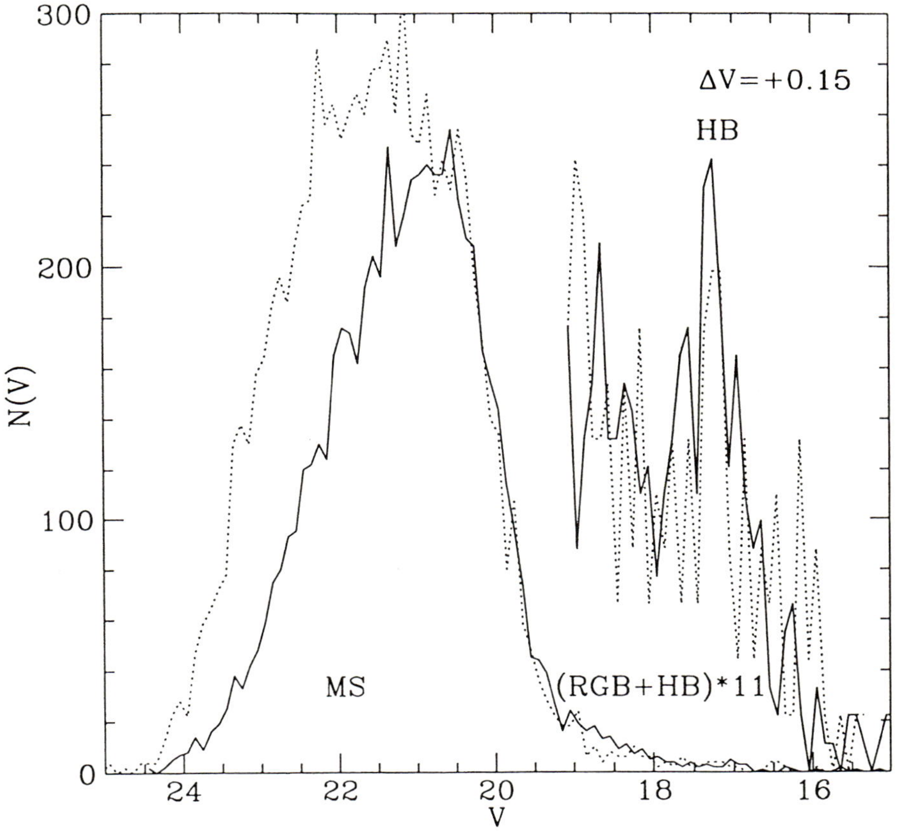

Fig. 3. The luminosity function (LF) of main sequence (MS) and red giant (RGB+HB) stars in Baade's Window (solid lines) and in NGC 6528 (WF2 field, dotted line). The cluster LF has been shifted by ΔV=+0.15 so as to bring into coincidence its HB peak (marked on the figure) with that of BW, and multiplied by a factor of 2 so as to normalize the two distributions at V=20.45, where both are reasonably complete. An identical scaling factor is obtained when normalizing the two distributions to the same number of RGB+HB stars brighter than V=19.45. For this display, the RGB+HB LFs have been multiplied by a factor 11, in order to avoid overlap with the LF of the disk stars in the spiral arm at 2 kpc, that span a similar luminosity range. The Baade's Window data are from Ortolani *et al.* (1995b) and have been obtained at the ESO/NTT telescope with ~ 0.4 arcsec seeing. (Adapted from Ortolani *et al.* 1995a.)

in an appreciably shallower slope of main sequence luminosity function. One can conclude that the cluster and Baade's Window data together strongly support the notion of a fast formation and chemical enrichment of the bulge as part of the spheroid formation process, early in the evolution of the Galaxy.

4 Conclusions

Having obtained from these HST and NTT data an old age for the galactic bulge, it may be tantalizing to generalize, and state that all bulges and spheroids are old, and formed dissipationally, early in the evolution of galaxies. This temptation should however be resisted, as we don't know for sure whether our own bulge is *prototypical or atypical* (Frogel 1990) among spiral bulges. As already mentioned, there are indeed hints that the next nearest bulge (M31), might be different from the galactic bulge in having a broader distribution of stellar ages (Rich *et al.* 1993).

Over the last ten years or so the idea of a late formation of spheroids had gained momentum and popularity. Several observational hints appeared to support it, and such picture was also perceived to be in nice agreement with the predictions of the CDM *standard* cosmological model. Yet, in more recent years some of this evidence has faded away, and new data have revitalized the *old*, traditional view of an early formation of spheroids, i.e., ellipticals, and the halo/bulge component of spirals. These new evidences include the tightness of the color-σ relation of cluster ellipticals (Bower *et al.* 1992), the tightness of their distribution about the so-called Fundamental Plane (Renzini & Ciotti 1993), the tightness of the $Mg_2 - \sigma$ relation for both local and moderate redshift ellipticals (Bender 1995), the population of bright red galaxies in clusters at high redshift ($z \sim 1$) observed from the ground (Aragon-Salamanca *et al.* 1993) and especially with HST (Dickinson 1995), and finally the high metallicity already present in very high redshift QSOs (Hamann & Ferland 1993). All these evidences argue for the bulk of stars in (at least most) elliptical galaxies having formed at high redshift ($z \gtrsim 2-3$, or even more), and having passively evolved thereafter (Renzini 1995).

One may argue that all this is not so surprizing after all, even in the frame of the standard CDM cosmology. Cluster elliptical galaxies and very high redshift QSOs may be forming out of 3-σ or 4-σ peaks of primordial density fluctuations, and as such being very rare, hence rather unrepresentative objects for the general population of galactic spheroids. Yet, our Milky Way is a spiral galaxy in a very loose group located rather away from-major density peaks in the distribution of galaxies. Nevertheless, her spheroidal component looks to be $\sim$ one Hubble time old, from the outer halo globular clusters all the way to the inner bulge. It seems we need a cosmological framework allowing a very quick build up of spheroids – no matter whether in clusters or in the field – early in the evolution of the universe.

Ackmowledgements. I am indebted to the team members Beatriz Barbuy, Eduardo Bica, Roberto Gilmozzi, Gianni Marconi, Sergio Ortolani (PI), and Mike Rich for their permission to present at this meeting our *HST* and *NTT* results.

References

Aragon-Salamanca, A., Ellis, R.S., Couch, W.J., Carter, D. 1993, MNRAS, 262, 764

Bender, R. 1995, in New Light on Galaxy Evolution, ed. R. Bender, R.D. Davies (Kluwer, Dordrecht), in press

Bower, R.G., Lucey, J.R. & Ellis, R.S. 1992, MNRAS, 254, 613

Carney, B. 1992, Mem. Soc. Astron. Ital. 63, 409

Cohen, J.G., Sleeper, C. 1994, AJ, 109, 242

Combes, F., Debbasch, Friedli, D., Pfenniger, D. 1990, A&A, 233, 82

Dickinson, M. 1995, in Fresh Views to Elliptical Galaxies, ed. A. Buzzoni, A. Renzini, A. Serrano, PASP Conf. Ser. , 283

DePoy, D.L., Terndrup, D.M., Frogel, J.A. 1993, AJ, 105, 2121

Eggen, O.J., Lynden-Bell, D., Sandage, A. 1962, ApJ, 136, 748

Frogel, J.A. 1988, ARA&A, 26, 51

Frogel, J.A. 1990, in Bulges of Galaxies, ed. B.J. Jarvis, D.M. Terndrup, ESO Conference and Workshop Ser. 35, 177

Gerhard, O., Binney, J. 1993, in Galactic Bulges, ed. H. Dejonghe, H.J. Habing (Kluwer, Dordrecht), p. 275

Hamann, F., Ferland, G. 1993, ApJ, 418, 11

Hasan, H., Pfenniger, D., Norman, C. 1993, ApJ, 409, 91

Hesser, J.E., Shawl, S.J., Meyer, J.E. 1986, PASP, 98, 403

Holzman, J.A., *et al.* 1993, AJ, 106, 1826

Larson, R.B. 1975, MNRAS, 173, 671

Larson, R.B. 1990, PASP, 102, 709

Lee, Y.-W. 1992, AJ, 104, 1780

McWilliam, A., Rich, R.M. 1994, ApJS, 91, 749

Minniti, D. 1995, AJ, 109, 1663

Ortolani, S., Renzini, A., Gilmozzi, R., Marconi, G., Barbuy, B., Bica, E., Rich, R.M., 1995a, Nat, submitted

Ortolani, S., Renzini, A., & Rich, R.M. 1995b, in preparation

Paczyński, B. *et al.* 1994, AJ, 107, 2060

Pfenniger, D. Friedli, D. 1991, A&A, 252, 75

Raha, N., Sellwood, J.A., James, R.A. Kahn, F.D. 1991, Nat, 352, 411

Renzini, A. 1991, in Observational Tests of Cosmological Inflation, ed. T. Shanks *et al.* (Kluwer, Dordrecht), p. 131

Renzini, A. 1993, in Galactic Bulges, ed. H. Dejonghe, H.J. Habing (Kluwer, Dordrecht), p. 151

Renzini, A. 1994, A&A, 285, L5

Renzini, A. 1995, in Stellar Populations, ed. G. Gilmore, P. van der Kruit (Kluwer, Dordrecht), p. 325

Renzini A., Ciotti, L. 1993, ApJ, 416, L49

Renzini, A., & Greggio, L. 1990, in Bulges of Galaxies, ed. Jarvis, B.J., Terndrup D.M, ESO Conference and Workshop Ser. 35, 47

Rich, R.M. 1985, Mem. Soc. Astron. Ital. 56, 23

Rich, R.M., Mould, J.R., Graham, J. 1993, AJ, 106, 2252
Rich, R.M., Mighell, K.J. 1995, ApJ, 439, 145
Sandage, A., Cacciari, C. 1990, ApJ, 350, 654
Stenmetz, M., Müller, E. 1994, A&A, 281, L97
Terndrup, D.M. 1988, AJ, 96, 884
Walker, A.R. 1992, ApJ, 390, L81
Whitelock, P., Feast, M., Catchpole, R. 1991, MNRAS, 248, 276

The Milky Way in the Near-IR, First Results of DENIS

Stéphanie Ruphy[1]

[1] Observatoire de Paris-Meudon, DESPA,
5, place Jules Janssen, 92190 Meudon, France

DENIS (Deep Near-Infrared Survey of the Southern Sky) (Epchtein et al, 1994) will provide, in the next years, the first digitized survey of the southern sky, with a 3σ detection limit of 18, 16 and 14, in the I, J and K_s bands respectively. The DENIS instrument has been in operation at the 1-meter telescope of the European Southern Observatory at La Silla in Chile, since September 1994, with the J and K_s channels, and has already provided data on large areas of the sky.

Deep star counts are obtained from these preliminary data, in various directions of our Galaxy. These star counts are compared to the results of models of stellar population synthesis for the Galaxy, in order to derive some constraints on the parameters of these models. Two models are involved in the star counts presented in this paper : the "Besançon" model (Bienaymé et al, 1987) which is a self consistent model of the Galaxy, and the "Sky" model (see Cohen, 1994, for reference) which is based on a realistic representation of the Galaxy.

At high and intermediate galactic latitudes, the agreement with the DENIS data is very satisfactory for the two models, as seen, for instance in the J band, in Fig. 1, whereas in the galactic plane, the comparison reveals some discrepencies that need to be investigated (for instance the controversial value of the scale height of the K giants can be constrained by these counts).

Furthermore, J-K histograms near the anticenter direction should allow to confirm the existence of a possible sharp cutoff of the thin disc at about 5.5 kpc from the Sun. This cutoff was first revealed by star counts in the visible range (Robin et al, 1992). In the near-infrared, it entails a deficiency of sources in the $0.6 \leq J\text{-}K \leq 1$, and $12 \leq K \leq 13.5$ intervals, due to the absence of the contribution of red giants beyond 5.5 kpc, as shown in Fig. 2, where a qualitative comparison of the DENIS data with the predictions given by the "Besançon" model seems to favour the existence of the cutoff. Thanks to larger sample with a better limit in completion, this very preliminary result should be soon confirmed.

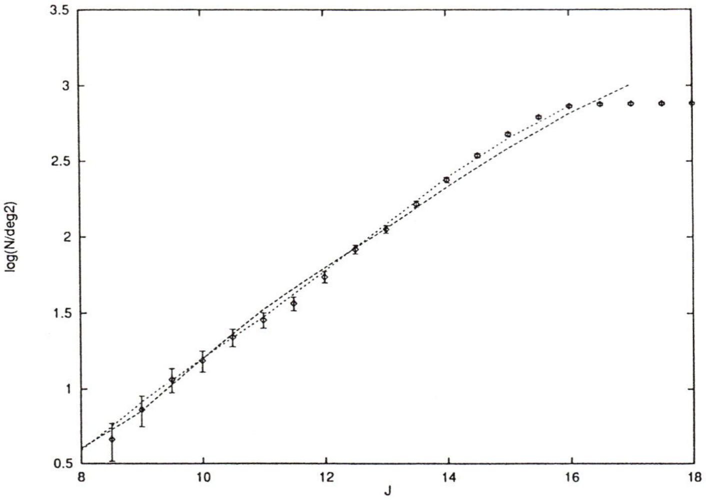

Fig. 1. Cumulative star count around l = 220° and b = −61°, on a 0.2 square degree field, in the J band, as compared to the model predictions (long dashes : "Besançon" model, short dashes : "Sky" model, diamonds with poissonian error bars : DENIS data)

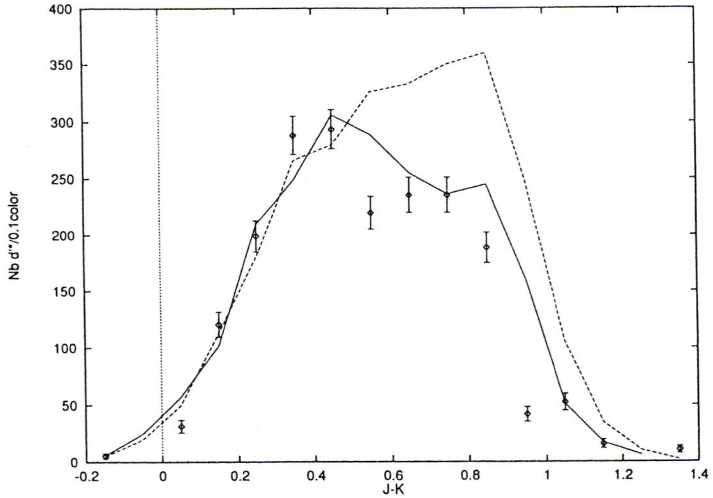

Fig. 2. J-K predicted distribution by the "Besançon"model, toward l = 212.8°, b = 3°, on a 0.2 square degree field , in the 12≤K≤13.5 interval : solid line, prediction with a cutoff of the thin disc at 5.5 kpc from the Sun ; dashes, prediction without cutoff ; diamonds, DENIS data with poissonian error bars.

References

Bienaymé O., Robin A. C., Crézé M., 1987, A&A, 180, 94.
Cohen M., 1994, Ap&SS, 217, 181.
Epchtein N. + 25 authors, 1994, Ap&SS, 217, 3.
Robin A. C., Crézé M., Mohan V., 1992, A&A, 265, 32.

HST Observations of the Ionizing Star Cluster NGC 595 in the Spiral Galaxy M33

Eliot M. Malumuth[1], William H. Waller[2] and Joel Wm. Parker[2]

[1]Astronomy Programs, Computer Sciences Corporation, LASP,
Code 681, Goddard Space Flight Center, Greenbelt MD. 20771, USA
[2]Hughes STX, LASP, Code 681, Goddard Space Flight Center,
Greenbelt MD. 20771, USA

Abstract. As part of a program to study the resolved and composite properties of the giant HII regions in the Local Group spiral galaxy M33, we have obtained, U, B, V, and UV images of the star cluster NGC 595 using the PC chip of the WFPC-2 camera aboard the HST. Photometric reductions were done using DAOPHOT-2. The U-B vs. B-V color-color diagram was used to determined the reddening of each star. The color-magnitude diagram is consistent with a cluster age of 4.5 Myr. Using the 4.5 Myr isochrone we derive the IMF slope for NGC595 to be −0.92 within a circular region of radius of 54 pc when corrected for an underlying disk population.

1 Introduction

The nearby, face-on Scd galaxy, M33 provides the ideal laboratory for testing the possibility of a link between metallicity and the Initial Mass Function (IMF). Claims of such a link have been based on observations of [OIII]/Hβ line ratios and other composite spectral indices. The inferred effective temperatures of the ionizing clusters are seen to increase with decreasing [O/H] abundance (Vilchez and Pagel 1988). This anticorrelation suggests that the IMF is biased toward hotter, higher mass stars in regions of lower metal abundance. At a distance of 0.84 Mpc M33 is the nearest galaxy possessing a large number of giant HII regions of differing metallicity.

NGC 595 is located $4'.25$ to the north-west of the nucleus of M33. Of the 5 clusters in this study it has the second lowest O/H abundance. Vilchez et al. (1988) find an O/H abundance of 2.75×10^{-4} or 0.36 solar.

2 Observations and Reductions

Images of NGC 595 were obtained on 1994 July 18 using the Hubble Space Telescope and the WFPC2 camera with the F170W, F336W, F439W, and F547M (hereafter UV, U, B, and V) filters. The exposure times were 700, 320, 360, and 200 seconds respectively. Each exposure was broken into 2 equal parts by the CR-split option to clean cosmic rays.

Photometric reductions were done using the DAOPHOT-2 package (Stetson, Davis, and Crabtree 1990). Absolute calibrations were done using the prescription of Holtzman *et al.* (1995). We found 561 stars on the V image, 345 stars on the B image, 272 stars on the U image, and 100 stars on the UV image. A total of 267 stars are common to the U, B, and V images while 83 were detected on all 4 images.

The U-B vs. B-V color-color diagram was used to determine the reddening for the 267 stars with U, B, and V photometry. The average reddening derived is $E_{B-V} = 0.36 \pm 0.28$.

Contours of Hα emission were used to define the cluster boundary as a circular region of radius of 54 pc.

3 The IMF

Models of the cluster evolution with Z=0.008 (.4 solar) and an age of 4.5 Myr were used to estimate the effective temperature, bolometric correction and initial mass.

The IMF, uncorrected for the presence of an underlying disk population was determined by counting the number of stars in 8 equal log mass bins between 10 and 51 $M_\odot$ and dividing by the size of the bin and the defined area of the cluster in kpc^2. The uncorrected slope is $\Gamma = -1.26 \pm 0.06$.

A background correction was made for the presence of an underlying disk population using the V-band data on the WFC CCD chips taken with the PC image of NGC 595. When the number of stars in the same area of the "background" are subtracted from each bin we get a corrected IMF slope of $\Gamma = -0.92 \pm 0.11$.

References

Holtzman, J.A. *et al.* (1995): The Photometric Performance and Calibration of WFPC2, Preprint.

Stetson, P.B., Davis, L.E., and Crabtree, D. R. (1990): Future Development of the DAOPHOT Crowded-field Photometry Package, ASP Conf. Ser. 8, CCDs in Astronomy, ed. G.H. Jacoby (San Francisco: ASP), 289.

Vilchez, J.M. and Pagel, B.E.J. (1988): On the Determination of Temperatures of Ionizing Stars in H II Regions, MNRAS, **231**, 257.

Vilchez, J.M., Pagel, B.E.J., Diaz, A.I., Terlevich, E., and Edmunds, M.G. (1988): The Chemical Composition Gradient Across M33, MNRAS, **235**, 633.

Discovery of the Galactic Bar Population

Y.K. Ng[1,5], G. Bertelli[2,3], C. Chiosi[2], and A. Bressan[4]

[1] Leiden Observatory, P.O. Box 9513, 2300 RA Leiden, the Netherlands [2] Department of Astronomy, Vicolo dell'Osservatorio 5, 35122 Padua, Italy [3] National Council of Research, CNR - GNA, Rome, Italy [4] Astronomical Observatory, Vicolo dell'Osservatorio 5, 35122 Padua, Italy [5] Present address: IAP, CNRS, 98 bis Boulevard Arago, F-75014 Paris, France

Abstract. The discovery of the Galactic Bar Population is reported from a detailed analysis of the OGLE star counts (Paczyński et al. 1994) in Baade's Window ($1^\circ\!.0, -3^\circ\!.9$). A better agreement between the observed and the simulated star counts is obtained through the introduction of the 'bar' population, which has an age in the range 8 – 9 Gyr and a metallicity between $0.005 < Z < 0.030$.

An analysis has been made of the OGLE star counts (Paczyński et al. 1994) with the HRD-GST (Hertzsprung-Russell Diagram Galactic Software Telescope; see Ng 1994 and Ng et al. 1995a for details). The results obtained from previous analysis with the HRD-GST are described in various papers, i.e. Bertelli et al. 1995ab, Ng 1994, and Ng et al. 1995ac. The results of the present analysis with (solid line) and without (dashed line) the Bar population are shown in figure 1. It shows a comparison between the observed colour counts from Baade's Window Colour-Magnitude Diagram (CMD) and the synthetic counts from Monte-Carlo (MC-) simulations with the HRD-GST. It shows that the observed counts are in better agreement with the simulations when the Bar population is included. See also Carraro & Ng (1995) for a discussion about the relation of this population with the observed age-metallicity relation of open clusters. For simplicity the stars from the bar are distributed inside a flattened spheroid. Figure 2 shows the synthetic CMD for the bar population. A complete description of the analysis is presented in a separate paper (Ng et al. 1995b), where a complete list of the populations' input specifications for the MC-simulations can be found.

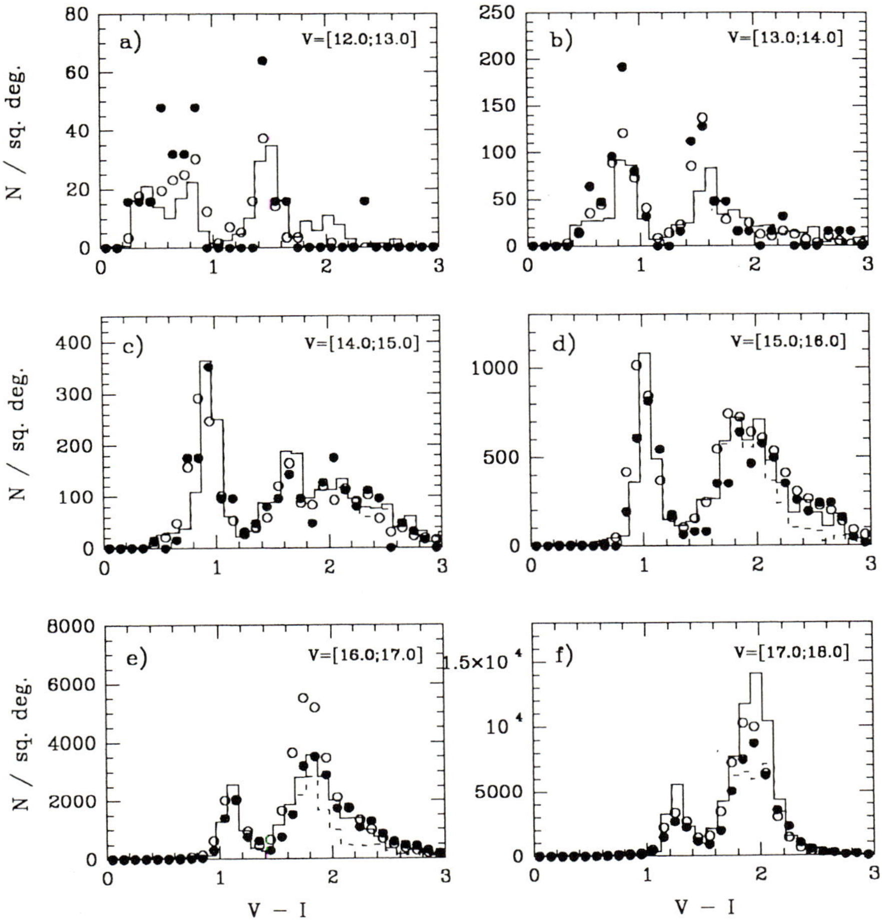

Fig. 1. The observed (filled dots for OGLE BW subfield #3; open dots for the whole OGLE BW field) and simulated (solid line with BAR population and the dotted line without the BAR population) colour distribution between $12 < V < 18$

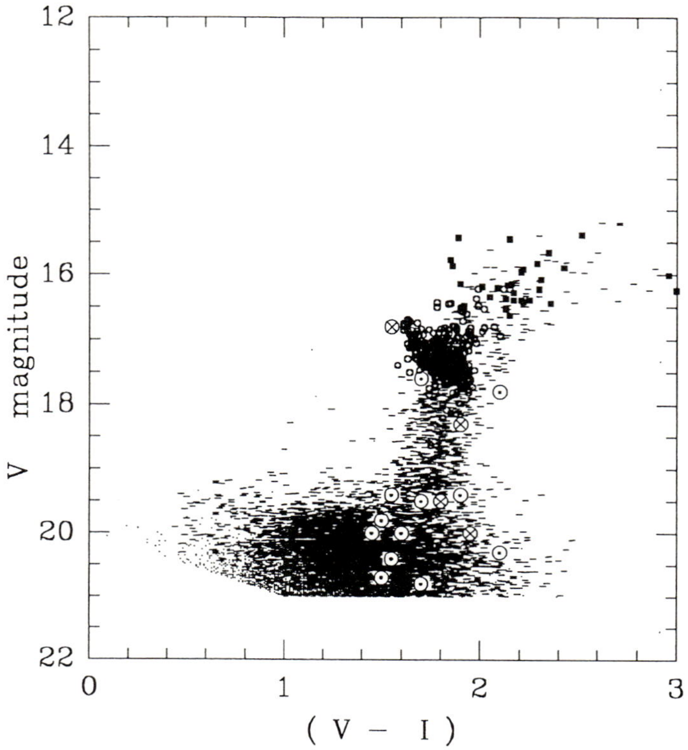

Figure 2

Acknowledgements. G. Bertelli, A. Bressan, and C. Chiosi acknowledge the financial support received from the Italian Ministry of University, Scientific Research and Technology and the Italian Space Agency (ASI). ANTARES, an astrophysics network funded by the HCM programme of the European Community, provided financial support for Ng's working visits to Padova & the research contribution for this workshop.

References

Bertelli, G., Bressan, A., Chiosi, C., Ng, Y.K., Ortolani, S., 1995a, A&A 301, 381

Bertelli, G., Bressan, A., Chiosi, C., Ng, Y.K., 1995b, A&A *in press*

Carraro, G., Ng, Y.K., 1995, IAU Symposium 171, 'New Light on Galaxy Evolution', 26 – 30 June 1995, Heidelberg (Germany)

Ng, Y.K., 1995, Ph.D. thesis, Leiden University, the Netherlands

Ng, Y.K., Bertelli, G., Bressan, A., Chiosi, C., Lub, J., 1995a, A&A 295, 655

Ng, Y.K., Bertelli, G., Chiosi, C., Bressan, A., 1995b, A&A *submitted*

Ng, Y.K., Bertelli, G., Bressan, A., Chiosi, C., 1995c, *in preparation*

Paczyński, B., Stanek, K.Z., Udalski, A., Szymański, M., Kałużny, J., Kubiak, M., Mateo, M., 1994, AJ 107, 2060

Integrated Infrared Colors of Star Clusters in the Galaxy and M31

A.Vallenari[1], C.Chiosi[2], R.Tantalo[2]

[1] Observatory of Padova, Vicolo Osservatorio 5, Padova, Italy

[2] Department of Astronomy, University of Padova, Vicolo Osservatorio 5, Padova, Italy

Abstract. We compare theoretical integrated infrared colors of stars clusters with observational data in the Milky Way and M31 to infer ranges of ages, masses and metallicities to be correlated to the other properties of the parent galaxies.

1. M_0-(V-K)$_0$ plane. Starting from the stellar isochrones by Bertelli et al (1994) we have constructed single stellar population models (SSP) and derived the integrated IR colours and magnitudes for star clusters. The behaviour of young and intermediate age clusters in the dereddened M_0-(V-K)$_0$ plan is affected by stochastic effects due to fluctuations in the number of AGB stars. This effect on the (V-K) colour is relevant for ages younger than a few Gyr and prevent any clear estimate of age, mass and metallicity derived from the location of a young cluster in this plane. At ages typical of globular clusters, stochastic effects of AGB stars are negligible. Therefore, the V-(V-K) plane can be used to derive plausible ranges of mass and metallicity for old clusters. In Fig.1 we plot the M_{V0}-(V-K)$_0$ data of clusters of M31 and the Galaxy. In the case of M31, 85 clusters are displayed (stars). The data are from Bonoli et al. (1992) and Battistini et al. (1987). The reddening of individual clusters when available is taken from Frogel et al. (1980). Otherwise the mean value of E(V-K)=0.28 is adopted. The data of Galactic clusters (squares) are from Burstein et al. (1984) and Brodie & Huchra (1990). The data are compared with the theoretical SSPs in the age range 10-19 Gyr. The solid lines are loci of constant mass of the SSP, the dotted lines are loci of constant metallicity. Although some degeneracy between mass and age is present, a certain degree of discrimination between age and mass is possible. Milky Way clusters have a maximum mass of about 5 $\times 10^6$ $M_\odot$. M31 clusters seem to be more massive. The majority of the clusters in the sample has metallicities ranging from Z=0.0004 to Z=0.008 with the exception of a few. As already noticed by Huchra et al. (1991), the M31 clusters are also slightly more metal rich than those of the Milky Way.

2. (J-K)-(V-K) plane. The trend of Galactic and M31 clusters in the (J-K)$_0$−(V-K)$_0$ plot follows a sequence of metallicity with little dependence on the age. Fig. 2 presents the (V-K)$_0$−(J-K)$_0$ plane of Galactic and M31 Globular Clusters. The lines of constant age and varying metallicity derived from SSP are also plotted. The solid line is for 19 Gyr, whereas the dashed line is for 8 Gyr. The metallicity increases from Z=0.0004 (bottom left) to

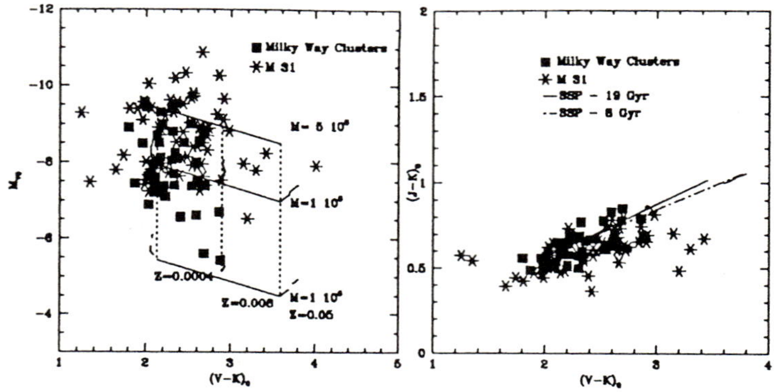

Fig. 1-2. Fig.1 presents Milky Way and M31 clusters in the plane M_{V0}- $(V\text{-}K)_0$. The lines are SSP in the age range 10-19 Gyr for different metallicities. Fig.2 presents the Milky Way and M31 clusters in the two colour plane, compared with theoretical SSP

Z=0.05 (top right). As already noticed by various authors, the clusters of M31 overlap the region filled by Galactic objects, indicating a similarity of properties as age, stellar content... The colours of the clusters in M31 seem to be more dispersed than those of Galactic clusters. Bonoli et al. (1987) estimate an error of about 0.25 in the (V-K) colour and at least 0.15 in (J-K) colour. Although part of the dispersion in the data can be accounted for by the observational errors, there might be another intrinsic source of scatter. The analysis of the two colour plane (Fig. 2) shows that the scatter in question cannot be ascribed to the age because ages as young as 1 Gyr should be invoked to account for it. Equally, stochastic effects cannot be invoked because in the age range typical of globular clusters (12-15 Gyr) these are found to be negligible. We suggest that, keeping constant all other relevant parameters, the scatter in the colours could result from different extensions of the HB passing from cluster to cluster and from the Milky Way to M31.

References

Battistini P., Bonoli F., Braccesi A., Federici L., Fusi-Pecci F., Marano B, Borngen F., 1987, A&AS 67, 447

Bertelli G., Bressan A., Chiosi C., Fagotto F., Nasi E. 1994, A&AS 106, 275

Bonoli, F., Delpino F., Federici L., Fusi-Pecci F., Longmore A.J., 1992, A&AS 96, 163

Brodie J.P., Huchra J.P., 1990, ApJ 362, 503

Burstein D., Faber S.M., Gaskell C.M., Krumm N., 1984, ApJ 287, 586

Frogel J.A., Persson S.E., Cohen J.G., 1980, ApJ 240, 785

Huchra J.P., Brodie J.P., Kent S.M., 1991, ApJ 370, 495

Pagel B.E.J., Simonson E.A., Terlevich, R. J., Edmunds M.G. 1992, MNRAS 255, 325

Persson S.E., Aaronson M., Cohen J.G., Frogel J.A., Matthews K., 1983, ApJ 266,105

III

DYNAMICAL STRUCTURE OF SPIRAL GALAXIES

Observational Evidence for a Compact Dark Mass at the Galactic Center

Eckart, A., Genzel, R., Krabbe, A., Hofmann, R., Tacconi-Garman, L.E., Kroker, H., Thatte, N., Weitzel, L.

Max Planck Institut für Extraterrestrische Physik, Giessenbachstrasse, D-85748 Garching b. München, Germany

Abstract. We present improved recent results from near-infrared high spatial resolution speckle observations with the MPE SHARP 1 camera at the NTT as well as imaging spectroscopic observations with the MPE 3D spectrometer at the ESO-MPIA 2.2m telescope, both at La Silla, Chile. The new K-band speckle maps reach K magnitudes of about 16 and resolve the previously found object at the position of the radio source Sgr A*(R) into a small cluster of compact sources. With one exception, their polarizations are similar to other sources in its vicinity and thus are probably caused by anisotropic foreground dust extinction in the Galactic plane. The Sgr A*(IR) complex does not exhibit any significant flux density variations at 2.2μm on time scales of minutes or years. We therefore interpret Sgr A*(IR) as a small local clustering of luminous stars ($M_K \approx$-3) near/at the position of the compact radio source. The radial velocity dispersion of 35 early and late type stars with distances of 1 to 12" from Sgr A* is 153$\pm$18 km/s. The new results favour the existence of a central dark mass of the order of $\approx 3\times10^6 M_\odot$ (density $\geq 10^{8.5}$ $M_\odot$ pc^{-3}, M/L$\geq$10 $L_\odot/M_\odot$) within 0.14 pc of the dynamic center.

High resolution near-infrared imaging with the MPE camera SHARP (0.05" pixels, 12.8" field of view; Hofmann et al. 1993) resolved the 2.2μm emission in the central parsec (at 8.5 kpc 1"=0.041 pc) into about 340 stars and gave the first evidence for an infrared counterpart (Sgr A*(IR)) of the compact radio source Sgr A* (R) (Eckart et al. 1992, 1993) which is the most likely candidate for a possible massive black hole at the center of our Galaxy (e.g.Lo 1989). Imaging spectroscopy revealed a small cluster of luminous and probably massive, blue supergiants (Allen, Hyland, and Hillier 1990, Krabbe et al. 1991). Sellgren et al. (1990) found evidence that the depth of the 2.3μm CO bandhead absorption decreases in the central 10", perhaps indicating a lack of late type stars there. Understanding better the properties of Sgr A*(IR) and its relation to Sgr A*(R), as well as determining the identification and distribution of different types of stars in the central parsec are of great importance for the analysis of the structure and evolution of the Galactic center stellar cluster.

For this purpose we have carried out new near-infrared speckle, as well as, imaging spectroscopic observations. The new speckle observations were carried out with the MPE infrared high resolution camera, SHARP (Hofmann et al. 1993) at the 3.5m New Technology Telescope (NTT) of the European Southern Observatory (ESO) in La Silla, Chile, on August 3-6, 1993, and April 21-29, 1994. The bulk of the spectroscopic observations was carried out with the MPE imaging spectrometer 3D (Weitzel 1994) in August 1994 at the ESO-MPIA 2.2m telescope on La Silla, Chile, in conjunction with our fast tip-tilt guider ROGUE (Thatte et al.1995). The new data now allow us to study the distribution, and properties of the stars in the central cluster. We can also determine properties of the possible counterpart of the compact radio source Sgr A* (R).

1 Properties of Sgr A*(IR)

Sgr A*(IR) is a cluster of compact sources: Previous SHARP K-band images had already shown that Sgr A*(IR) is extended ≈0.5" east-west (Eckart et al. 1993, 1994) or multiple. This structure has recently been confirmed by Close et al. (private communication) via 0.2" K-band imaging using the tip-tilt FASTTREC I camera (Close & McCarthy 1994) at the Steward Observatory 2.3m telescope at Kitt Peak. In good agreement with previous maps the our new diffraction limited speckle images of the central parsec now clearly reveal that its spatial extent is due to the superposition of at least half a dozen individual compact components within ≈0.5" of the nominal position of Sgr A*(R) (Figs.1) with K≈14 to 15. Their projected distances from each other are of the order of our resolution (8×10^{-3}pc or about 1700 A.U. at 8.5 kpc) so that the structure of Sgr A*(IR) is confusion limited at ≈0.1" resolution.

Sgr A*(IR) is not variable: In the course of about three years (August 1991 to April 1994) we did not detect any long-term flux variability greater than $\pm0.25^m$ (3 σ) of the total flux of Sgr A*(IR) in the K-band ($m_K\approx12$ in a 1" aperture). To test for short-term variability we examined the relative fluxes of IRS 7, IRS 16C, and Sgr A*(IR) in one of our best (1 second exposures; 0.3" seeing) data sets from March 1992. On the raw, image motion compensated frames, we integrated the fluxes over 0.5"x0.5" apertures centered on IRS 7, IRS 16C, and Sgr A*(IR). We then binned the data into intervals of 20 seconds each. The only residual variations over the course of 60 minutes are <15% for the Sgr A*(IR)/IRS 7 ratio and <7% for the IRS 16C/IRS 7 ratio.

Sgr A*(IR) has the same colors as IRS 16 and has no strong spectral features: The J-K colors of Sgr A*(IR) as well as the central and northern sources of the IRS 16 cluster are very similar (J-K=5.2) and - most likely due to local extinction variations in the HII region Sgr A West - about 0.5 magnitudes bluer than most other sources that surround the central few arcseconds. This statement holds with the exception of the foreground stars Irr1 (≈1" WNW of IRS 16C) and Irr2 (e.g. Rosa et al. 1992). There is no exceptional color gradient across Sgr A*(IR) in our 0.2" J-K color map and no significant signal at Sgr A*(IR) in the 2 spectral lines we investigated.

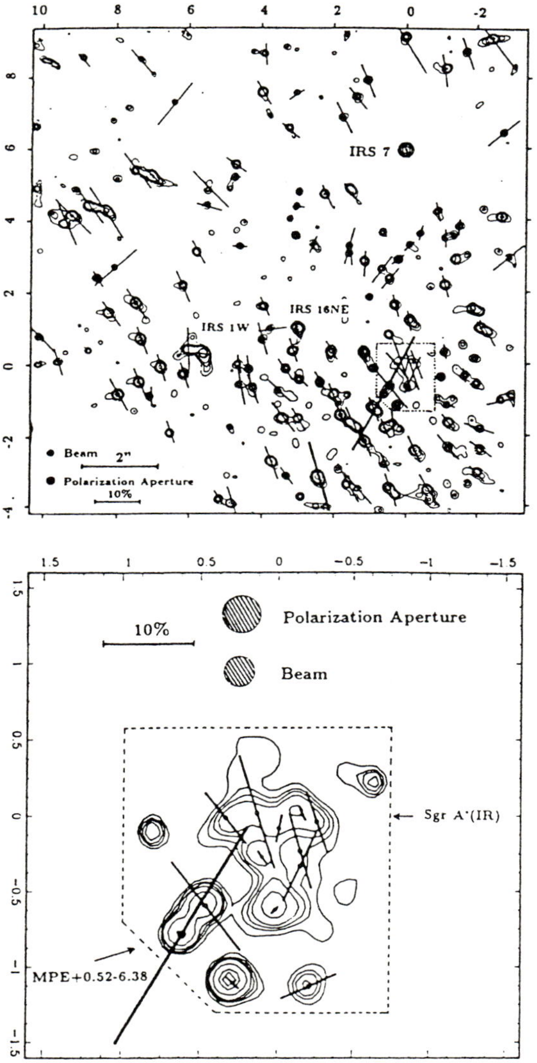

Fig. 1. Top: 0.2" resolution total intensity images as obtained from the polarization data with the polarization vectors superimposed. Two sources show a significantly larger degree of polarization: The source MPE+0.52-6.38 (the numbers indicate the offset from IRS 7 in arcseconds) with 29% at -28° and IRS 21 (MPE-8.57+2.20)with 17% at 16° The mean flux-weighted polarization of all sources in the field of view is 4% at 25°. Typical uncertainties are ±0.5% and 5°. **Bottom:** 0.2" resolution zoom of the central 2"×2" containing Sgr A* and the highly polarized source MPE+0.52-6.38.

Sgr A*(IR) has no large intrinsic polarization: At the obtained resolution and sensitivity the degree and angle of the K-band polarizations of the Sgr A*(IR) sources are similar to those of sources in its vicinity and thus are probably caused by anisotropic absorption by aligned interstellar dust grains along the line of sight to the Galactic Center (Lebofsky et al. 1982, Knacke and Capps 1977). An upper limit to the intrinsic polarization of Sgr A*(IR) is about 2%. There is one strongly polarized source at the periphery of the Sgr A*(IR) complex 1"±0.2" SE of the nominal position of Sgr A*(R): MPE+0.52-6.38 (numbers indicate offset from IRS 7 in arcseconds) with 29% at -28°. Its color is similar to sources in its vicinity. Its large polarization is therefore unlikely due to local anisotropic absorption but more likely caused by scattering.

Sgr A*(IR) is likely a central stellar condensation and Sgr A*(R) is dark: Given our new findings only the combination of its position and the large surface density of objects make Sgr A*(IR) now look peculiar with respect to the surrounding stars. We conclude that Sgr A*(IR) is most likely a local clustering of moderately luminous stars with $M_K \approx -3$. These could either be early type main sequence (B0 V: $\approx 10^{4.4}$ $L_\odot$) or late type (K2-4 III: $\approx 10^{2.3}$ $L_\odot$) giant stars. To answer which - if any - of the half a dozen sources in Sgr A*(IR) can be identified with Sgr A*(R) the present ±0.2" uncertainty of the relative positioning of the radio and infrared frames has to be substantially improved (e.g Eckart et al.1993, Yusef-Zadeh and Melia 1992). If the 1 to 3×10^6 $M_\odot$ central mass concentration within 5" (0.2 pc) of Sgr A* (Genzel, Hollenbach and Townes 1994) refers to Sgr A*(R), its mass to luminosity ratio presently is at least ≈ 100 $L_\odot / M_\odot$.

Number Density Distribution: Following Eckart et al. (1993) we computed the radial distribution of number surface density averaged in 0.3" wide annuli centered on Sgr A*(R) for 600 sources with $6.8 < m_K < 15$ and 264 sources with $6.8 < m_K < 13.8$. As in Eckart et al. (1993, 1994) the number distribution is centered within ±1" (3σ) on Sgr A*(R) and sources with K<13.8 at a distance ≤10" from the centroid follow an isothermal distribution with a core radius of ≈0.2 pc (5"). Including the faintest sources (K<15) indicates a somewhat flatter distribution (core radius ≈0.3pc) between 2" and 10". However, a proper correction for crowding in the central few arcseconds has not been done which would increase the central surface density. At the same time Sgr A*(IR) now shows up as a central excess (10 sources per arcsec^2) above the best fitting isothermal distribution (3 to 4 sources per arcsec^2). This excess is presently only significant at the 2σ level, above that of the flat (isothermal) distribution, however. Sgr A*(IR) may thus be a central cusp in the overall stellar distribution but substantial improvements in both angular resolution (confusion) and sensitivity are required in order to verify this possibility with better statistical significance.

2 The Stellar Content of the Central Parsec

Two sources show a significantly larger degree of polarization compared to the other cluster members: The source MPE+0.52-6.38 (see above) and IRS 21 (MPE-8.57+2.20) with 17% at 16°. IRS 21 has a very red 2μm spectrum without strong spectral lines (Krabbe et al. 1995), suggesting it is heavily absorbed or has a disk. Polarizations as large as observed can be obtained via scattering of the radiation of a central source by dense asymmetrically distributed dust. IRS 21 may therefore be a young massive star that has recently formed in the central part of the Sgr A(West) gas filaments.

¿From our speckle line maps (Eckart et al.1995) it is possible, for the first time, to identify individual early type stars (HeI line emission) and late type stars (CO absorption) in the central few arcseconds. The He I line to continuum map in clearly reveals about 15 He I emission line stars (see also Krabbe et al. 1991, 1994, Genzel, Hollenbach and Townes 1994). The CO 2.3μm line to continuum map shows for the first time about 15 candidates for being CO absorption stars with $K \leq 12.5$ in the inner 10". Such stars could either be solar mass, late M giants or intermediate mass (2-8 $M_\odot$) AGB stars. About 20 objects with $K \leq 12.5$ which do not show significant features in the He I or CO line may either be main sequence O stars, or B/A/F supergiants, or embedded young stars. Since we have detected a substantial number of fairly bright CO absorption sources within the central 10" the drop in CO bandhead strength (Sellgren et al. 1990) is likely caused not by the disappearance of CO sources but by the additional presence of bright early type stars, including the He I emission line stars. This conclusion gains additional support from the fact that the bandhead strength as a function of radius as mapped by Sellgren et al. (1990) is similar to the variation of the ratio of the faint star ($11 \leq K \leq 14$) to total ($K \leq 14.5$) SHARP surface brightness or the variation of surface brightness of stars with CO absorption to total surface brightness as derived by Allen (1994).

3 Velocity Dispersion and Mass

The stellar velocities measured with 3D are presented in Krabbe et al. 1995 together with CGS4 and FAST data. These observations allow to derive velocity dispersions for early- and late-type stars. The overall rotation of the stellar cluster appears to be small ($\leq$70 km/s) and has been neglected here. Correcting the apparent distances of the stars from the center of the cluster for projection effects and subtracting velocity uncertainties in squares from the observed velocities one can obtain the intrinsic velocity dispersion of the cluster and calculate the inclosed masses from the virial theorem and the Bahcall-Tremaine (1981) estimator. Fig.2 shows the smaller and more conservative virial masses along with other mass estimates from the literature. Our new results present a very compelling case for the presence of

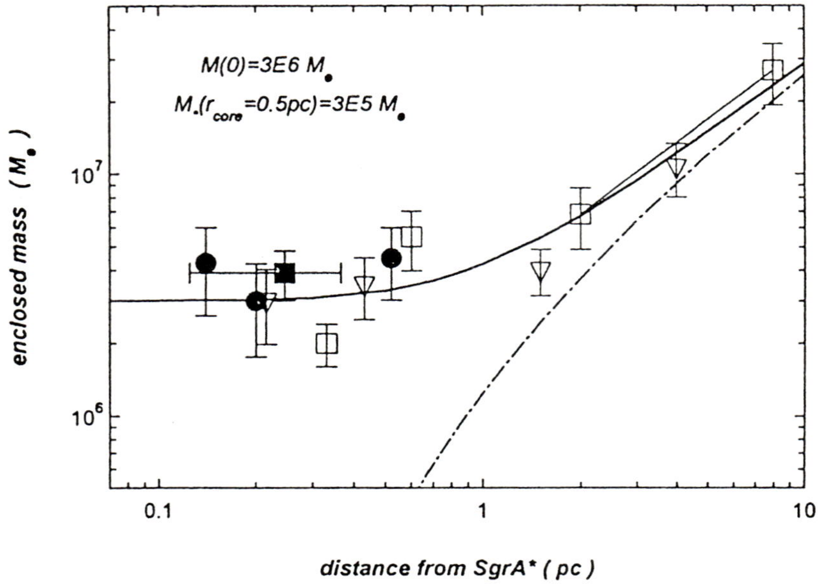

Fig. 2. Enclosed mass as a function of (true) distance from Sgr A*. Our new mass estimates derived from early and late type star velocity dispersions are marked by filled circles and a filled black rectangle (with 1 σ error bars). Open squares (stellar velocity data) and triangles (gas dynamics) represent several mass estimates taken from the literature (see references in Krabbe et al. 1995). The dashed curve shows the mass distribution of an isothermal cluster with an assumed core radius of 0.5pc and a mass of $3\times10^5 M_\odot$ within this radius. The continuous curve is a model containing a $3\times10^6 M_\odot$ central mass plus the isothermal cluster.

a (2-4)$\times10^6$ $M_\odot$ mass concentration. The model plotted in Fig.2 contains 3×10^6 $M_\odot$ within 0.14 pc with a mass density of $10^{8.5}$ $M_\odot$ pc^{-3} and a M/L ratio of at least 10 and probably as much as 100 (Eckart et al. 1995). The velocity dispersions and masses within 0.25-0.39 pc derived with the two estimators and the combination of 35 stellar velocities within 11.7" are now between 4.3 and 4.7 σ above values that can be explained by the stellar cluster alone. The velocity dispersions of the massive early type stars agree well with those of the intermediate-mass late-type stars . This fact strengthens the case for a dominant central mass at the center of the galaxy. The statistical significance of the dark mass concentration is now quite convincing and depends little on velocity errors and methods for deriving the masses. The only remaining caveat is the possibility that stellar orbits with large radial anisotropies dominate. This will be tested with the ongoing MPE proper-motion experiment (Eckart et al. 1995). Positions of individual brighter stars can be measured to much better than the 50mas pixel scale of the SHARP I camera. The data is being collected once or twice a year starting in August 1991. If a 3×10^6 $M_\odot$ black hole is present at the position of Sgr A* the stars of the IRS 16 complex that are at $\leq$ 2" ($\leq$ 0.08 pc) from it must have on average proper motions of $\geq$ 7 milliarcsec/year, or a $\geq$0.4 of the diffraction limited beam of the NTT in 6 years (taking into account the motions in both spatial coordinates).

References

Allen, D.A., 1994, "The Nuclei of Normal Galaxies", eds. R.Genzel & A.I.Harris Kluwer (Dordrecht), p.293

Allen,D.A., Hyland,A.R. and Hillier,D.J., 1990, M.N.R.A.S 244, 706

Bahcall, J.N., Tremaine, S., 1981, Ap.J. 244, 805

Close, L.M., McCarthy, D.W., 1994, P.A.S.P, 106, 77

Eckart, A., Genzel, R., Krabbe, A., Hofmann, R., van der Werf, P.P., Drapatz, S., 1992, Nature 335, 526

Eckart, A., Genzel, R., Hofmann, Sams, B.J., Tacconi-Garman, L.E., 1993, Ap.J.Let. 407, L77

Eckart, A., Genzel, R., Hofmann, Sams, B.J., Tacconi-Garman, L.E., Cruzalebes, P., 1994, "The Nuclei of Normal Galaxies", eds. R.Genzel & A.I.Harris Kluwer (Dordrecht), p.305

Eckart, A., Genzel, R., Hofmann, R., Sams, B.J., Tacconi-Garman, L.E., 1995, Ap.J.Let. 445, L23

Genzel, R., Hollenbach, D.J. and Townes, C.H., 1994, Rep.Progr.Phys. 57, 417

Hofmann, R., Blietz, M., Duhoux, P., Eckart, A., Krabbe, A. and Rotaciuc, V., 1993, in Progress in Telescope and Instrumentation Technologies, ed. M. H. Ulrich, ESO Report, 42, 617

Knacke, R.F., and Cabbs, R.W., 1977, Ap.J. 216, 271

Krabbe,A., Genzel,R., Drapatz,S. and Rotaciuc,V., 1991, Ap.J.Let. 382, L19

Krabbe, A., Blietz, M., Drapatz, S., Eckart, A., Genzel, R., Hofmann, R., Lutz, D., Najarro, F., Sams, B., Tacconi-Garman, L., van der Werf, P., 1994, "High Resolution Near-Infrared Imaging", in "Infrared Astronomy with Arrays (3)", I.McLean (ed.), (Kluwer:Dordrecht), p.484

Krabbe, A., Genzel, R., Eckart, A., Najarro, F., Lutz, D., Cameron, M., Kroker, H., Tacconi-Garman, L.E., Thatte, N., Weitzel, L., Drapatz, S., Geballe, T., Sternberg, A., Kudritzki, R.-P., 1995, Ap.J.Let. 447, L95

Lebofsky, M.J., Rieke, G.H., Deshpande, M.R., Kemp, J.C., 1982, Ap.J.Let. 263, 672

Lo, K.Y., 1989, "The Center of the Galaxy", (ed.) Morris, M., (Kluwer:Dortrecht), p.527

Rosa,M.R., Zinnecker,H., Moneti,A. and Melnick,J., 1992, A&A 257, 515

Sellgren, K., McGinn, M.T., Becklin, E.E., Hall, D.,N.,B., 1990, Ap.J. 359, 112

Thatte, N., et al. 1995, in prep.

Weitzel, L., 1994, PhD Thesis, Ludwig-Maximilian-Universität München

Applications of a Self-Consistent Model for the Galactic Bar

HongSheng Zhao

Max-Planck-Institute für Astrophysik, 85740 Garching, Germany

Abstract. A stable self-consistent stellar dynamical model for the Galactic bar is constructed from about 500 numerically computed orbits with an extension of the Schwarzschild technique. The model fits the *COBE* found asymmetric boxy light distribution and the observed stellar kinematics of the bulge. The model is also consistent with the non-circular motions of the HI and CO velocity maps of the inner Galaxy, and the microlensing event rate towards the bulge. The technique used here can also be applied to interpret light and velocity data of external systems.

1.1 Introduction

It is now widely known that our Galaxy has a central bar with its near end on the positive Galactic longitude side. Two strong evidences for this picture come from the asymmetric light distribution found by the *COBE* team (Weiland et al. 1994) and the non-circular motion of the HI and CO clouds (Binney et al. 1991). Interestingly the large microlensing optical depth towards the bulge found by the OGLE and MACHO teams (Alcock et al. 1995, Udalski et al. 1994) is also in agreement with most lenses being in the near side of a massive bar in the center, pointing nearly towards us (Zhao et al. 1995). While the observations can directly rule out existing oblate rotator models (Kent 1992), it is much harder to device a bar model that fits the observations qualitatively. The traditional N-body approach to make a bar from an unstable disk is often not good enough for quantitative interpretations although some remarkable progress has been made in this direction (Fux et al. 1995).

In this talk, I show an equilibrium model for the stellar bar that is made particularly to fit the *COBE* light distribution and the stellar kinematics of the Galactic bar. The basic technique is as Schwarzschild (1979), but we have implemented many necessary technical modifications (see the original model in Zhao 1994 PhD thesis and an improved model in Zhao 1995a). The model is made by populating 500 orbits in a fixed bar potential. The mass on each orbit is determined to fit the observations of the light and kinematics in the least square sense and also to reproduce the potential self-consistently. In particular, Figure 1 shows that the model fits the asymmetric boxy shape of the observed light distribution (Weiland et al. 1994, Dwek et al 1995) up to 10° from the center. The model also predicts a solid body rotation field and radial fall off of the dispersion, which is consistent with stellar radial velocity data of the bulge (see de Zeeuw 1993).

The self-consistent model can directly tell us how different orbit families are populated. The model bar's mass is divided between explicitly integrated direct orbits and some orbits (which I call collective-orbits) with an implicit isotropic distribution function of $f = f(E_J)$. Figure 2 shows the composition of the orbits. Figure 3 shows the relative fraction of the major orbits. We find that although the dominant orbits in the model are still the direct boxy orbits (60% in mass), which are responsible for both the boxy contours in the *COBE* map and the rapid rotation of the bulge, the rest of the mass is in the collective-orbits, which implicitly contain 2/3 chaotic orbits and 1/3 retrograde orbits. As one does not expect significant amount of retrograde or chaotic orbits be populated during the formation of the bar from the disk, we argue that the large fraction seen in the model poses a possible challenge to this canonical scenario.

Other results of the model are shown elsewhere (Zhao 1995a,b), which include 1) testing the stability of the constructed bar model, 2) fitting the gas longitude-velocity diagrams, 3) fitting the microlensing data of the bulge. The model is found to be stable for at least 1 Gigayear despite the fact that it has a nucleus with 5% mass of the bar. The model also can fit the HI and CO $l-v$ diagrams of the inner Galaxy with the non-intersecting closed orbits of the potential, which is based on the *COBE* light distribution and has been used for constructing the stellar model. The enlongated bar is also consistent with the rate of microlensing events detected by the MACHO and OGLE teams towards the bulge. The typically long duration of these events (about 20 days) also argues against stellar mass functions with a large fraction of brown dwarfs. In summary, while the range of plausible models remains to be explored, the current model is consistent with a variety of recent observations of the Galactic bar. The model results will be made electronicly available to the community in the near future.

This work is a bulk part of the PhD thesis work done in Columbia University. I thank David Spergel and Michael Rich for advices and supports during this work.

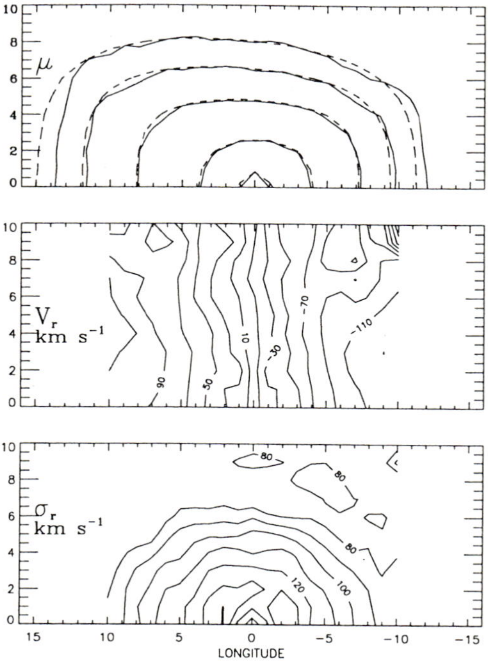

Fig. 1. From top to bottom the solid contours are the surface density (spaced with one magnitude interval), line-of-sight velocity V_r and dispersion σ_r maps of the self-consistent bar model in Galactic coordinates. Also shown in dashed contours is the reprojected surface density of the Dwek et al. (1995) volume density model, which was derived from the *COBE* map.

References

Alcock, C. et al. 1995, ApJ, 445, 133

Binney, J.J. et al. 1991, MNRAS, 252, 210.

de Zeeuw, T. 1993, in *Galactic Bulges*, Eds. H. Dejonghe and H.J. Habing (Kluwer Academic Publ.: Netherlands), p 191.

Dwek, E. et al. 1995, ApJ 445, 716

Fux, R., et al. 1995, in IAU Symp. 169 "Unsolved Problems of the Milky Way", ed. L. Blitz

Kent, S.M., 1992, ApJ 387, 181

Schwarzschild, M. 1979, ApJ, 232, 236.

Udalski, A., et al. 1994, Acta Astronomica, 44, 165

Weiland, J. et al. 1994, ApJ, 425, L81.

Zhao, H.S. 1994, Ph.D. thesis, Columbia University, New York

Zhao, H.S. 1995a, submitted to Monthly Notices

Zhao, H.S. 1995b, in "Barred Galaxies", IAU Colloquium 157, ed. R. Buta et al.

Zhao, H.S., Spergel, D.N., & Rich, R.M. 1995, ApJ, 440, L13,

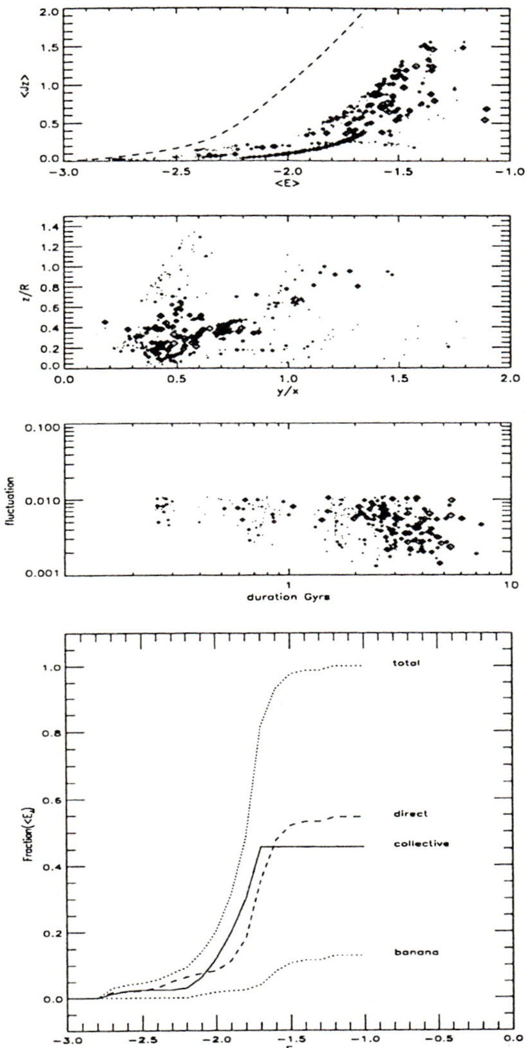

Fig. 2. The upper three panels shows the time averaged energy vs angular momentum distribution of the orbits, the axis ratio distribution, and the level of time dependency of the regular orbits at the end of the orbit integration. Each isolated dot indicates an orbit assigned very low but non-zero weight; dots with diamond and plus symbols of increasing size indicate orbits with increasing weights. The dashed line in the top panel corresponds to the dynamical boundary. The bottom panel shows the fractions of various types of orbits in the self-consistent model with the energy lower than E_J. The typical radius of an orbit is related to the Jacobi energy E_J so that $E_J = -2.2$ and -1.8 correspond to 0.5 kpc and 1 kpc respectively. Note that most of the mass is in the regular direct boxy orbits.

Towards a Three-Dimensional Model of the Galaxy

David N. Spergel[1], Sangeeta Malhotra[1] and Leo Blitz[2]

[1]Princeton University Observatory, Princeton, NJ 08544 USA,
[2] University of Maryland, College Park, MD 20742 USA

Abstract. We present preliminary results of our modelling of the J,K,L and M band emission detected by the DIRBE instrument on the COBE satellite. Our model includes the contribution of Zodiacal light and a three-dimensional model for the distribution of dust based upon DIRBE observations of dust emission at 240 microns. Our approach enables us to construct dust removed maps of our Galaxy that we will use to model the structure of the bar, the spiral structure of Galaxy and the shape of the disk.

1 Introduction

In the past few years, infrared observations have altered our view of our Galaxy. They suggest that the Milky Way is a significantly smaller Galaxy than our companion, M31, (Kent et al. 1991) and imply that we live in a barred Galaxy (Blitz and Spergel 1991, Weiland et al. 1994, Dwek et al. 1995).

Because the Galaxy is much more transparent at infrared wavelengths, observations at 1 - 10 microns are an important probe of the distribution of stars in our Galaxy. Emission at J, K and L bands, is believed to primarily from giant stars, hence, it is tracing the underlying stellar mass distribution. While this hypothesis needs to be tested, particularly given the detection of the strong spiral arms seen in external galaxies at K band (Rix & Zaritsky 1995, Rhoads 1995a, Rhoads 1995b), we will assume that it as our working hypothesis in constructing a model for our Galaxy.

Kent, Dame & Fazio (1991) used the Spacelab IRT data (at 2.4 microns) to construct an axisymmetric model for the galactic disk and bulge. In their analysis, they used the HI and CO observations to trace the gas distribution and then assumed a constant dust/CO and dust/HI ratio so that they could model the dust absorbtion at 2.4 microns. They found that the galactic disk (in the infrared) has a much smaller scale length (3.0 kpc) then had been previously estimated using visual and radio observations. This small scale length is consistent with recent HST M dwarf star counts (Bahcall et al. 1995). Kent (1992) constructed a self-consistent oblate rotator model for the bulge based upon the IRT data and observations of stellar kinematics.

Infrared observations soon challenged this axisymmetric galaxy model. Blitz & Spergel (1991) analyzed the Matsumoto et al. (1982) 2.2 μm balloon data and showed that the bulge appeared non-axisymmetric. Binney et al. (1991) showed that the gas motions in the inner Galaxy implied that the Galaxy is barred. Observations by the DIRBE experiment at 1.25, 2.2, 3.5 and 4.9 μm provided definitive evidence for the Galaxy being barred (Weiland et al. 1994). This photometric evidence has been complemented by analysis of star counts (Nakada et al. 1991; Whitelock & Catchpole 1992; Weinberg 1992; Stanek et al. 1994); the excess of gravitational microlensing events in the direction of the bulge (Paczynski et al. 1994; Zhao, Spergel & Rich 1995) the presence of an off-center hotspot in the 1.8 MeV COMPTEL sky map (Chen, Gehrels & Wiehl 1994) and the detection of vertex deviation in Baade's window (Zhao, Spergel & Rich 1994). All of these observations are consistent with the Milky Way being a barred Galaxy (see Blitz 1993 and Gerhard 1995 for recent reviews).

Dwek et al. (1995) constructed parameterized three-dimensional models for the bar as fits to the COBE DIRBE data. Following Hauser (1993), they used a simple empirical model to model the contribution of Zodiac emission, which is important at high galactic latitude and at 4.9 μm. Dwek et al. (1995) modelled the dust absorption as entirely foreground. This assumption forced them to exclude all of the data at $|b| < 3^o$ at 1.25 and 2.2 μm and at $|b| < 2^o$ at 3.5 and 4.9 μm. While the details of their fit was sensitive to the functional form of their parameterization, they generally found that the data was best fit by a bar with axis ratios (1:0.33:0.22) at an angle of 20 ± 10°.

2 Our Analysis

We are working on constructing an improved model for the infrared emission. We model the dust emission as distributed along the line-of-sight and are plan to use a non-parametric approach to modelling the shape of the bar. These changes allow us to model the disk shape (which requires using data below 3°) and allows us to construct an improved model for the shape of the bar.

The data used for this modelling are the all sky DIRBE maps at solar elongation angle $\epsilon = 90^o$ released by the DIRBE team. In most cases the observations were made at solar elongation angles near 90 degrees and the $\epsilon = 90^o$ maps were constructed by the DIRBE team by interpolation/extrapolation (cf. DIRBE explanatory supplement 1995). The contribution of the zodiacal dust to NIR light was modelled and subtracted by fitting the analytic form

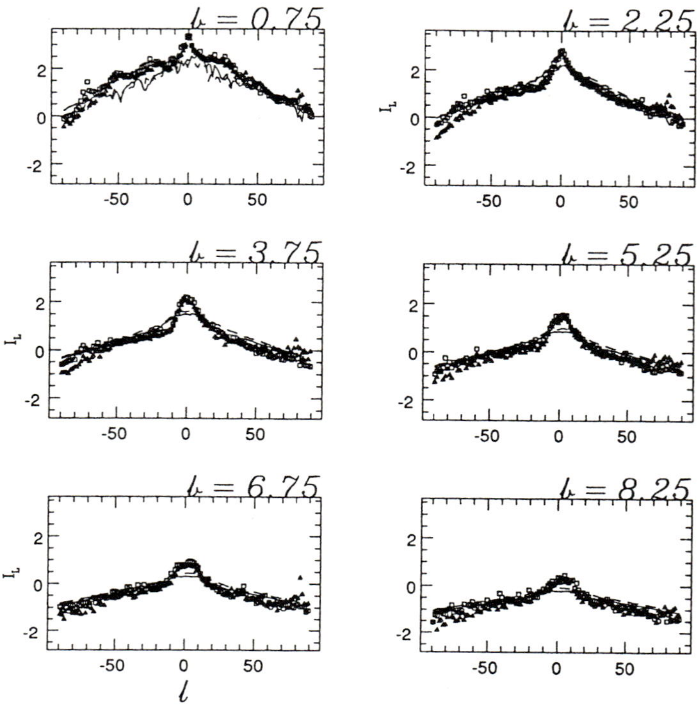

Fig. 1. This plot compares the L band data with the results of the model. The triangle are the data for $+b$ and the squares are the data for $-b$. The solid line is the model for $+b$ and the dashed line is the data for $-b$.

given by Hauser(1993). At each ecliptic longitude bin the ecliptic latitude profile of the emission shows a very sharp lower cutoff; the upward scatter of points is due to Galactic contribution. We fit a lower envelope to this emission, excluding the emission at galactic longitude $b < 25^o$. The zodiacal emission is weak in these bands, so the residuals from the subtraction are estimated to be a few % of Galactic emission. The maps obtained after subtracting zodiacal emission were median filtered to remove point sources.

After removing the zodiac emission using the procedures outlined in Hauser (1993) and Weiland et al. (1994), the next step in the modelling program is to correct the infrared observations for dust absorption.

We use the COBE 240 micron observations as a tracer of the spatial distribution of the dust. Sodroski et al. (1993) found that the dust emission at 240 microns was optical thin throughout the Galaxy and that the dust temperature varied slowly from 22 K in the galactic center to 17 K in the outer Galaxy. The 240 micron emission is probably a better tracer of the dust distribution than the CO and HI emission as the H_2/CO ratio appears to vary by nearly an order-of-magnitude between the galactic center and the outer Galaxy (Blitz et al. 1985, Sodroski et al. 1994).

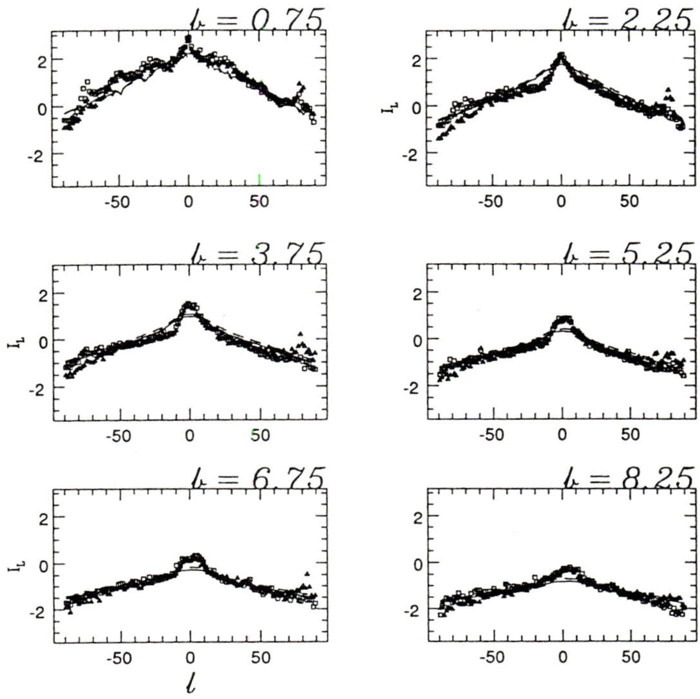

Fig. 2. This plot compares the M band data with the results of the model. The triangle are the data for $+b$ and the squares are the data for $-b$. The solid line is the model for $+b$ and the dashed line is the data for $-b$.

We model the 240 micron emission by first fitting an axisymmetric dust model to the data. We assume that the dust layer has a scale height that increases linearly with r:

$$\rho_d(r, z) = \rho_{dust} * \exp(-|r|/r_{dust}) * \exp(-|z|/z_{dust})$$

where $z_{dust} = \beta * r + z_{min}$. This model was a significantly better fit to the data than a constant scale height model. The linearly increasing scale height is consistent with the gas having a constant azimuthal velocity dispersion, and the Galaxy having a flat rotation curve with a constant ratio of azimuthal to radial epicyclic frequencies. The best least square fit to the data was $\rho_{dust} = 1100.$, $r_{dust} = 3.82$ kpc, $\beta = 0.02$ and $z_{min} = 0.05$ kpc. This dust distribution is consistent with the vertical distribution of HI (Malhotra 1995).

We then improve on the axisymmetric model by rescaling it it along each line of sight to fit the 240 emission data:

$$\rho_{disk}(r, z, phi) = \rho_d(r, z) \frac{I_{observed}(l, b)}{I_{axi}(l, b)}$$

where $I_{observed}$ is the observed intensity and I_{axi} is the intensity predicted by the axisymmetric model. This rescaling has the effect of adding extra dust in the direction of nearby spiral arms. This approach significantly improved our fit to the observed reddening.

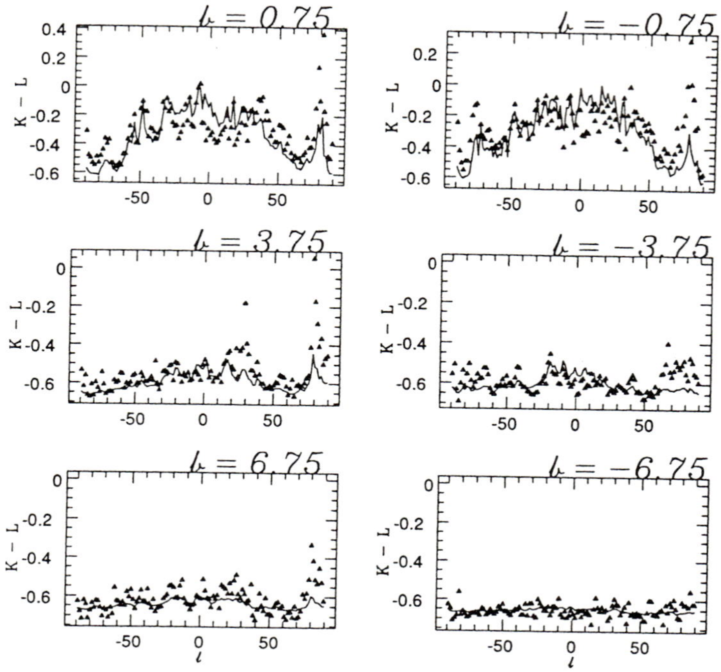

Fig. 3. This plot compares the K-L reddening predicted from modelling the 240 μm emission (solid line) with the observed reddening (triangles).

We then simultaneously fit the emission at K,L, and M bands using our dust model together with a multiparameter model for the disk and bulge:

$$I_\nu(l,b) = \int_0^\infty ds \rho_\nu(s) \exp(-\tau_\nu(s))$$

where $\rho_\nu(s) = C_\nu(\rho_{disk}(r,z) + \rho_{bulge}(x,y,z)]$ and τ_ν is the optical depth at each wavelength. In estimating the optical depth, there was only one free parameter, A_V/I_{240}, the ratio of V band extinction to the intensity measured at 240 microns. The Rieke & Lebofsky (1985) values for A_K/A_V, A_L/A_V and A_M/A_V were used in the analysis. In the analysis, we used the Dwek et al. (1995) G2 parameterization for the bulge and modelled the disk stellar distribution as

$$\rho_{disk}(r,z) = \exp(-r/r_{disk}) * \mathrm{sech}^2(z/z_{disk})$$

The model has six parameters to fit the three color data at 1800 positions: the overall normalizations in K, L and M, the disk scale length and scale height and the ratio of the extinction in V to the emission at 240 microns. The best fit parameters were found using a conjugate gradient minimization scheme (Press et al. 1991) to minimize the least square fit to the K, L and M magnitudes. In order to avoid local minima, the minimization was repeated with 10 different starting positions. The best fit model parameters are $r_{disk} =$ 2.8 kpc,$z_{disk} = 0.276$ pc and $A_V/I_{240} = 0.0215$. Note that the stellar disk appears to be significantly more centrally concentrated than the dust layer. This small scale length is consistent with Bahcall et al. (1995) analysis of the M dwarfs detected by HST and with Kent et al. (1991) analysis of their IRT data.

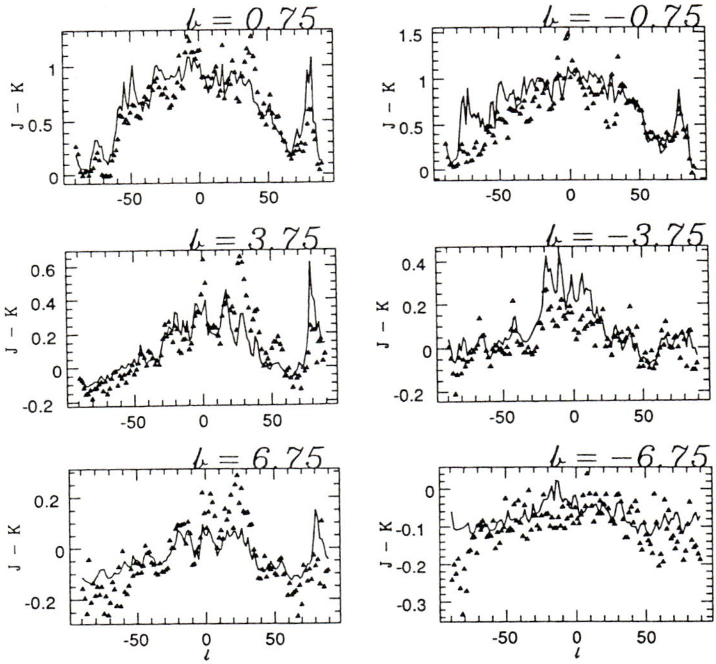

Fig. 4. This plot compares the J-K reddening predicted from modelling the 240 μm emission (solid line) with the observed reddening (triangles).

Figure 1 compares the model to the L band data. Outside of the bulge region, the model is a good fit to the data. The discrepancy in the bulge regions suggests that the Dwek et al. (1995) model, which was fit to data only outside of $|b| > 3$ is not a good fit to the central region of the bulge. The Dwek model is clearly not steep enough in the inner Galaxy. The model also underestimate the L band luminosity at b = 0.75. This may be due to our representing the stellar distribution as a single population. The light in

L band is likely a combination of a thin supergiant population and a thicker giant population.

Figure 2 compares the model to the M band observations. The model fit was somewhat improved by assuming a slightly higher A_M/A_V than suggested by Reike and Liebofsky: $A_M/A_V = 0.04$ rather than $A_M/A_V = 0.028$.

Figure 3 compares the predicted reddening in K-L with the observed reddening.

As a check on the dust modelling, we have calculated the predicted J-K reddening as a function of position. The results of this comparison is shown in figure 4. There is only one free parameter in the fit: the unreddened J-K color of the starlight. Note that the J band data was not used in the fit and that the J band reddening is much larger than the reddening in K,L and M, thus, the good agreement between model and data is reassuring.

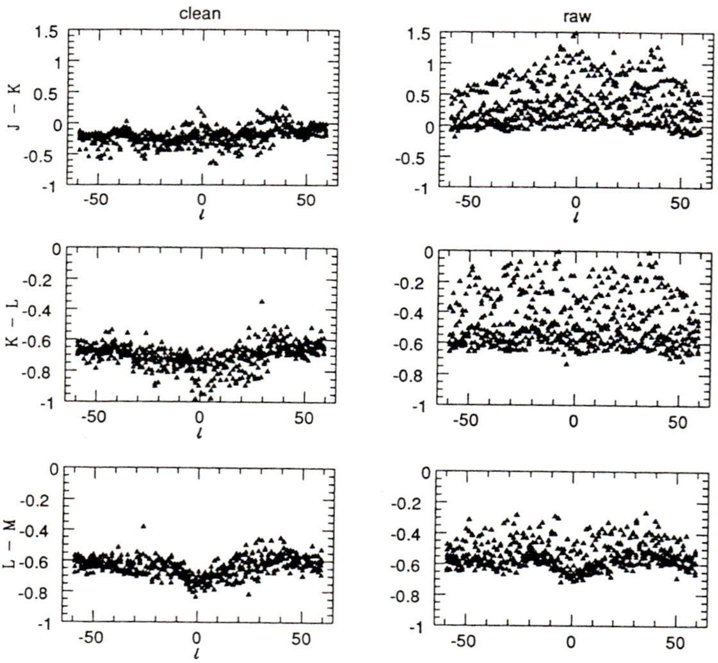

Fig. 5. Figure 5 shows this comparison before and after dust correction for $|b| < 6$. In K - L, the dispersion in color is 0.076 magnitudes.

Having fit a model for the stars and dust, we can now predict the reddening along each line of sight in the model. This enables us to rescale the observations to produce a "dust-free" galactic map that we will use in our analysis of the structure of the disk and bulge:

$$I_{\text{"dust-free"}}(l, b) = I_{DIRBE}(l, b) * I_{dust-freemodel}(l, b)/I_{model}(l, b)$$

where $I_{model}(l,b) = \int ds\rho_*(s)\exp(-\tau_s)$ and $I_{dust-freemodel}(l,b) = \int ds\rho_*(s)$.

These dust removed maps can be computed for each of the four infrared bands. If the dust corrections have been properly done and if there is no variation in the intrinsic stellar color, then the color corrected maps should have uniform color. This final check on our analysis is shown in Figure 5, which compares the dispersion in color before and after dust correction.

In K - L, the dispersion in color is 0.076 magnitudes. This is probably a reasonable guess on the uncertainty in L. In M band, there is a systematic effect where the pixels towards $l = 0$ are somewhat overcorrected (0.05 - 0.1 magnitudes). It is not clear whether this systematic effect is due to zodiac subtraction or due to errors in the dust modelling.

Our approach, like that of previous workers, ignores the contribution of scattered light. This is potentially an important contribution at infrared wavelengths (de Jong 1995). We have checked the contribution of scattered light by performing a Monte-Carlo integration of the equations of radiative transfer. In this calculation, we assume the stellar and dust distribution derived from our best-fit analysis. However, we now include the effects of dust scattering. Dust scattering properties ($\sigma_{scatt}/\sigma_{ext}$, average scattering angle) were based on figure 3 in Kim, Martin and Hendry (1994) and we used a Henyey-Greenstein function to model the scattering cross-section as function of angle. The Monte-Carlo integration was done both for the Kim et al. values for $\sigma_{scatt}/\sigma_{ext}$ and with scattering "turned off" ($\sigma_{scatt}/\sigma_{ext}$). We found that scattering contributed a maximum of 10% of the emission at J band and less than 5% of the emission at K, L and M bands. While we do hope to include scattering in a later version of the least-square fit analysis, we do not think that it will alter the results.

3 Future Work

These infrared maps can be used to trace the spiral structure in the inner Galaxy. There are clear peaks seen in the infrared data when we are viewing the spiral arms along tangents. The location of these tangents, together with the location of the tangents in CO data (Dame et al. 1987) should enable us to construct an improved model for the spiral structure of the Galaxy (Spergel et al. 1996).

We also plan on constructing an improved model for the galactic bulge. The deviations shown in figures 1 and 2 suggest that there is significant room for improvement over the Dwek et al. (1995) model at low b. Binney & Gerhard (1995) have developed a non-parametric algorithm for deprojecting observations of the Galactic bulge. Their Richardson-Lucy (Richardson 1972; Lucy 1974) deprojection algorithm is based on the assumption that the bulge is eight-fold symmetric. This assumption is likely to be valid as galactic bars are thought to have triaxial symmetry. This algorithm has been tested against noisy data and it successfully recovers the bar structure. In collaboration with Binney and Gerhard, we plan to apply this approach to our dust-free maps

(Binney et al. 1996). We also plan to extend their algorithm by adding a regularization scheme (see Press et al. 1991) that ensures that the inverted density distribution is not only positive definite, but also smooth.

Acknowledgements We would like to thank James Binney, Bruce Draine, Ortwin Gerhard, and James Rhoads for helpful discussions. This work is supported by NSF grant AST 91-17388 and by a NASA ADP grant to LB and DNS. DNS would like to thank Arcetri Observatory and the Department of Theoretical Physics at Oxford for their hospitality while this work was completed. We thank the organisers of this workshop for organizing an interesting and topical workshop.

References

Arendt, R.G., et al. 1994, ApJ, 425, L81.
Bahcall, J.N., Flynn, R., & Gould, A. 1995, IAS preprint.
Bernstein, G.M., et al. 1994, A.J., 107, 1962.
Binney, J.J., Blitz, L., Gerhard, O.E., & Spergel, D.N., 1996, in preparation.
Binney, J.J., Gerhard, O.E., Stark, A.A., Bally, J., Uchida, K.I. 1991, MNRAS, 252, 210.
Binney, J.J. & Gerhard, O.E. 1995, submitted to MNRAS.
Blitz, L. & Spergel, D.N. 1991, ApJ, 379, 631.
Blitz, L. 1993, in 'Back to the Galaxy', S.S. Holt & F. Verter, eds., AIP, New York, p 98.
Blitz, L., Bloemen, J.B.G.M., Herman, W. & Bania, T.M. 1985, A & A, 143, 267.
Chen, W., Gehrels, N. & Diehl,R. 1995, ApJ, 44, L57.
Dame, T., et al. 1987, ApJ, 322, 706.
de Jong, R. 1995, Ph.D. thesis.
Dwek, E. et al. 1995, ApJ, 445, 716.
Gerhard, O.E., 1995 in IAU Symposium 169, Unsolved Problems in the Milky Way, ed. L. Blitz, Kluwer, Dordrecht.
Hauser, M.G. 1993, in Back to the Galaxy, ed. S.S. Holt & F. Verter (New York: AIP), 201.
Kent, S.M., 1992, ApJ, 387, 181.
Kent, S.M., Dame, T.M. & Fazio, G. 1991, ApJ, 378, 131.
Kim, S-H., Martin, P.G., & Hendry, P.D. 1994, ApJ, 422, 164.
Lucy, L.B. 1974, AJ, 79, 745.
Malhotra, S. 1995, ApJ 448, 138.
Matsumoto, T. et al. 1982 in The Galactic Center, ed. G. Riegler & R. Blandford(New York: AIP), 48.
Nakada, Y. et al. 1991, Nature, 353, 140.
Paczynski, B. et al. 1994, ApJ, 435, L113.
Press, W.H., et al. 1991, Numerical Recipes, Cambridge University Press.
Reach, W.T., et al. 1995, SISSA preprint, to appear in Ap.J. (Sept. 20, 1995).
Rhoads, J.E. 1995a, in "The Formation of the Milky Way" workshop proceedings, Cambridge University Press, eds. E.J.Alfaro and A.J. Delgado.
Rhoads, J.E. 1995b, this conference.
Richardson, W.H. 1972, J.Opt. Soc. Am., 62, 55.

Rieke, G.H. & Lebofksy, R.M. 1985. ApJ, 288, 618.
Rix, H.-W., & Zaritsky, D. 1995 submitted.
Sodroski, T.J., et al. 1993, ApJ 428, 638.
Sodroski, T.J., et al. 1994, COBE preprint 94/14.
Spergel, D.N., Malhotra, S. & Blitz, L., 1996, in preparation.
Stanek, K.Z., et al. 1994, ApJ, 429, L73.
Weiland, J., et al. 1994, ApJ, 425, L81.
Whitelock, P. & Catchpole, R. 1992, in The Center, Bulge and Disk of the Milky Way, ed. L. Blitz (Dordrecht: Kluwer), 103.
Weinberg, M.D. 1992, ApJ, 384, 81.
Zhao, H., Spergel, D.N. & Rich, R.M. 1994, A.J., 108, 2154.
Zhao, H., Spergel, D.N. & Rich, R.M. 1995, ApJ, 440, L13.

On the Deprojection of Axisymmetric and Triaxial Stellar Systems

Ortwin Gerhard

Astronomisches Institut, Universität Basel, Venusstr. 7, CH-4102 Binningen

Abstract. The deprojection problem for transparent stellar systems is discussed. An axisymmetric model galaxy can be made disky or boxy without altering its projected image by adding a 'konus density'. The range of invisible konus densities increases as the inclination of the galaxy decreases from edge-on to face-on.

The deprojection of triaxial systems is indeterminate not only because of the well-known global changes of the axial ratios and viewing angles which leave the image unchanged. There also exist seven families of 'funnel densities' corresponding to invisible disky or boxy structures with respect to the three principal planes and four other, inclined planes. In the triaxial case, the uncertainties in the deprojection are important for all projection directions.

1 Introduction

Galaxies are much more transparent at near-infrared wavelengths than in visible light. Despite possible contributions from young supergiants and certain interstellar lines, the NIR emissivity is the most faithful tracer we have of the distribution of old stars, and thus of the distribution of stellar mass (see the papers by Krabbe, Rhoads, Silva and others in these proceedings).

However, even for a transparent system of stars we can in actual observations only measure the two-dimensional surface brightness distribution. How much can we learn from such measurements about the unprojected distribution of light? This is an old problem, of course, but one that will be increasingly relevant for the study of spiral galaxies, now that NIR photometry is becoming routine. In this paper I want to discuss some recent, perhaps surprising, results on this old problem.

To make progress on deprojection generally requires an assumption on the *three-dimensional* light distribution $\rho(\boldsymbol{x})$ which cannot be justified by the surface brightness data, $I(\boldsymbol{x}')$. Here $\boldsymbol{x}$ and $\boldsymbol{x}'$ denote positions in space and on the sky, respectively. Such an assumption on the symmetry or functional form of $\rho(\boldsymbol{x})$ may be based on a priori information, or may simply be made to make the problem tractable, and prevents arbitrary shifting of light along the various lines-of-sight. It is clear that only when $\rho(\boldsymbol{x})$ is so constrained, and shifting of light down lines-of-sight is thereby restricted to coherent patterns, can one hope that deprojection be unique or nearly unique.

A priori information on the shape of a single object viewed from just one direction can come from a statistical study of objects that appear similar and are observed from random directions. For example, the globular clusters around the Galaxy are all nearly round on the sky, and one can therefore assume that a given cluster must be nearly spherical. The luminosity density of a spherically symmetric system is uniquely determined from the surface brightness distribution by the well-known Abel transform

$$\rho(r) = -\frac{1}{\pi}\int_r^\infty \frac{\mathrm{d}I}{\mathrm{d}R}\frac{dR}{(R^2-r^2)^{1/2}}, \tag{1}$$

where r and R denote intrinsic and projected radius, respectively.

This is an exceptional case, however. The fact that the isophotes of elliptical galaxies are nearly elliptical is consistent with, but does not imply spheroidal isodensities (Stark 1977). Indeed, we now believe from kinematic data that ellipticals are generally triaxial, although many may not be far from axisymmetric (Franx, Illingworth & deZeeuw 1991).

2 Axisymmetric bodies

Since the kinematic data indicate that many ellipticals and bulges may at least be approximately axisymmetric, it is desirable to know how well the three-dimensional luminosity density $\rho(r,\theta)$ of an axisymmetric galaxy can be recovered from its projected surface brightness $I(x',y')$, given an assumed inclination angle i between the galaxy's symmetry axis (the z-axis) and the line of sight to the observer. While iterative algorithms have been successfully used to recover distributions $\rho(r,\theta)$ for a large number of galaxies with i assumed different from 90° (e.g., Binney, Davies & Illingworth 1990; van der Marel 1991; Dehnen 1995), until recently the uniqueness of the resulting model galaxies has remained unclear.

On the one hand, Rybicki (1986) gave a simple argument based on the 'Fourier slice theorem' that, for $i \neq 90°$, any given surface brightness distribution, $I(x,y)$ must be the projection of an infinite number of luminosity densities $\rho(r,\theta)$. On the other hand, Palmer (1994) proved that the relationship between I and ρ is unique for $i \neq 0$ provided that the density $\rho(R,\theta)$ is 'band-limited': that is that its expansion in Legendre polynomials $P_l(\cos\theta)$,

$$\rho(r,\theta) = \sum_{l=0}^{L} \rho_l(r) P_l(\cos\theta), \tag{2}$$

contains only a finite number of terms. Since any plausible luminosity density can be approximated to sufficient accuracy by a band-limited density, one might think that for practical purposes the relationship between I and ρ is one-to-one for all $i \neq 0°$.

However, this turns out not to be true. In a recent paper Gerhard & Binney (1995; GB) showed by explicit construction that, for $i \neq 90°$, an infinite number of physically plausible luminosity densities *are* compatible with a given surface-brightness distribution. In particular, they showed that one may choose between disky and boxy luminosity densities for the same photometric data. This is of some astrophysical importance because, from isophote analysis, weak disks and box-shaped structures in elliptical galaxies are common (e.g., Bender *et al.* 1989).

From any two space-densities which, for fixed i, project to the same surface brightness distribution, a difference density distribution can be constructed which projects to zero brightness. GB referred to such an invisible density distribution as a 'konus' density, because konus densities are constructed from functions that vanish outside a 'cone of ignorance' with opening angle $90° - i$ in Fourier space. Rybicki (1986) first showed by use of the Fourier slice theorem that such functions leave no trace in the projected image of an axisymmetric system and, therefore, that the deprojection of axisymmetric systems is in principle indeterminate. By explicit construction, GB demonstrated that konus densities can be found that are everywhere non-singular and compact in the sense that they fall off sufficiently rapidly with distance from the center. This latter property is essential because a konus density must necessarily be positive in some parts of space and negative in others, and after adding a konus density to a visible one the resulting total density must be everywhere non-negative.

Fig. 1 shows an example of a konus density; this particular one projects to exactly zero, everywhere on the sky, for all inclination angles $i \leq 65°$. Fig. 2 shows a spheroidal and a disky luminosity density which project to exactly the same surface brightness distribution. The latter was obtained from the former by adding a multiple of the konus density in Fig. 1.

The konus density of Fig. 1 is a member of an analytic three-parameter family with regular spatial densities that asymptotically fall off rapidly to infinity ($\propto R^{-3}$ on the major axis, and $\propto z^{-5}$ on the minor axis). Konus densities from this family have the special property that at large R and small z they tend to constant-scale-height disks.

In a subsequent paper, Kochanek & Rybicki (1995) have obtained new families of konus densities with a technique based on the concept of 'semikonus' functions. These are functions that vanish outside a half-cone rather than a cone in Fourier space, and they have the property that the corresponding complex space densities are closed under multiplication (the products remain semikonus densities). By taking the real part, one can thereby obtain new konus densities. One interesting property of the konus densities constructed

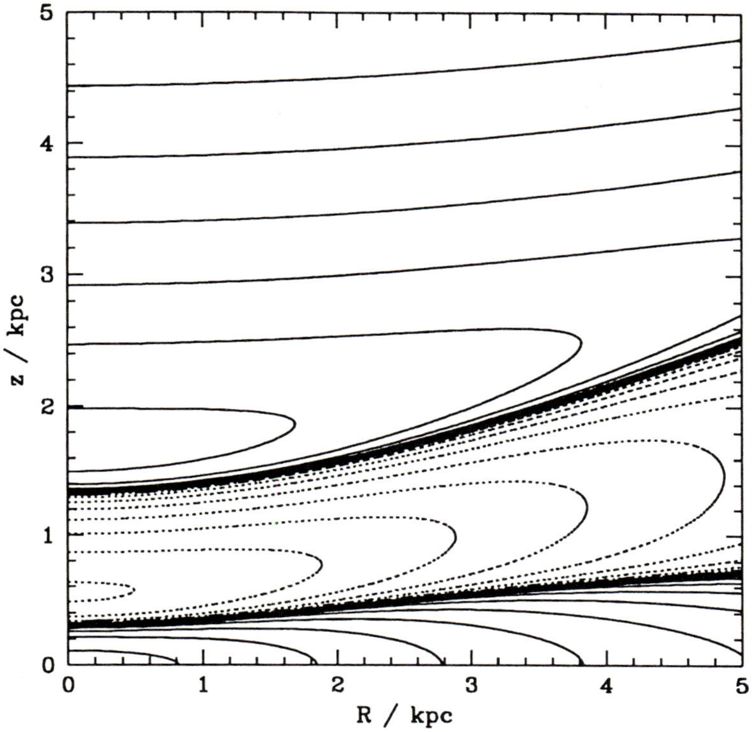

Fig. 1. Contour plots in the meridional plane of a konus density from the family of GB, with $i = 65°$, $\alpha = 1$, $\beta = 2$. Contours are uniformly spaced in $\log(|\rho|)$ and for negative density are dotted.

by Kochanek & Rybicki is that they can have many sign alterations (there are only two in Fig. 1). These solutions also have some special asymptotic properties (but different ones from those of the family described by GB), which appear to be due to the assumptions made in order to find analytic solutions.

Thus from the results of these papers it appears that the range of physically plausible konus densities is large, in that neither the radial fall-off nor the angular structure of konus densities are strongly constrained.

If oblate density distributions cannot be recovered uniquely from their projections for $i \neq 0$, how is this compatible with Palmer's (1994) result, and how does it affect the deprojection of galaxy images? There are several aspects to this, which are discussed in more detail in GB. (i) Konus densities have infinitely many terms in their Legendre expansions, so that the

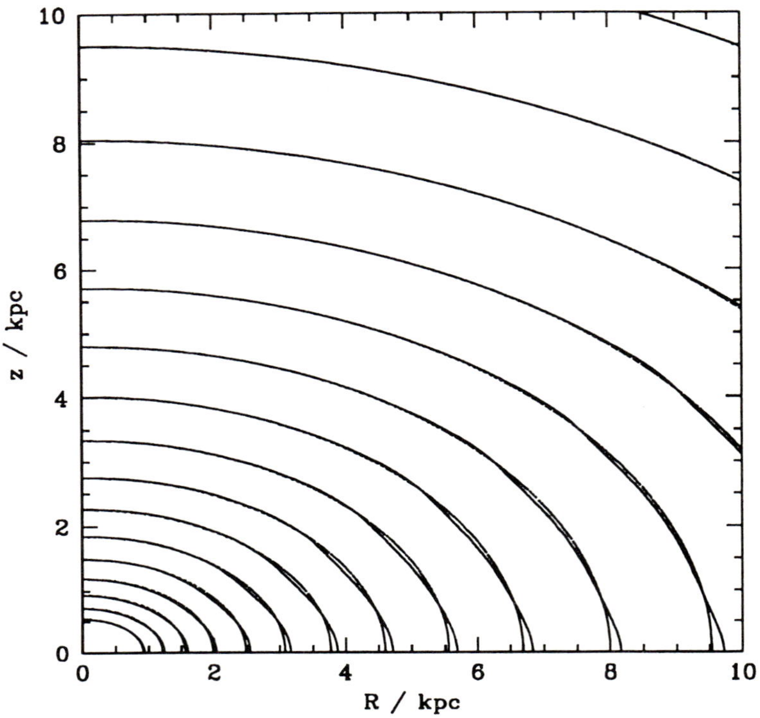

Fig. 2. A disky system with an elliptical surface brightness distribution. The figure shows contours in the meridional plane for two deprojections of the same surface brightness data. The dotted elliptical contours correspond to the density $\rho(m) = \rho_0 m^{-1}(m + m_c)^{-2}$, where m is a spheroidal radius with axis ratio 0.6, and $m_c = 2$ is a break radius parameter. The full contours show the sum of this density and the konus density of Fig. 1 with a plausible positive amplitude f.

assumption that the three-dimensional density is band-limited at some order L [eq. (2)] excludes them from consideration. Nonetheless, konus densities can in principle be recovered by Palmer's inversion algorithm by a suitable limiting procedure.

(ii) Palmer's algorithm can be used to deproject surface brightness data which do not derive from a band-limited density distribution, provided, however, that the coefficients of an angular Fourier expansion of the data on the sky fall off 'sufficiently' rapidly. This condition is relatively weak for near-edge-on inclinations, but becomes very strong for near-face-on inclinations. In the latter case, the addition of one small coefficient I_{L+2} can change significantly the recovered density at all lower orders. Hence the range of surface brightness distributions that can be successfully deprojected diminishes as i decreases.

(iii) With decreasing i the deprojected density becomes more and more sensitive to noise in the data. The density distribution obtained by truncating the infinite Legendre expansion of a konus density at high order, projects to a faint surface brightness distribution. When this pattern of surface brightness is present in smoothed noisy data, a truncated konus density of large amplitude is generated when the data are deprojected. GB showed the example of a deprojected image of NGC 2300 to illustrate this.

In summary: As the inclination of an axisymmetric galaxy decreases from edge-on to face-on, the range of invisible konus densities, the resulting uncertainty in the deprojection, and the sensitivity to noise all increase. The fact that by adding konus densities to a model galaxy it can be made disky or boxy without changing the projected image, implies that disk-to-bulge ratios are not well-determined from photometry unless the disk is strong or the system is seen precisely edge-on.

3 Triaxial bodies

From the work of Contopoulos (1956) and Stark (1977) it is well-known that an ellipsoidal stellar system projects to an image whose surface density contours are similar ellipses, for all possible choices of the direction of projection. Let this direction be specified by two angles (θ, ϕ) in the galaxy-intrinsic frame. If these two angles, the axial ratio of the projected ellipses q, and a further parameter ψ are known, then the two intrinsic axis ratios are fixed. (ψ measures the misalignment of the principal axes of the image with the projection of the intrinsic minor axis on the sky). Furthermore, it is then possible by the inversion of an Abel equation similar to eq. (1) to determine $\rho(x)$ from $I(x, y)$; the formulae and a summary of this subject are given, e.g., in Gerhard (1994).

In reality, often only q can be measured, while the direction of projection and hence the intrinsic axial ratios of the ellipsoidal system remain unknown. There are then many choices of these parameters which lead to the same image. It is likely (and generally assumed) that a similar ambiguity occurs also for triaxially symmetric, but not precisely ellipsoidal systems. Although all elliptical galaxies have nearly elliptical isophotes, and it is therefore reasonable to assume that any one of them has nearly elliptical isophotes in all projections and must then be 'nearly ellipsoidal', the deviations from ellipsoidalness will generally be much larger than the measured deviations of the isophotes from ellipses, because of the ill-posed nature of the deprojection problem (e.g., Merritt 1993).

Besides these global ambiguities there are also other, more localized ambiguities in the deprojection of triaxial stellar systems. These are related to the existence of invisible densities similar to the konus densities in axisymmetric systems (paper in preparation). To see this, consider again projection as an operation in Fourier space and recall the Fourier slice theorem, which says that the Fourier transform of the projected surface density $I(x', y')$ is the two-dimensional slice $A(k'_x, k'_y, 0)$ of the Fourier transform $A(\boldsymbol{k})$ of the density $\rho(\boldsymbol{x})$.

Hence from the surface brightness distribution obtained by projecting a triaxially symmetric stellar system along an arbitrary direction $\hat{\boldsymbol{k}}$, we can obtain the density's Fourier transform on a plane in $\boldsymbol{k}$-space perpendicular to $\hat{\boldsymbol{k}}$ and containing the origin $\boldsymbol{k} = 0$. Because of the assumed triaxial symmetry, however, the images seen from eight symmetric directions must be identical. The observed image therefore provides information on *four* planes in Fourier space through $\boldsymbol{k} = 0$, because the two Fourier planes perpendicular to any pair of directions $\hat{\boldsymbol{k}}$ and $-\hat{\boldsymbol{k}}$ are identical. The four planes may be defined by normal vectors $(\pm\hat{k}_x, \pm\hat{k}_y, \hat{k}_z)$, say.

Fig. 3 shows a perspective view of these four planes in Fourier space for a general direction of projection. Also shown for better visualization is the upper half of an ellipsoidal surface, which depicts the orientation of the triaxial ellipsoid in the corresponding real space. Here the triad of unit vectors $(\boldsymbol{e}_x, \boldsymbol{e}_y, \boldsymbol{e}_z) = (\boldsymbol{k}_x, \boldsymbol{k}_y, \boldsymbol{k}_z)$ is chosen such that the longest axis of the ellipsoid points upwards, the intermediate axis to the right, and the shortest axis into the figure. One sees that the four planes divide Fourier space into a number of pyramid funnels bounded by either four or three infinite planar triangles; no funnels with more or fewer sides are possible. 4-funnels arise around the principal axis directions of the triaxial system, 3-funnels separate the 4-funnels and arise around directions intermediate in longitude and latitude. The term 'n-funnel' here denotes an n-pyramid funnel including its point-symmetric counterpart with respect to the origin. By symmetry, there are then three 4-funnels and four 3-funnels, which are each composed of two 'semi-funnels', respectively.

The sizes of the respective funnels are determined by the two angles (θ, ϕ) defining the direction of projection. Special cases occur when the system is viewed along a direction $\hat{\boldsymbol{k}}$ within a principal plane or along a principal axis. Then the number of independent Fourier planes shrinks to two and one, respectively. In the case of two planes, the intersection gives rise to four wedges, each of which can be thought of as a semi-4-funnel that has opened to infinity along one principal axis direction.

In each of the 4-funnels or 3-funnels the Fourier density is completely unconstrained by the projected image; these funnels are 'funnels of ignorance' much like the 'cone of ignorance' in an axisymmetric system. Fourier densities that vanish outside a particular funnel or semi-funnel are invisible, as well as any combination of Fourier densities from the seven independent funnels. The space density corresponding to any such Fourier density may be called a 'funnel density' or 'semi-funnel density'.

The global ambiguities in the deprojection of a triaxial stellar system, which, for example, allow changing the axial ratios of a triaxial ellisoid such

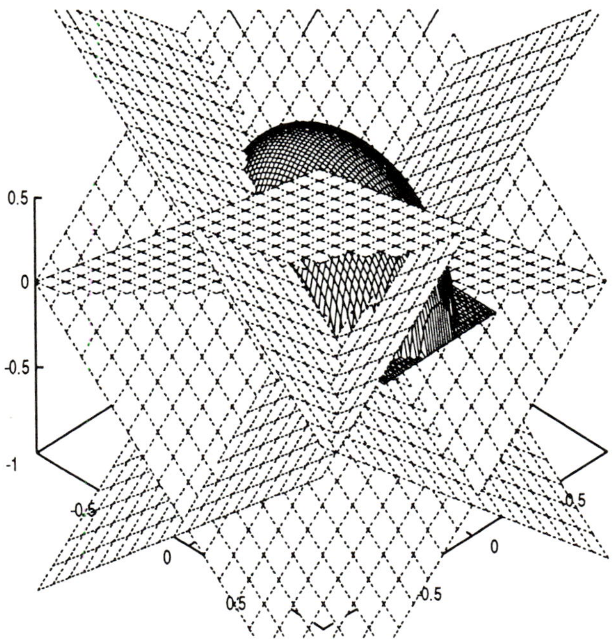

Fig. 3. The four planes in Fourier space on which the Fourier transform of the density of a triaxial stellar system is known from its projected image, for a general direction of projection. These planes divide Fourier space into three 4-funnels and four 3-funnels which are point-symmetric with respect to the origin $\boldsymbol{k} = 0$. Also shown is the upper half of an ellipsoidal surface with long axis upwards, intermediate axis to the right, and short axis into the figure, and a small portion of the plane containing the ellipsoid's intermediate and short axes. See text.

that the image remains unchanged, will correspond to global changes of the Fourier density in all funnels, with the only constraint that it must remain constant on the appropriate four Fourier planes (cf. Fig. 3).

More localized changes in the three-dimensional light distribution that leave the image unchanged can be achieved by adding individual funnel densities to a given triaxial density distribution. Particular examples for such funnel densities are provided by the axisymmetric konus densities. For given projection angles (θ, ϕ), axisymmetric cones with opening angles up to some maximum value $\alpha_{\max}(\theta, \phi)$, will fit completely into a particular 3- or 4-funnel. Thus funnel densities can be generated which equal the konus densities such as shown in Fig. 1, in the appropriate axisymmetric coordinate system specified by the konus direction. E.g., if the Fourier density corresponding to

Fig. 1 is fitted into the 4-funnel along the major-axis direction, the corresponding funnel density, when added to an ellipsoidal system, will generate disky or boxy structures with respect to the principal plane containing the two shortest axes.

The number of possible funnel structures is much richer, however: (i) The funnel densities need not be axisymmetric; (ii) funnel densities from the four 3-funnels can give rise to inclined invisible disks; and (iii) it is likely that by combining Fourier densities in several funnels invisible structures intermediate between the global axis ratio changes and the disky-to-boxy invisible densities may be constructed. Moreover, as the projection angles (θ, ϕ) vary, some funnels increase in size and others decrease in size. In particular, contrary to the axisymmtric case, there are always funnels with large volume since all of Fourier space but four planes of measure zero remains unspecified. This implies that substantial disks of appropriate orientation may be hidden in a triaxial galaxy for almost any direction of projection.

Alltogether these results suggest that the class of triaxial densities that projects to a given image could be very large indeed.

References

Bender R., Surma P., Döbereiner S., Möllenhoff C., Madejsky R., 1989, A&A, 217, 35

Binney J.J., Davies R.L., Illingworth G.D., 1990, ApJ, 361, 78

Contopoulos G., 1956, Zs. Ap. 39, 126

Dehnen W., 1995, MNRAS, 274, 919

Franx M., Illingworth G.D., de Zeeuw T., 1991, ApJ, 383, 112

Gerhard O.E., 1994, in Galactic dynamics and N-body simulations, Lect. Notes in Physics 433, Contopoulos G., Spyrou N.K., Vlahos L., eds, Springer, Berlin, 191

Gerhard O.E., Binney J.J., 1995, MNRAS, 00, 00

Kochanek C.S., Rybicki G.B., 1995, preprint

Merritt D., 1993, in Structure, Dynamics, and Chemical Evolution of Elliptical Galaxies, Danziger I.J., Zeilinger W.W., Kjär K., eds, ESO, Garching, 275

Palmer P.L., 1994, MNRAS, 266, 697

Rybicki, G.B., 1986, in Structure and dynamics of elliptical galaxies, IAU Symp. 127, Zeeuw P.T., ed, Kluwer, Dordrecht, 397

van der Marel R.P., 1991, MNRAS, 253, 710

Stark A.A., 1977, ApJ, 213, 368

Near IR Observations and Dynamics of Barred Galaxies

E. Athanassoula[1]

[1] Observatoire de Marseille, 2 Place Le Verrier,
F-13248 Marseille Cedex 04, France

Abstract. We discuss some of the analytical barred potentials used e.g. for studies of gas flow or orbital structure in barred spirals, and mention their relative merits. Potentials obtained directly from observation are more realistic, but their calculation relies on two major assumptions, i.e. the mass-to-light ratio is constant over the face of a galaxy, and the disc and bar have a constant vertical thickness. Changing the vertical density profile does not seem to influence much the results, provided the same value for the projected light and for $< z^2 >$ are kept. Changing the scale-height, on the other hand, changes the amplitude of the Fourier components of the potential.

The most interesting result, however, comes from comparing potentials obtained from images at different wavelength bands. If near-IR rather than B band photometry is used, considerably more dark matter is needed to account for the observed rotation curves.

Using the potential calculated from a near-IR image of NGC 4314, we discuss both symmetric and asymmetric periodic orbits in the bar region of that galaxy, and in particular the various 3/1 and 4/1 families.

The shape of bars, at least in early type strongly barred galaxies, is quite rectangular-like, in particular near the outer parts of the bar. This could trigger a substantial amount of ergodicity. We discuss the various types of orbits that can account for this shape, presenting the relative merits of each solution.

Finally, some suggestions are made for future work with near-IR photometry of barred and, more generally, disc galaxies.

1 Potentials

Any response calculation, such as used in e.g. the gas flow or orbital structure studies in barred disc galaxies, needs a realistic expression for the potential. Potentials of the type

$$\phi(r,\theta) = \phi_o(r) + \phi_2(r)cos(2\theta) + \phi_4(r)cos(4\theta) + ...$$

have been often used, with ad hoc functions for $\phi_m(r), m = 0, 2, 4...$ A better choice is to use a given axisymmetric background together with a Ferrers's (1877) ellipsoid, i.e. an ellipsoid with a density distribution of the type

$$\rho = \rho_o(1 - \frac{x^2}{a^2} - \frac{y^2}{b^2} - \frac{z^2}{c^2})^n \qquad \text{for} \qquad \frac{x^2}{a^2} + \frac{y^2}{b^2} + \frac{z^2}{c^2} < 1$$

and

$$\rho = 0 \qquad \text{elsewhere}$$

This density distribution is reasonable (Athanassoula 1992) and contains parameters with physical meaning, such as the central volume density, semi-major axis etc. It is thus well suited for studying the dependence of a given effect on a physical parameter, e.g. the amount of gas inflow in a barred galaxy as a function of the axial ratio of the bar. In principle the values of these parameters for a given galaxy can be obtained directly from the observations. In practice the easiest quantity to obtain is the axial ratio b/a, although in galaxies with strong bulges even this is not trivial. Furthermore, it varies with radius (e.g. Athanassoula et al. 1990), so that it is not directly convertible to a unique value necessary for Ferrers's formula. To determine the central concentration of the bar, or its mass, the light distribution needs to be decomposed into an axisymmetric and a barred component. This is not so easy since the two components are not rigid, as in the simple Ferrers's model, but respond to each other. For example a disc does not stay axisymmetric in the presence of a bar, and a given star or orbit does not belong uniquely to one of these components, but can change allegiance with time.

In order to stay close to observations, is thus seems better to forego the decomposition stage altogether, and to calculate the potential directly from the projected light distribution. In practice this presents two problems. First, it is necessary to know the mass-to-light ratio (M/L) as a function of radius, and, since the object is not axisymmetric, azimuthal angle. Since this quantity is not observable directly, the simplest procedure is to assume a constant value. Until the advent of IR observation it was thought that a constant value pertains even to the data in the B band (e.g. Schweizer 1976 or Wevers 1984 for radial distributions, while Duval & Athanassoula (1983) showed that the $(M/L)_B$ of the bar and lens in NGC 5383 have neighbouring values). The advent of IR observations has shown that this is only a rough approximation, as will be discussed below. Since the H and K bands, and to a lesser extent the J band, are less affected by dust than the B-band, and since the young stars do not seem to contribute much to these wavelengths (see other contributions in this volume), it is now hoped that the M/L = constant approximation applies to IR photometry, so as to obtain a good description of the real projected mass distribution.

A second, more serious, problem is geometry, since disc galaxies are three dimensional objects observed in a two dimensional projection on the plane of the sky. We can obtain a three dimensional picture only by making certain hypotheses, guided by observations of galaxies seen from different orientations. Thus to a zero-th order approximation, bulges are taken to be spherical and discs are considered as thin objects of constant scaleheight with a vertical distribution that can be well represented by an exponential, a *sech* or a $sech^2$ function (Van der Kruit & Searle 1981a,b; 1982a,b; Wainscoat, Freeman & Hyland 1989; Barnaby & Thronson 1992)

The situation is less clear for bars, which are difficult to recognise in edge-on galaxies. Early observations (Kormendy 1982, Tsikoudi 1980, Wakamatsu & Hamabe 1984, de Carvalho & da Costa 1987, Hamabe & Wakamatsu 1989) of edge-on SB0 galaxies were used to infer that bars are thin objects. On the other hand theoretical studies and numerical simulations (Combes & Sanders 1981, Combes et al. 1990, Raha et al. 1991) have revealed the existence of vertical instabilities and the formation of thicker bars. Such bars do not have a constant thickness, but have a peanut-like shape, thus further complicating the deprojection problem. No quantitative theoretical study has so far given us sufficient information to allow us, after assuming the bar is thick and peanut-shaped, to deproject it. Therefore, given our present limited knowledge, assuming a spherical bulge and a disc and bar of constant thickness may still be considered a workable approach, although it is not clear at this stage what exactly the effect of this approximation is (cf. section 4).

2 Results from near-IR imaging of a few barred galaxies

The Ohio State University (hereafter OSU) survey will provide a library of photometrically calibrated images of 200 to 300 galaxies from wavelength bands spanning from 0.4 to 2.2 μm. So far four barred galaxies, NGC 1097, 3351, 4314 and 7479 have been analysed and published. Thus BVRJHK photometry is available for NGC 1097, 3351 and 7479, and JHK for NGC 4314. To these can be added NGC 1365, for which the OSU survey has obtained a J frame. The types of these galaxies range from SBa to SBc, and they have widely ranging morphologies, including asymmetries, spirals, inner rings, dust lanes, nuclear rings and inner bars. Based on this small sample, Alice Quillen and I (Quillen & Athanassoula, in preparation; see also Quillen 1995 and this volume) have calculated potentials using the method outlined by Quillen, Frogel & González (1994, hereafter QFG), and tested how they depend on various free parameters and unknown quantities.

We considered three different vertical density profiles, following the exponential, *sech* and $sech^2$ laws correspondingly, and normalised so that their integrals were equal to 1. The results change very little from one case to another, provided that they have the same value for $< z^2 >$. Changing the scale height does not have any effect on the phases of the various Fourier components, but does affect their amplitudes, the highest amplitude corresponding, as expected, to the flatter density distribution. The relative effect of changing the scaleheight depends on how much structure the galaxy has in scales of the order of the considered scaleheight.

The most interesting result has been obtained by comparing potentials calculated from images of different wavelengths. The galaxies in our small sample become bluer in their outer parts. Hence, a comparison of the axisymmetric component of the potential calculated from images in different wavelength bands shows that the missing mass problem is aggravated in the near-IR. Thus halos obtained by comparing rotation curves and disc near-IR profiles need to be more massive and more centrally concentrated than those obtained from B photometry. This is in agreement with results of Peletier et al. (1994), De Jong & Van der Kruit (1994), and Peletier & Balcells (this volume), who find that the radial scalelengths of disc galaxies are larger in B than in the IR, but nevertheless needs to be verified with a larger sample. Sizeable differences are seen also in both the amplitudes and the phases of the nonaxisymmetric components.

3 Orbits in the bar region of NGC 4314

NGC 4314 is well suited for work on orbits because it is seen nearly face-on and has a small size bulge with an extent less than 7" (Benedict et al. 1992, 1993), corresponding to 340 pc for a distance of 10 Mpc. Thus uncertainties in the inclination and the form of the bulge will not influence much the potential calculated from the observations. Furthermore it has remarkably constant near-infrared colours in the bar region (QFG). The potential of this galaxy was calculated by QFG and a polynomial was fit to the Fourier components (see QFG and Quillen this volume, for more details). Patsis, Athanassoula & Quillen (this volume and in preparation) have used two versions of this potential to calculate orbits. In one version (hereafter "C" model) only the axisymmetric and the $cos(2\theta)$ and $cos(4\theta)$ terms are retained, while for the second (hereafter "T" model) the axisymmetric and the first three even Fourier coefficients were considered. In both cases the adopted value of the pattern speed is that given by QFG. The search for periodic orbits was not confined, as in many other studies, to those orbits that are symmetric with respect to the bar major axis. This was, of course, mandatory for the "T" model, but enabled us also to find several families of asymmetric orbits in the "C" model. For reasons of continuity with previous work on orbital structure (see e.g. Contopoulos & Grosbøl 1989 for a review) we will continue discussing orbits in terms of their position on the characteristic diagram - i.e. the (x, E_J) plane, where x is taken along the bar minor axis and E_J is the Jacobi constant - although in the case of asymmetric orbits a point on this diagram does not uniquely define an orbit.

Complex dynamical structure is found in the parts of the characteristic diagram that correspond to very small radii. Although this is of considerable theoretical interest, it may not be relevant to the orbital structure in NGC 4314, since for radii less than 7" the bulge contribution will be nonnegligible, so we will not discuss it further here.

At the 3/1 resonance region we found the standard 3/1 family, which, for the "C" model, starts perpendicularly off the $x = 0$ axis and is, as for many other models, unstable. For the "T" model these orbits do not start off exactly perpendicular and have a small stable section on the characteristic diagram. More important though, the search for asymmetric periodic orbits resulted in finding, for both the "C" and the "T" model, another family, which has one side roughly parallel to the bar minor axis and is stable. A possible application of this family will be discussed in the next section.

The region of the 4/1 resonance shows a lot of complexity compared to previously studied models. Previous work showed the existence of two types of 4/1 orbits around the ultra harmonic resonance (hereafter UHR): rectangular and diamond shaped. For NGC 4314 we also find these two families of 4/1 orbits but their layout on the characteristic diagram is more complex than either of the two types of UHRs discussed by Contopoulos (1988). This will be discussed in more detail elsewhere (Patsis, Athanassoula and Quillen, in preparation). For here it suffices to say that in the case of the "C" model the diamond shaped family is stable and branches off the x_1 family, while the rectangular shaped family is mainly unstable, with only a short stable section, corresponding to a very small interval of values for the Jacobi constant. For the "T" model we find stable 4/1 orbits which are a continuation of the x_1 family and some skew and roughly-rectangular-like orbits, with a short stable section on the characteristic diagram, corresponding again to a very short Jacobi constant interval. The shapes of these orbits are, nevertheless, very reminiscent of the shape of the outer bar isophotes in NGC 4314. Some implications of the existence and stability of the 4/1 families will be discussed in the next section.

4 The shape of bars in the disc plane

Athanassoula et al. (1990) studied the shapes of bars in the symmetry plane of the disc, using a small sample of eleven SB0 and one SBa galaxies with strong bars. They introduced generalised ellipses, i.e. forms described by the equation

$$\left(\frac{|x|}{a}\right)^c + \left(\frac{|y|}{b}\right)^c = 1,$$

which for $c = 2$ are ellipses, for $c < 2$ lozenges and for $c > 2$ have a rectangular-like shape. Fitting such shapes to the isophotes in the bar region, they conclude that the bar shapes are more rectangular-like than elliptical like, and that they can be well represented with generalised ellipses. The shape parameter c has a maximum near the end of the bar, with values, for the sample considered, in the range between 2.7 and 5.3, i.e. in all cases well above 2. The ellipticity of the isophotes decreases with radius with a maximum of the order of 0.8, corresponding to an axial ratio a/b of the order of 5.

Athanassoula (1990) finds that surfaces of section calculated in potentials with rectangular-like isodensity contours, or, equivalently, potentials with relatively large $m = 4$ components, show a lot of ergodicity, contrary to potentials with elliptical-like isodensity contours or relatively low $m = 4$ components. She explains this with the help of Chirikov's criterion (Chirikov 1969). Indeed for $m = 2$ there can be within and around the bar region at best two major primary resonances : corotation and inner Lindblad resonance, while for $m = 4$ there are three : inner Lindblad, inner ultraharmonic and corotation. Since the latter two are usually not far from each other, resonance overlap can be achieved and ergodicity can set in for relatively low potential amplitudes, contrary to the case for $m = 2$.

What type(s) of orbits could account for the rectangular-like shape of observed bars? One would intuitively have expected that real bars would have stable rectangular-like 4/1 orbits which are able to trap regular quasi-periodic orbits around them. It thus comes as a surprise that the, admittedly fragmentary and perhaps even questionable, evidence seems to point to the opposite conclusion.

Assuming infinitesimally thin density distributions and using the approximate analytical formulae given by Contopoulos (1988) it is possible to test the type of the UHR for the sample of early type galaxies of Athanassoula et al. (1990). Indications are that in all cases the UHRs are of type 1, and the rectangular-like orbits unstable. This result is rather disconcerting, although it can be criticised on at least two accounts. The density distribution in real galaxies is not infinitesimally thin, and the approximate analytical formulae of Contopoulos (1988) use the epicyclic approximation and are ill suited for numerical applications. To forego the last problem we have calculated periodic orbits in the potential of NGC 936, but the results still show that the rectangular-like orbits are unstable. Because of the uncertainties introduced by our assumption of infinitesimal vertical thickness we tried various alternatives for the potential in that galaxy. Thus we tried several values of the corotation radius between R_{bar} and 2.1 R_{bar}, we halved, doubled and quadrupled the $m = 0$ component, neglected the components $m = 6$ and 8, or the components $m = 4$, 6 and 8. Even so, the rectangular-like orbits were always unstable.

As mentioned in the previous section the potential of NGC 4314 has been calculated with the hypothesis of a constant thickness (QFG) and the bulge is sufficiently small not to affect the potential around the UHR. Hence the scepticisms voiced above for the case of NGC 936 do not hold in this case, and thus the structure of the UHR is of particular interest.

We find that both "C" and "T" models have only a very restricted region each on their characteristic diagram that corresponds to rectangular-like and stable periodic orbits. Of course the distribution function could be such that these regions, albeit restricted, play an important role in the structure of the outer parts of the bar. This can not be ruled out until proper studies of the distribution function have been made. Furthermore the short extent of this region could be attributed to other sources of uncertainty. E.g. noise in the data might affect the quality of the polynomial fit. Although this may be a concern further out in the galaxy, it should not be important at the radius of the UHR. Other criticisms can be found. For example we have assumed a constant thickness for the bar and disc, which, as discussed in the first section, may be wrong. Nevertheless, in view of the discussion in the last couple of paragraphs it is necessary to seek alternative orbits which can account for the rectangular-like shape. This question has already been considered by Athanassoula (1991) and we will here only extend somewhat that discussion.

- As mentioned in section 3, for NGC 4314 one of the two branches of 3/1 periodic orbits is stable. Thus pairs of 3/1 orbits with reflection symmetry (i.e. a given 3/1 orbit and its mirror image with respect to the bar minor axis), or orbits trapped around them, could enhance the bar rectangularity provided they extend sufficiently far in the region where the bar rectangularity is greatest.
- Orbits trapped around the x_1 family. These can be rectangular-like even though the corresponding x_1 periodic orbit is quite elongated and even has pointed ends. Good examples can be seen in Figure 12 of Athanassoula (1991) and in Figure 3-18 of Binney & Tremaine (1987). Unfortunately their radial extent is rather short and thus they should not contribute to the rectangularity in the outer parts of the bar where it is most needed.
- Semiergodic orbits can be trapped by cantori for very long times (Petrou 1984) and, at least in the case of NGC 936, have the right shape and extent. Whether this is a more general phenomenon and whether it could account for the rectangular-like shape of the bar remains to be seen. It is worth mentioning, however, that an important contribution of the $m = 4$ component in that region should induce a substantial amount of ergodicity.

Thus the situation is far from clear, and the main question of what orbits can account for the rectangular-like shape observed at least in early type strongly barred galaxies, remains largely unanswered. Future work in this area should concentrate in two directions. On the one hand we should see how far we should push all hypotheses inherent in the potential calculations, and particularly those concerning the three-dimensional geometry of the galaxy, in order to obtain UHRs of type 2. On the other hand a larger sample should be analysed in order to study further the behaviour of the relevant periodic and trapped orbits, as well as the semi-ergodic orbits trapped by cantori.

5 Some suggestions for future work

(together with A. Bosma)

The advent of surveys with homogeneous photometric material, as the OSU survey, opens the way for a number of interesting analyses on barred, and more generally on disc galaxies.

In particular many aspects of the dark matter problem should be rethought. Provided that it is established that near-IR M/L ratios vary little with radius and that the profiles are not much affected by dust, radial luminosity profiles can be used for rotation curve decompositions, as done so far with blue or red photometry by e.g. Kent (1986, 1987), Athanassoula, Bosma & Papaioannou (1987), or Begeman, Broeils & Sanders (1991). As seen in section 2, IR observations should lead to a larger amount and a different radial distribution of dark matter than what has been obtained so far with B photometry. Thus issues like whether dwarfs are more dark matter dominated than giants, whether the amount and central concentration of the halo depend on galaxy type and luminosity, whether low surface brightness giants are dark matter dominated, or whether the central concentration of the halo influences the presence or absence of warps (Bosma 1991), can be addressed anew. Still in the dark matter problem, IR observations could help establish whether faint luminous halos exist and how important they are (Sackett et al. 1994, Casali & James 1995).

More specifically about barred galaxies it would be interesting to see whether the boxy shape ($c > 2$ in the generalised ellipse equation) found for early type galaxies is equally strong in near-IR as in B, and to extend this work to later types. Calculating potentials from IR photometry will allow us to address whether barred galaxies have inner Lindblad resonances or not (Athanassoula 1991), whether their Langrangian points are stable, what orbital populations make the rectangular-like shape (provided this persists in the IR), and how various properties of the bar, like its length, axial ratio, shape, mass or radial profile, relate to global properties of the galaxy, like its magnitude, type or rotation.

Several other calculations, like gas flows, applying the Tremaine-Weinberg (1985) method to calculate bar pattern speeds, or calculations of orbital structures or distribution functions, are ready to use realistic bar potentials.

Near-IR photometry can be used also for studying "bars within bars" (Shaw et al. 1995, Quillen et al. 1995 and Friedli et al. in preparation) and their relations to AGNs, as well as for finding the fraction of spirals that are barred. It should also be used for searching for hidden bars in apparently nonbarred galaxies with rings (Athanassoula 1995, Athanassoula et al. 1995)

The above long list, which is only a fraction of the work that can be made with IR photometry, clearly illustrates the use of such observations to the dynamics of barred galaxies.

Acknowledgements. The work on the orbital structure of NGC 4314 (section 3) is done in collaboration with Panos Patsis and Alice Quillen, and the study of the potentials from IR images (section 2) in collaboration with Alice Quillen. I thank them both, as well as Albert Bosma, for many interesting discussions. The Ohio State University galaxy survey is being supported in part by NSF grant AST 92-17716.

References

Athanassoula, E. (1990): *Annals of the New York Ac. of Sc* **596**, 181

Athanassoula, E. (1991): in *Dynamics of Disc Galaxies*, ed. B. Sundelius, Göteborg, Sweden, 149

Athanassoula, E. (1992): MNRAS **259**, 328

Athanassoula, E. (1995): in *Barred Galaxies*, eds. R. Buta, B. Elmegreen and D. A. Crocker, in press

Athanassoula, E., Bosma, A., Papaioannou, S. (1987): A&A **179**, 23

Athanassoula, E., Bosma, A., Guivarch, B., Verdes-Montenegro, L. (1995): in *New Light on Galaxy Evolution*, eds R. Bender and R. Davies, Kluwer Academic Publ., Dordrecht, in press

Athanassoula, E., Morin, S., Wozniak, H., Puy, D., Pierce, M.J., Lombard, J., Bosma, A. (1990): MNRAS **245**, 130

Barnaby, D., Thronson, H.A. (1992): AJ **103**, 41

Begeman, K.G., Broeils, A.H., Sanders, R.H. (1991) MNRAS **249**, 523

Benedict, G.F., Higdon, J.L., Tollestrup, E.V., Hahn, J.M., Harvey, P.M. (1992): AJ **103**, 757

Benedict, G.F., Higdon, J.L., Jefferys, W.H., Duncombe, R., Hemenway, P.D., Shelus, P.J., Whipple, A.L., Nelay, E., Story, D., McArthur, B., McCartney, J., Franz, O.G., Fredrick, L.W., Van Altena, Wm.F. (1993): AJ **105**, 1369 (1992): AJ **103**, 757

Binney, J., Tremaine, S. (1987): *Galactic Dynamics*, Princeton Univ. Press, Princeton

Bosma, A. (1991): in *Warped Disks and inclined Rings around Galaxies* eds S. Casertano, P. Sackett and F.H. Briggs, Cambridge University Press, p. 181

Casali M.M., James P.A. (1995): MNRAS **274**, 265

Chirikov, B.V. (1969): Nuclear Physics Institute of the Siberian Section of the USSR Academy of Sciences, Report 267

Combes, F., Sanders, R.H. (1981): A&A **96**, 164

Combes, F., Debbash, F., Friedly, F., Pfenniger, D. (1990): A&A **233**, 82

Contopoulos, G. (1988): A&A **201**, 44

Contopoulos, G., Grosbøl, P. (1989): AAR **1**, 261

De Carvallo, R.R., da Costa, L.N. (1984): A&A **171**, 66

De Jong, R.S., Van der Kruit, P.C. (1994): *Infrared Astronomy with Arrays* ed. I. Mc Lean, Kluwer Pub, Dordrecht, p. 123

Duval, M.F., Athanassoula, E. (1983): A&A bf 121, 297

Ferrers, N.M. (1877): Quart. J. Pure Appl. Math. **14**, 1

Hamabe, M. Wakamatsu, K. (1989): ApJ **339**, 783

Kent, S. (1986): AJ **91**, 1301

Kent, S. (1987): AJ **93**, 816

Kormendy, J. (1982): in *Morphology and Dynamics of Galaxies*, eds. L. Martinet and M. Mayor, Geneva Obs., p. 113

Peletier, F.F., Valentijn, E.A., Moorwood, A.F.M., Frendling, W. (1994): A&A **292**, 369

Petrou, M. (1984): MNRAS **211**, 283

Quillen, A.C., Frogel, J.A., González, R.A.(1994): ApJ **437**, 162 (QFG)

Quillen, A.C., Frogel, J.A., Kucinski, L.E., Tendrup, D.M. (1995): AJ **110**, 156

Quillen, A.C. (1995): in *Barred Galaxies*, eds. R. Buta, B. Elmegreen and D. A. Crocker, in press

Raha, N., Sellwood, J.A., James, R.A., Kahn, F.D. (1991): Nature **352**, 411

Sackett P.D, Morrison H.L., Harding P., Boroson T.A. (1994): Nature **370**, 441

Schweizer, F. (1976): ApJS **31**, 313

Shaw M., Axon D, Probst R. and Gatley I. (1995) MNRAS **274**, 369

Tremaine, S., Weinberg, M.D. (1985): ApJ **282**, L5

Tsikoudi, V. (1980): ApJS **43**, 365

Van der Kruit, P.C., Searle, L. (1981a): A&A **95**, 105

Van der Kruit, P.C., Searle, L. (1981b): A&A **95**, 116

Van der Kruit, P.C., Searle, L. (1982a): A&A **110**, 61

Van der Kruit, P.C., Searle, L. (1982b): A&A **110**, 79

Wainscoat, R.J., Freeman, K.C., Hyland, A.R. (1989): ApJ **337**, 163

Wakamatsu, K., Hamabe, M. (1984): ApJS **56**, 253

Wevers, B.M.H.R., (1984): Ph. D. Thesis, Groningen

Studying Galaxy Dynamics with IR Images

Alice C. Quillen[1]

[1] Ohio State University, Astronomy Dept., 174 West 18th Avenue, Columbus, OH 43210, USA

Abstract. Detailed studies of stellar and gas dynamics has been carried out primarily in model galaxies which are not good approximations to the shapes of real galaxies (Athanassoula 1991). The high quality of recently available near-IR images of galaxies makes it possible to make realistic mass models of the matter that is traced by starlight. The gravitational potential derived from the mass model of a galaxy can be used to conduct stellar and gas dynamical studies in the galaxy, and the results can be compared to observed velocity fields and other observations of these galaxies. In this report we discuss how accurately the gravitational potential can be estimated based on an infrared image of the galaxy. We describe two uses of such an estimate of a gravitational potential: stellar orbits integrated in the barred galaxy NGC 4314 and an estimate of the gas inflow rate along the bar in NGC 7479.

1 Observations

Galaxies look different in the near-IR (J,H,K) than in the optical (B,V,R), bands (see Figures 1 and 2). Note that the near-IR images are smoother or have less small scale structure than the optical images. This is real since the images shown are presented at the same spatial resolution. The optical and infrared wavelengths have different sensitivity to extinction from dust and are dominated by light from different stellar populations.

Extinction from dust is significantly lower in the near-IR $A_{K(2.2\mu\mathrm{m})} \sim 0.1 A_{V(0.55\mu\mathrm{m})}$ than at optical wavelengths. Dust typically has a lower vertical scale height than the stars in a galaxy so that when dust is concentrated in sharp features in the plane of the galaxy such as along spiral arms or along a bar, a sharp absorption feature can be seen.

The lower infrared extinction can also makes it possible to detect underlying features such as stellar rings, spiral arms, bars, and nuclei that may be difficult to see in optical images– this is particularly true in the central regions of galaxies where there may be a high column depth of dust (for example see Figure 2 showing the nuclear spiral and bar in NGC 1097 which is apparent in the near-IR).

Near-IR images of galaxies detect light primarily from cool giants and dwarfs that contribute a major fraction of the bolometric luminosity of a galaxy. Particularly in spiral galaxies, these stars are much better tracers of the mass distribution of the galaxy than are the bluer, hotter stars (Frogel 1988). This older population is typically more evenly distributed than young blue stars which can be concentrated in linear features following spiral arms, rings or bars. This older population also should have a higher vertical scale height and larger velocity dispersion than newly formed stars, so that the galaxy appears smoother when viewed in the near-IR.

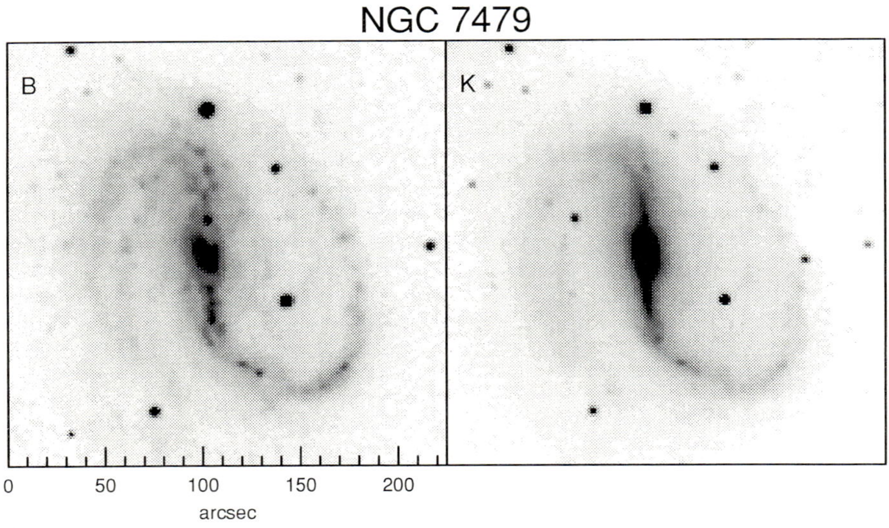

Fig. 1. B and K band images of NGC 7479.

We can make the assumption that near-IR images are a good tracer of mass (note that in cases of high recent starbursts, supergiants can contribute at 2.2μm (K band) so in these situations K band light may not be an ideal tracer of mass). Recent comparison of IR generated to observed Hα rotation curves (Heraudeau et al. 1995) confirms that in the central regions of late-type galaxies (central few exponential scale lengths) IR maximal disk models are usually successful at reproducing rotation curves.

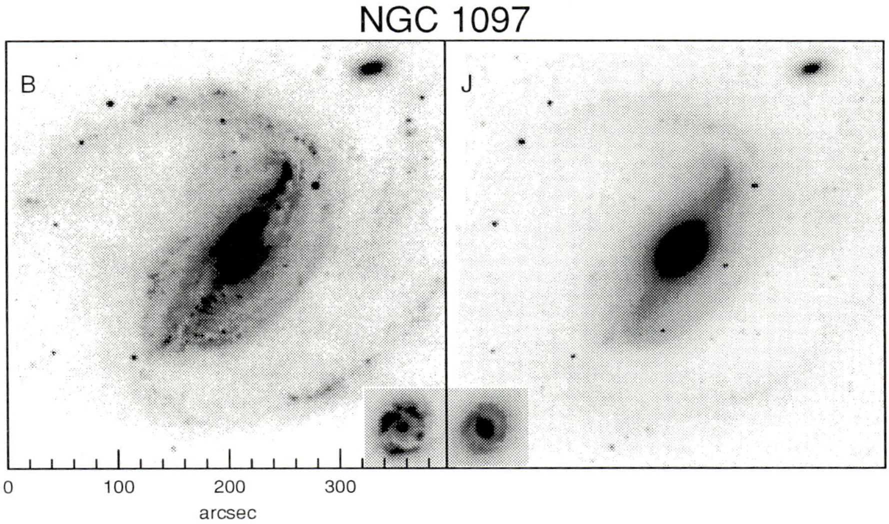

Fig. 2. B and J band images of NGC 1097. Images in the lower central part of the figure are B and J images of the nuclear ring and are at 1/2 scale of the larger images.

2 Estimating the Gravitational Potential from Images

The major force on gas and stars in galaxies is the gravitational potential. An estimate of the gravitational potential in a galaxy can be used for a variety of studies, for example:

- Orbits in barred galaxies
  - Kent & Glaudell (1989) stellar orbits in the barred galaxy NGC 936
  - Quillen et al. (1994a) stellar orbits in the barred galaxy NGC 4314
  - Patsis et al. (1995) a study of orbital families in NGC 4314
- Estimating the torque on the gas and the gas inflow rate
  - Quillen et al. (1994b) in NGC 7479
- Estimating the torque on the stars and the stellar heating rate
  - Gnedin, Goodman & Rhoads (1995) in M100
- Using the potential as a basis for fluid simulations
  - Florida, Paris and Maryland groups all working on the simulation of the fluid dynamics of the ISM in NGC 7479 with SPH, sticky particle and hydro fluid codes respectively.
  - Lindblad, Lindblad & Athanassoula (1995) hydro simulation of the HI in NGC 1365

The gravitational potential is given by the convolution of the the mass (or light) density with the function $1/r$. Our images are observed on a grid so that a quick way to estimate the potential is by using an FFT to do the convolution and no assumption about the shape of the galaxy is required. However, our galaxy images are two-dimensional not three-dimensional, so some assumption about of the vertical structure of the galaxy is needed. Observations of edge-on galaxies find that scale heights of disks are constant as a function of radius (see references cited in Quillen et al. 1994a, in future it would be useful if the vertical structures of 3D N-body simulations are described in detail, so that we can determine how bad an assumption this is). If we assume that the scale height is constant across the disk of the galaxy we can estimate the potential by convolving our images with a function $g(r)$ that is

$$g(r) = \int_{-\infty}^{\infty} \frac{\rho_z(z)dz}{\sqrt{r^2+z^2}}$$

where $\rho_z(z)$ is describes the vertical density distribution of the disk. The function $g(r)$ is equivalent to $1/r$ for $r \gg h$ where h is the vertical scale height of the disk, and levels off for $r < h$, so that the potential will have little structure at scales less than the vertical scale height h.

How well can one estimate the potential? The following things can affect the estimate: the vertical scale height, the choice of vertical density function, the wavelength of image used to trace mass, the bulge-disk decomposition, and projection effects caused by the viewing angle of the galaxy (not discussed here). In this section we discuss the decomposition of the potential into its Fourier components or moments so that each moment can be described in terms of an angle and a magnitude.

2.1 Vertical scale height and choice of vertical density function

Studies of edge-on disks in the IR find that for a galaxy light density $\propto e^{-R/h_d} e^{-z/h}$ that the ratio of scale lengths $q_0 \equiv h/h_d \sim 1/12$. We used values for the vertical scale height determined from this relation as a starting point for our estimate of the potential. We note that it is likely that vertical scale lengths are correlated with Hubble type and that galaxies of an individual Hubble type should have a range of q_0 (Fouqué et al 1990). We note that values for q_0 given by Fouqué et al. (1990) were much larger than those found in optical studies of edge-on disks which were larger than those found from IR studies of edge-on disks (see references in Quillen et al. 1994a). Hopefully modeling of the dust and multi-band statistical surveys will be able to resolve these uncertainties.

We estimated the gravitational potential in a few barred galaxies to determine the effect of varying the vertical scale height. We found that doubling the scale height typically affected the strength of the moments at the 10-15% level and had little effect upon the angles of the moments. We also found that the potential is insensitive (moments are affected less than 1-2%) to the choice of vertical density function for functions with the same $< z^2 >$ (Quillen & Athanassoula 1995).

2.2 Wavelength of Image

Because of young stars and absorption from dust we found a large difference in the strength of the moments as well as the angles between potentials generated from optical images and potentials generated from infrared images. Late-type galaxies usually have optical to infrared radial color gradients in the sense that they become bluer with increasing radius (deJong 1995). This means that IR generated maximal disk rotation curves drop faster than optically generated ones implying that more dark matter is needed at smaller radii than previously found from optical studies. I and (Heraudeau et al. 1995) find from a dozen or so galaxies that K band mass-to-light ratios range from $1.0 - 1.4 M_{\odot}/L_{K\odot}$.

Dwarfs can have color gradients in the sense that the galaxy becomes redder with increasing radius (e.g. in NGC 1705 Quillen et al. 1995a) so that dwarf mass models may not necessarily need to be as halo dominated as previously found. Dwarfs also typically have significantly lower V-K colors than larger galaxies so that infrared mass-to-light ratios for dwarf maximal disk models may be even higher than previously expected. These results are somewhat premature; clearly more careful dynamical and multi-band studies of dwarfs need to be done.

3 Orbits in NGC 4314 and NGC 7479

NGC 4314 is a nice galaxy in which to integrate orbits because it has a small bulge (making it unnecessary to do a bulge-disk decomposition), and it is close to face-on so that no correction for the inclination is necessary. The near-IR colors are constant almost everywhere, so we do not need to correct for changes in the mass-to-light ratio. NGC 4314's high symmetry about the origin, lack of gas, and weak spiral arms imply that the bar is close to stable. NGC 4314 also has prominent peaks in intensity near the end of its bar (which we refer to as knobby ends to the bar) which we would like to understand (see Figure 3).

Two parameters affect the shape of closed orbits inside corotation: h, the vertical scale height and Ω_b the bar pattern speed. We found that if we placed the corotation radius near the end of the bar, that there was an $m = 4$ Inner Lindblad resonance near the location of the knobby features near the end of the bar. The smaller the scale height, the more structure is present in the potential so that h affects ellipticity of orbits almost everywhere, whereas changing Ω_b moves the location of the resonances and causes larges changes in the shapes of the orbits only near the resonances. Because of this we found that we could constrain both parameters.

We found that closed orbits within the $m = 4$ resonance were lozenge shaped and had pronounced peaks near the end of the bar with low speeds (implying high density) nearest the $m = 4$ resonance. This implies that the knobby features at the ends of the bar are caused by the shape of these most peaked lozenge shaped orbits (see Figure 3).

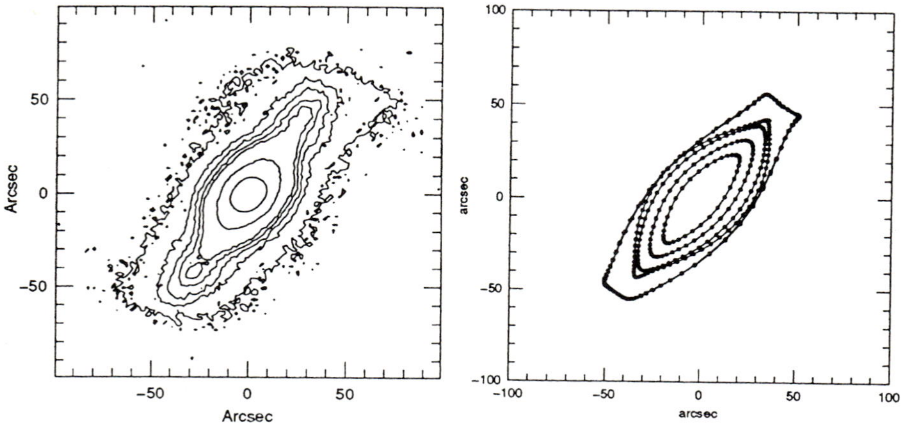

Fig. 3. The left hand side shows contours of the K (2.2μm) image of the face-on barred galaxy NGC 4314. The right hand side shows prograde closed orbits in the frame in which the bar is stationary. These orbits are predicted from the gravitational potential which was estimated from the infrared image. Points are plotted at equal time intervals in a single orbit. The density of the galaxy should increase where the velocity decreases.

We also found that the spiral arms began very close to the location of the $m = 4$ resonance. Studies of other galaxies should determine if this is true in general.

It would be interesting to consider the ellipticity of orbits as a function of radius in late type bars as compared to early type bars. I suspect that the reason early type bars have shoulders in their profile along the major axis is because the orbits have large changes in ellipticity as a function of radius. Late type bars should have a large separation between resonances, and orbits with constant ellipticity as a function of radius so that they have exponential profiles along their major axis. Such a comparison could be done either by searching for closed orbits with numerical integration (as with NGC 4314) or by carefull study of color gradients. In a galaxy with a color gradient, it is possible to constrain the ellipticity of the orbits by comparing the ellipticity of the isohpotes with the ellipticity of the iso-color contours (Quillen et al. 1995b).

I also tried integrating close orbits in NGC 7479 which is highly asymmetric. It was much more difficult to find the closed orbits because of the lack of symmetry in this galaxy. The m=1,3,5 moments increase the number of resonances and also cause bifurcations in the closed orbit families (Patsis 1995). I found it difficult to find orbits that looked like the galaxy for a range of pattern speeds and vertical scale heights. Possible explanations are that I had bad estimates for the potential, or that the system is evolving quickly. Alternatively ergodic orbits may conspire to support to the pattern so that there are few closed orbits that resemble the galaxy.

4 Torque on the gas and the gas inflow rate along the bar in NGC 7479

NGC 7479 is a strongly barred galaxy with star formation occuring along its bar. Its highly asymmetrical spiral arms suggest that it is in a transient or quickly evolving state. In fact, the appearance of the galaxy is similar to the galaxy displayed in the minor merger simulation of Mihos & Hernquist (1994) during the time when gas inflow occurs. In NGC 7479 the molecular gas (traced in CO emission) is coincident with emission in Hα and lies in a linear feature, running along, but slightly offset set from the center of the bar as seen in the near-IR images (see Figure 4). This implies that there is a torque on the gas. Hydro-simulations of gas flow in barred galaxies find that the gas in these linear features is stationary in the frame in which the bar is still, so that that the angular momentum in the gas can be approximated as $l \sim r^2 \Omega_b$ and the torque $dl/dt \sim 2r\Omega_b dr/dt$. The gas inflow rate can be estimated as $\frac{dr}{dt} = \frac{d\Phi/d\theta}{2r\Omega_b}$, so that an estimate of the azimuthal derivative of the gravitational potential at the location of the gas emission can be used to estimate the gas inflow rate. We note that inflow is difficult to determine from velocities observed in the gas because they show large streaming motions. Using our estimate for the gravitational potential we derive a gas inflow rate in NGC 7479 of 10-20 km/s which corresponds to a mass inflow rate of 4 $M_\odot$ per year. We note that this estimate is roughly consistent with that observed in Mihos and Hernquist's' SPH simulation but is somewhat higher than that observed in Athanassoula's hydro simulations.

We eagerly away results of simulations (listed above) to see what values they find are consistent with the observations. In particular we hope that these simulations can constrain what kind of ISM might be consistent with the observations. High inflow rates are not inconsistent with SPH simulations but are large compared to those predicted by sticky particle and hydro simulations. We suggest that the low dissipation required for ring formation is not necessarily inconsistent with high inflow rates along bars if the form of dissipation required to simulate the flow is dependent upon the sheer and compression observed in these shocks.

This work was done in collaboration with the OSU galaxy survey group (R. W. Pogge, D. L. DePoy, J. A. Frogel, R. A. González, S. V. Ramírez, D. M. Terndrup, L. E. Kuchinski, G. Tiede, K. Sellgren, R. Davies), and also

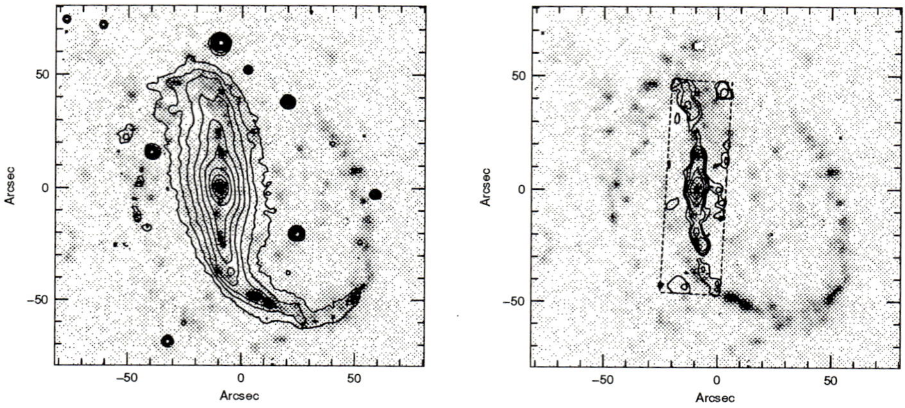

Fig. 4. The left hand side shows an Hα emission gray scale and K (2.2μm) contour overlay of the asymmetric barred galaxy NGC 7479. Note that the Hα is offset from the center of the bar. The right hand side shows an Hα emission gray scale and CO(1-0) emission contour overlay. Note that the $H\alpha$ and CO are almost coincident. The offset of these emission features from the center of the bar imply that there is a torque on the gas. This torque, estimated from the gravitational potential (derived from the infrared image) implies a gas inflow rate along the bar of 10–20 km/s.

J. D. Kenney, and E. Athanassoula. I am grateful to the Observatoire de Marseille for its hospitality and support during Spring 1995.

References

Athanassoula, E., *Bars: Orbit, Gas Flows and Observational Input", 1991, "Dynamics of Disc Galaxies*, ed. B. Sundelius, Göteburg, Sweden, p. 149
De Jong, R. S. 1987, Ph.D. Thesis, Rijksuniversiteit Groningen
Frogel, J. A. 1988, ARA&A, 26, 51
Fouqué, P., Bottinelli, L., Gouguenheim, L., & Paturel, G. 1990, ApJ, 349, 1
Gnedin, O., Goodman, J. & Rhoads, J. ESO/MPA Workshop on Spiral Galaxies in the near IR, ESO Garching b. Munchen, 7-9 June, 1995
Heraudeau, Ph., Simien, F., Mamon G., ESO/MPA Workshop on Spiral Galaxies in the near IR, ESO Garching b. Munchen, 7-9 June, 1995
Kent, S. M., & Glaudell, G. 1989, AJ, 98, 1588
Lindblad, P. A. B., Lindblad, P. O. & Athanassoula, E. 1995, this conference
Mihos, J. C., & Hernquist, L. 1994, ApJ, 425, L13
Quillen, A. C., Frogel, J. A. & González, R. A. 1994a, ApJ, 437, 162
Quillen, A. C., Frogel, J. A., Kenney, J. D, Pogge, R. W., & DePoy, D. L. 1994b, ApJ, 441, 549

Quillen, A. C., Ramírez, S. V., & Frogel, J. A. 1995a, AJ, 110, 205
Quillen, A. C., Ramírez, S. V., & Frogel, J. A. 1995b, submitted to ApJL
Quillen, A. C., & Athanassoula, E. 1995, in preparation
Patsis, P. ESO/MPA Workshop on Spiral Galaxies in the near IR, ESO Garching b. Munchen, 7-9 June, 1995

Dynamical Evolution of Disk Galaxies

J A Sellwood

Rutgers University, Department of Physics and Astronomy, PO Box 849, Piscataway NJ 08855, USA

Abstract. We now believe disk galaxies to be rapidly evolving objects. My aim here is to sketch a few of the main mechanisms that must have caused galaxies to change since their formation and to question whether we may still be underestimating these effects. In the first part, I discuss why the very high amplitude spiral density waves implied by the fascinating new IR data present problems for dynamicists. In the second part I focus on the secular evolution of barred galaxies, and pay particular attention to the dynamical friction problem that is now becoming quite serious.

1 Spiral arms

Spiral density waves provide the only possible form of "viscous" angular momentum transport in stellar discs. The formula for the gravity torque caused by a spiral perturbation in a galaxy appears to have first been written down by Lynden-Bell & Kalnajs (1972). They showed that the gravitational couple across an infinite cylindrical surface parallel to the spin axis of a galaxy is

$$C_z = \frac{1}{4\pi G} \int_S f_R f_\phi dS,$$

where f_R & f_ϕ are the radial and azimuthal parts of the perturbation forces that arise from the spiral density variations. Since the couple depends on their product, its magnitude rises directly as the square of the amplitude of a perturbation of constant shape. For a fixed amplitude, the couple also peaks for open spiral waves with pitch angles around 45°, dropping to zero both for rings and for bars.

The open and intense patterns seen in the near IR (e.g., Rix & Zaritsky 1995), if they indeed reflect a mass contrast of approximately the same magnitude, imply a much larger rate of angular momentum transport than was previously expected (e.g. Bertin 1983). (See also the poster paper by Gnedin

& Goodman, this meeting.) While the consequent more rapid secular changes are of great interest, the high rate of angular momentum transport presents us with a significant puzzle, since it must also strongly "heat" the stellar disk (Lynden-Bell & Kalnajs; Carlberg & Sellwood 1986) – i.e., the random motions of disk stars about circular orbits will increase rapidly.

A rapid heating rate presents two problems: First, it is impossible for the stars to organize themselves into strong spiral patterns if the Toomre Q parameter is large. Since it is unattractive to argue that the observed strong spiral phase is a short-lived episode in the evolution of every galaxy, we must invoke an efficient cooling scheme to balance the rapid heating. Second, it is hard to reconcile with the observed variations in velocity dispersion with age in the Solar neighbourhood, which can be accounted for by spirals of modest amplitude.

Sellwood & Carlberg (1984) argued that a modest rate of infall of fresh gas had to be invoked to maintain the responsiveness of the system. (See also Toomre 1990.) In our picture, gas settles to nearly circular orbits in the disk and then forms stars with small peculiar velocities. Infall cools more effectively than just dissipation in the gas of a closed system, and on-going star formation keeps the gas and stars dynamically coupled to each other (c.f., Carlberg & Freedman 1985). The required infall rate is a strong function of the activity amplitude to be maintained, and would appear to be uncomfortably large for the Rix & Zaritsky sample.

Carlberg & Sellwood found that the comparatively mild spiral patterns in their simulations could account for the observed velocity dispersion in the disk of the Milky Way. In a further study, Jenkins & Binney (1990) concluded that mild spiral patterns, with potential variations corresponding to some (9–13 $\mathrm{km\,s^{-1}})^2$, are all that is required to account for the local stellar kinematics, though stronger, more open spirals could be allowed provided the old disk stars took little part.

The new IR data therefore present a real puzzle and, from a dynamicist's point of view at least, it would be much better if the mass contrast in the arms were not as large as the IR surface brightness variations suggest.

2 Bar pattern speeds

It is highly likely that bars in real galaxies are formed through the global bar instability of stellar disks discovered in some of the first large N-body simulations (e.g., Hohl 1971). The instability leads to a rapidly rotating bar, in the sense that co-rotation lies not far outside the end of the bar. While there is still plenty of scope for more detailed comparisons, there is a broad correspondence between observed photometric and kinematic properties of bars and those in N-body simulations (e.g., Sparke & Sellwood 1987).

A more careful definition of a rapidly rotating bar is one in which the semi-major axis of the bar, a_{B}, is not much less than the distance, D_{L}, from the bar centre to the Lagrange point on the bar major axis. Merrifield & Kuijken (1995) have reported the only reasonably precise direct estimate of a pattern speed in a barred galaxy; they deduce that $D_L/a_B = 1.4 \pm 0.3$ in the SB0 galaxy NGC 936. Moreover, there is indirect evidence for similar pattern speeds from the locations and shapes of dust lanes along bars, which correspond most closely to shocks seen in hydrodynamical simulations when $1.0 < D_L/a_B < 1.4$ (Athanassoula 1992, Weiner et al. 1995).

If this were the end of the story, we might be justified in feeling that we were on the right track towards developing an understanding of barred galaxies. Unfortunately, bars in galaxies also evolve over time in a number of ways, one of which leads to a theoretical prediction that bars should have much lower pattern speeds.

After a bar has formed, it can continue to grow in length by trapping more disk stars through interactions with slower spiral patterns (Sellwood 1981). It is also expected to buckle out of the plane (Raha et al. 1991) and form a thicker peanut-shaped object (Combes & Sanders 1980, Combes et al. 1990). It will be braked through dynamical friction against the bulge/halo (Sellwood 1980, Weinberg 1985). Moreover, shocks within the bar will quickly cause any gas in the bar region to lose angular momentum and settle towards the centre (e.g., Friedli & Benz 1993). If enough gas can be driven into a very dense central concentration, the bar may self-destruct (e.g., Norman et al. 1996). I do not discuss all these inevitable secular changes here, but focus on the puzzle presented by dynamical friction.

3 Dynamical friction

A bar rotating within a massive distribution of bulge stars and halo particles must experience dynamical friction. Even though the halo particles are expected to have have large random velocities, the perturbations to their orbits caused by the motion of the bar will be such as to produce slightly overdense regions lagging the bar, and the gravitational attraction between the overdense regions and the bar will remove angular momentum from the bar and torque up the halo. My first rather crude 3-D simulations with a population of halo particles (Sellwood 1980) revealed quite a large torque, but were not run for long enough to determine the final fate of the system. Weinberg (1985), using a semi-analytic approach, estimated for a set of reasonable parameters that the bar loses some 20% of its angular momentum in each rotation period!

Recently, Debattista & myself (in preparation) have returned to this problem equipped with much more powerful computers, and similar work is also being carried out by Athanassoula. In models integrated for 40 initial bar rotation periods, we find that fully one third of the angular momentum of the initial disk is transferred to the halo before the torque drops to almost zero. Interestingly, the halo in this "final" state has not been spun up sufficiently to co-rotate with the bar, but many particles appear to be locked in resonance with it.

This huge angular momentum loss causes the bar to slow down dramatically; the pattern speed of the bar drops to about one fifth of its value at the time of its formation, roughly in proportion to the fraction of angular momentum lost from the bar region. As shown in Figure 1, a_B grows only slowly over this time, while D_L more than trebles. The "final" state therefore is one in which $D_L/a_B > 2$ – a value quite outside the range allowed by the observational evidence summarized above.

Our simulations are therefore demonstrating that Weinberg's alarming prediction was at least qualitatively correct, though we have yet to make a detailed quantitative comparison. The total mass of our bulge/halo is merely twice that of the disk and it is not at all extensive; its half-mass radius is about equal to the outer radius of the initial disk. Since increasing the halo density and/or making it more extensive would increase the torque, our modest parameters are probably causing us to *underestimate* the drag on the bar. There therefore seems little room to escape from a clear contradiction between theory and observation.

Our simulations have shown that bars are in fact quite astonishingly robust. Thus a possible solution that the bar could not survive such fierce braking is ruled out by our fully self-consistent models.

We cannot yet definitively exclude the possibility that bars in real galaxies grow in length as they are braked by the halo; the disk in our models was not extensive enough to permit continued strong spiral activity. However, a_B would have to nearly double to keep D_L/a_B in the required range, which seems extreme.

Other explanations of the disagreement, such as bars in real galaxies are weak, or halos are either not very massive or co-rotate with the bar, etc., also seem unattractive. One still more wildly speculative solution is that bars in real galaxies are transient features that do not survive long enough to be slowed down; to sustain this hypothesis, we require bars to reform frequently in order to account for the observed numbers of bars. The problems presented by this idea may be even more extreme than that which motivated it, but it is fun to explore what other advantages it might offer.

4 Transient bars?

We have already found that bars are robust, but they can be destroyed in two possible ways. One obvious way would be to hit a bar hard enough with an intruder; Athanassoula (1995) is exploring whether this can be achieved without also puffing up the disk to an excessive extent (e.g., Tóth & Ostriker 1992). Many groups (e.g., Norman et al. and references therein) have developed a second idea that gas inflow might build up a sufficient central mass concentration to destabilize the loop orbits that support the bar.

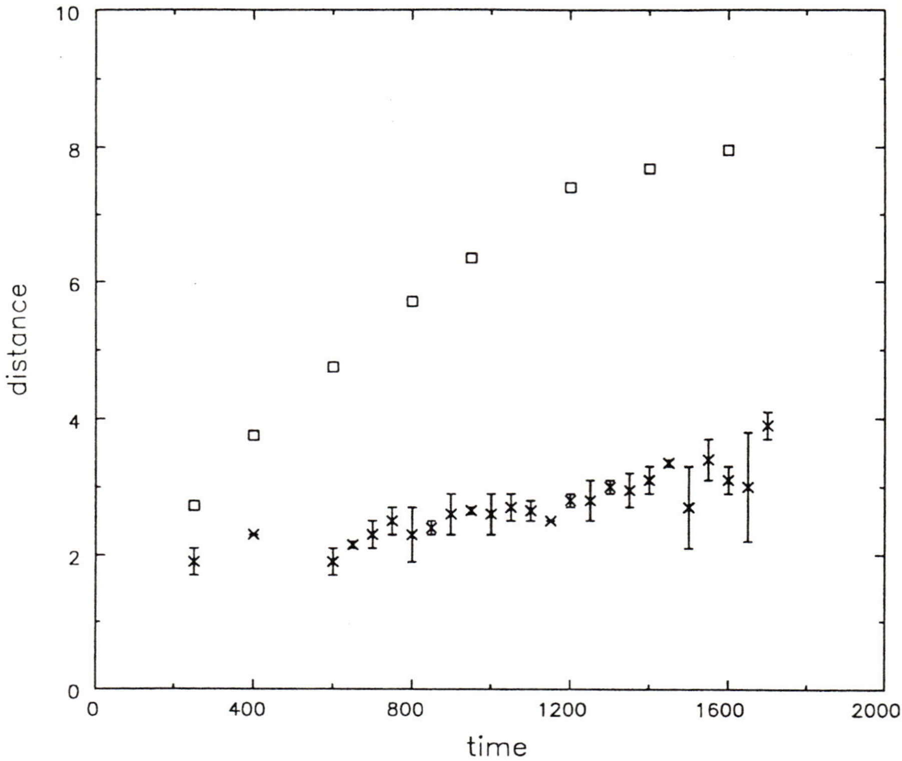

Fig. 1. The time evolution of a fully self-consistent disk-halo barred galaxy model. The abscissae are dynamical times and the total evolution corresponds to some 40 bar rotation periods. Open symbols show the distance from the bar centre to the major axis Lagrange point ($D_{\rm L}$), crosses show the semi-major axis of the bar ($a_{\rm B}$).

If we are to use either of these ideas to account for the absence of slowly rotating bars in galaxies, then they must operate quite quickly since the period before halo braking becomes substantial is quite short (see Figure 1). Then in order to account for the observed 30% frequency of bars in galaxies (e.g., Sellwood & Wilkinson 1993), we would need a new bar to form within the same galaxy. Such an occurrence would seem unlikely for two reasons:

(1) Disk stars will have been stirred into quite eccentric orbits by the formation, and then the destruction of the first bar. The system will require significant cooling before the disk becomes dynamically responsive enough to form a second bar.

(2) The central mass concentration that caused the bar to dissolve will itself have a stabilizing effect (Toomre 1981) making a second collective linear instability still less likely.

The cooling demands of item (1) are substantial, but I noted above that we may already be forced by IR observations of spiral arm amplitudes to infer a much higher infall rate than we had hitherto assumed. Item (2) is not insuperable either: if the disk is responsive, it can still be unstable to perturbations of finite amplitude (Sellwood 1989), such as could be caused by a passing companion galaxy. Both these arguments need to be stretched, perhaps unreasonably, but the idea of a second bar cannot yet be excluded.

If the above two serious difficulties can be overcome to allow a second bar to form in the same galaxy, do we have to strain credibility still further by requiring a truly recurrent cycle of bar formation and destruction? A bar phase lasting some 30% of any one cycle might naïvely seem called for, to account for the observed bar frequency. But perhaps we do not need to invoke such an extreme hypothesis, since our objective here is to account for the absence of slow bars. If the first bar torques up the halo sufficiently before being destroyed, it may be possible that the friction force on the second bar could be feeble enough to evade the pattern speed problem. Such an idea is largely wishful thinking at this stage, though there could be something to it: we have tried experiments in which we altered the azimuthal coordinates of halo particles at random, after the torque had dropped to a low value; "scrambling" the halo in this way led to only a mild new torque that decayed quickly and caused little fresh braking.

The picture would then be that a first bar formed early in the life of almost every galaxy, it quickly torqued up the inner bulge/halo before being destroyed – thereby adding yet more rapidly rotating material to the bulge (c.f. Norman et al.). The disk is rejuvenated by continued substantial infall which allows some 30–50% of galaxies to reform a bar, perhaps triggered by a passing companion. The frequency of bars, for which no convincing explanation has yet been given, has already been noted to be higher in galaxy groups and clusters (Elmegreen et al. 1990). The second generation bars must experience a much weaker torque from the halo, which could be possible if it has been torqued up sufficiently by the first bar.

A side advantage of this wildly speculative scenario is that the first generation bars will be driving gas into their centres at about the right epoch to account for the age dependence of the QSO luminosity function.

5 Conclusions

Many different arguments indicate that the structure of a disk galaxy evolves; the mere existence of spiral arms *requires* them to change over time. Most dynamicists used to think that secular changes are slow and largely insignificant, but this complacency is becoming difficult to sustain.

The near IR observations seem to be telling us that spiral patterns have much higher amplitudes than thought hitherto, presenting new challenges to the theory of disk dynamics. Bars evolve in many ways after their formation, which may lead to the addition of inner disk stars to bulges (e.g., Norman et al). One further secular effect caused by dynamical friction is only now being addressed by adequate N-body simulations which indicate that there are limited ways to escape from the pattern speed problem first clearly stated by Weinberg. These ideas may require a fundamental revision to our understanding of how galaxies evolve.

Acknowledgments

This work was supported by NSF grant AST 93/18617 and NASA Theory grant NAG 5-2803.

References

Athanassoula, E. (1992): The existence and shapes of dust lanes in galactic bars. MNRAS **259**, 345

Athanassoula, E. (1995) In *Barred Galaxies*, IAU Colloquium **157**, Eds. R. Buta, B. F. Elmegreen, D. A. Crocker, (to appear).

Bertin, G. (1983): On the dynamical evolution of spiral galaxies. A&A **127**, 145

Carlberg, R. G., Freedman, W. L. (1985): Dissipative models of spiral galaxies. ApJ **298**, 486

Carlberg, R. G., Sellwood, J. A. (1986): Dynamical evolution in galactic disks. ApJ **292**, 79

Combes, F., Sanders, R. H. (1980): Formation and properties of persisting stellar bars. A&A **96**, 164

Combes, F., Debbasch, F., Friedli, D., Pfenniger, D. (1990): Box and peanut shapes generated by stellar bars. A&A **233**, 82

Elmegreen, D. M., Bellin, A. D., Elmegreen, B. G. (1990): Statistical evidence that galaxy companions trigger bars and change the spiral Hubble type. ApJ **364**, 415

Friedli, D., Benz, W. (1993): Secular evolution of isolated barred galaxies. A&A **268**, 65

Hohl, F. (1971): Numerical experiments with a disk of stars. ApJ **168**, 343

Jenkins, A., Binney, J. (1990): Spiral heating of galactic discs. MNRAS **245**, 305

Lynden-Bell, D., Kalnajs, A. J. (1972): On the generating mechanism for spiral structure. MNRAS **157**, 1

Merrifield, M. R., Kuijken, K. (1995) The pattern speed of the bar in NGC 936. MNRAS **274**, 933

Norman, C. A., Sellwood, J. A., Hasan, H. (1996): Bar Dissolution and bulge formation. ApJ (to appear)

Raha, N., Sellwood, J. A., James, R. A., Kahn, F. D. (1991): A dynamical instability of bars in disk galaxies. Nature **352**, 411

Rix, H., Zaritsky, D. (1995): Nonaxisymmetric Structures in the Stellar Disks of Galaxies. ApJ **447**, 82

Sellwood, J. A. (1980): Galaxy models with live halos. A&A **89**, 296

Sellwood, J. A. (1981): Bar instability and rotation curves. A&A **99**, 362

Sellwood, J. A. (1989): Meta-stability in galactic discs. MNRAS **238**, 115

Sellwood, J. A., Carlberg, R.G. (1984): Spiral instabilities provoked by accretion and star formation. ApJ **282**, 61

Sellwood & Wilkinson 1993): Dynamics of Barred Galaxies. Rep. Prog. Phys. **56**, 173

Shaw, M. A., (1987) The nature of 'box' and 'peanut' shaped galactic bulges. MNRAS **229**, 691

Sparke, L. S., Sellwood, J. A. (1987) Dissection of an N-body bar. MNRAS **225**, 653

Toomre, A. (1981): What amplifies the spirals? In *Structure and Evolution of Normal Galaxies*, Eds. S. M. Fall & D. Lynden-Bell, Cambridge Univ Press, p. 111

Toomre, A. (1990): Gas-hungry Sc spirals. In *Dynamics & Interactions of Galaxies*, Ed. R. Wielen, Springer-Verlag:Berlin, Heidelberg, p. 292

Tóth, G., Ostriker, J. P. (1992): Galactic disks, infall, and the global value of Omega. ApJ **389**, 5

Weinberg, M. D. (1985): Evolution of barred galaxies by dynamical friction. MNRAS **213**, 451

Weiner, B., Sellwood, J. A., Williams, T. B. (1995): In *Barred Galaxies*, IAU Colloquium **157**, Eds. R. Buta, B. F. Elmegreen, D. A. Crocker, (to appear).

Disk Dynamics Based on Near-IR Photometry[2]

Preben Grosbøl[1] and Panos Patsis[2]

[1]European Southern Observatory, Karl-Schwarzschild-Str. 2, D-85748 Garching
[2]Max-Planck-Institute für Astronomie, Königstuhl 17, D-69117 Heidelberg

Abstract. Surface photometry of disk galaxies at 2 μm gives a much better indication of the mass distribution in their disk than that obtained from visual bands which are dominated by dust extinction and population effects. Perturbations in the disk of such systems (*e.g.* spiral patterns) can yield information on their dynamic state. Procedures for extracting and analyzing parameters of the spiral pattern in ordinary galaxies are discussed using K′ band photometry. General quantities for strength and shape of these perturbations are related to the possible location of major resonances in the galaxies. Finally, dust features located on (B-K′) maps are compared to the perturbation derived from near-IR.

1 Introduction

The detailed mass distribution or potential of a galaxy is important for the understanding of the dynamics of its disk. Rotation curves provide a good estimate of the general radial variation of the total potential but do not revile the relative distribution of matter between different components in the galaxy (*i.e.* bulge, disk and halo) nor its high spatial frequencies. Surface photometry maps suggest, on the other hand, the relative variation of luminous matter also at small scales. Major problems in deriving projected column densities from the observed luminosity distribution are attenuation of light due to dust and population effects which change the effective mass-to-light ratio. In addition assumptions on the three dimensional structure of matter must be made to estimate the potential.

The effects of dust and population variations are much larger at wavelength $\lambda < 1\mu$m that in the near infrared region (NIR). To derive quantities

[2] Based on observations collected at the European Southern Observatory, La Silla, Chile

related to the intrinsic mass distribution, it is essential to use a bandpass where such effects are as small as possible since corrections are very uncertain. The low surface brightness of galactic disks requires the use of relative broad bands where dust and population effects are very difficult to distinguish. The attenuation due to dust depends on the geometry and structure of the interstellar medium (Witt *et al.* 1992). Further, the problem becomes even more complex for the study of spiral structure due to the likely spatial correlation between population changes and properties of the interstellar medium. This suggests that the K′ filter (Wainscoat and Cowie 1992) is best suited for such studies having the longest mean wavelength still not significantly effected by thermal background emission. Although the 2.1 μm bandpass is much less effected by dust and Population I objects (Rix and Rieke 1993) than bands at shorter wavelength, these effects still remain a concern as the contribution from red super giants and OB associations (*e.g.* Brγ emission) is difficult to estimate.

This paper presents K′ surface photometry of 5 Sb-Sc galaxies and considers the possible implication on the disk dynamics of these systems. The observations and extraction of intrinsic parameters for the galaxies are described in the two next sections while the interpretation and conclusions are given in two last sections.

2 Observations and reductions

Five ordinary spiral galaxies were selected for the study of disk dynamics and spiral structure as listed in Table 1. Their Hubble type and systemic velocity taken from RSA are given while diameter D_{25} and projection parameters (PA,I) listed are from RC3. The physical scale is derived using a Hubble constant H_0=75 km s^{-1}Mpc^{-1}. The galaxies include both flocculent and grand design spirals (Elmegreen and Elmegreen 1982) and were chosen to have intermediate inclination angles allowing kinematic data to be obtained.

Table 1. Parameters for spiral galaxies observed

Galaxy	Type	v_0	D_{25}	PA	I	Scale
NGC 3223	Sb(s)I-II	2619 km/s	4.′1	135°	53°	220 pc/″
NGC 5085	Sc(r)I-II	1720 km/s	3.′4	38°	29°	140 pc/″
NGC 5247	Sc(s)I-II	1143 km/s	5.′6	20°	29°	95 pc/″
NGC 5861	Sc(s)II	1725 km/s	3.′0	150°	57°	245 pc/″
NGC 7083	Sb(s)I-II	2951 km/s	3.′9	5°	53°	250 pc/″

The K′-band observations were made with the 2.2m ESO/MPI telescope at La Silla in May 1993 using the IRAC-2A instrument with a 256 × 256 NICMOS-3 detector. A pixel scale of 0.″5 was used making it necessary to map the galaxies with mosaics consisting of 2-6 fields depending on their size and sky projection. A stack of 5 frames with 3″ relative offsets was made for each field with interleaving sky exposures. All frames were the mean of 90 2 s integrations yielding a total exposure time of 900 s on both object and sky. Visual maps in BVI were also obtained at the 2.2m with EFOSC2. The average seeing during the observations was 1″.

The object frames were sky subtracted using the mean of two nearby sky exposures from which stellar images were removed. After flat field correction, the stacks were averaged rejecting all outliers compared to the median of the 5 frames. This removed all bad pixels since the relative offsets were larger that the largest blemish. The individual fields were combined using stars in the overlap regions. The final surface photometry maps were calibrated using the aperture measurements in K by de Vaucouleurs and Longo (1988), however, no color term was applied. The transformation agreed with that derived from standard stars suggesting an uncertainty of 0.1 mag. A signal-to-noise ratio of 10 for a single pixel was reached at a level of K′≈ 17.6 mag/arcsec^2.

3 Extraction of intrinsic parameters

It is necessary to make a number of assumptions on the major structural components and their geometry to derive intrinsic parameters from the sky projected intensity maps. In the following, the luminous matter in a galaxy is modeled as a spherical bulge and a thin plane disk in which bar or spiral perturbations may exist. Since the K′ surface brightness mainly maps the distribution of Population II stars this is a reasonable assumption considering that only the inner disk, unlikely to be effected by warps, can be observed.

Table 2. Sky projection and disk parameters for galaxies

	disk		θ_2 phase		disk			
Galaxy	PA	I	PA	I	range	K'_0	h	α
NGC 3223	128.°2	46.°8	127.°1	44.°4	20–45″	16.5	23.″9	-1.28
NGC 5085	44.°0	40.°0	60.°0	47.°7	30–55″	17.7	50.″5	-0.76
NGC 5247	16.°9	27.°4	31.°0	24.°7	50–90″	18.7	65.″8	-0.97
NGC 5861	152.°2	56.°4	152.°7	55.°0	16–55″	17.1	25.″8	-1.11
NGC 7083	4.°5	54.°8	8.°1	50.°6	16–40″	16.2	14.″4	-1.87

The three standard procedures to determine the sky projection parameters for a spiral galaxy are to fit: a) a detail velocity map assuming mainly circular rotation in its disk, b) a 2 dimensional exponential disk to its outer parts avoiding regions where warps or asymmetries are seen, and c) the radial phase variation of the spiral arms (Danver 1942) assuming a smooth intrinsic variation. Due to lack of detailed velocity information, only the two latter methods could be applied. The accurate determination of the sky projection is important for identification of weak bars or oval distortions in the central region of the galaxies. Both methods were applied to provide an independent estimates with the results listed in Table 2. The formal error is of the order of 2° in both cases. The estimates agree well except for NGC 5085 and NGC 5247 which both have strong spiral perturbations in their outer disks. The projection parameters derived from the phase variation of the two armed spiral components were adopted since they are less effected by such perturbations. For NGC 5085, long slit spectra were available for PA = 24° and 90° suggesting that the major axis of this galaxy was located between these two direction in agreement with the adopted PA. The projection parameters were also used to correct the visual maps.

The intrinsic parameters of the galaxies were obtained by first removing all obvious foreground stars and then transforming the intensity maps into polar coordinates in the plan of the disk with origin at the central K′ intensity peak and using the projection parameters derived. Both direct and polar representations of NGC 3223 and NGC 5861 are shown in Fig. 1 where the intensities relative to the axisymmetric disk are given for the polar maps. The azimuthal intensity variation was Fourier transformed yielding amplitudes a_m and phases θ_m of the mth harmonics as function of radius. The mean value a_0 gives the axisymmetric intensity profile while *e.g.* the two armed spiral pattern is associated to the $m = 2$ component.

4 Discussion

All five galaxies show grand design two armed spirals over the entire disk in K′ whereas NGC 3223, NGC 5085 and NGC 7083 have partly filamentary, multiple-armed patterns on B images. This clearly indicates that the distribution of dust to a large extend is decoupled from that of the old stellar population as found by Block and Wainscoat (1991) and Block *et al.* (1994). Thus, the morphology of spiral galaxies determined from maps in visual bands may suggest some properties of the interstellar media but is only marginally related to the mass distribution and perturbations in the disk.

4.1 Axisymmetric profile

The radial K′ profile and (B-K′) color index of NGC 5085 and NGC 7083 are shown in Fig. 2 as examples of galaxies with shallow and steep color gradients. The radial region in which an exponential disk could be identified as the main component was estimated visually. This range and the fitted values for the disk are listed in Table 2 where K'_0 and h are the central intensity and scale length of the exponential disk, respectively. A power law with exponent α was also fitted to the disks (see Table 2). It was not possible to determine which of the two models (*i.e.* power law or exponential) were the better due to the small radial range. The surface brightness of the disks starts to decrease more rapidly just outside the range specified where also the spiral

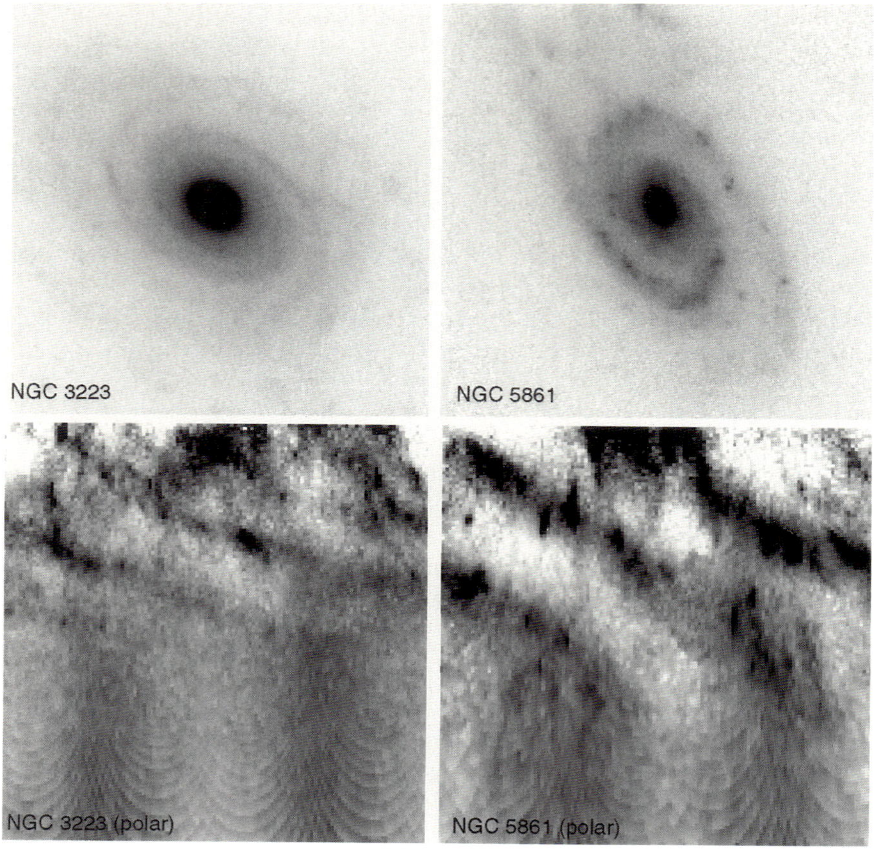

Fig. 1. K′ surface brightness maps in Cartesian and polar θ-ln(r) coordinates for NGC 3223 and NGC 5861 (foreground stars removed). The polar maps show relative intensities normalized to the mean radial profile

pattern ends. The break can be seen for all the galaxies but is strongest for NGC 5085 and NGC 5247. The transition from an inner to an outer disk is better observed using a power law representation in which case the exponent changes abruptly by a factor of two or more. The central regions have (B-K′)≈ 4.5 mag while the disks have slightly lower values in the inner parts and become bluer outwards. NGC 7083 has the strongest color gradient reaching (B-K′)≈ 2.5 mag in its outer disk region. A combination of dust, population and metallicity effects may give raise to these gradients (de Jong 1995).

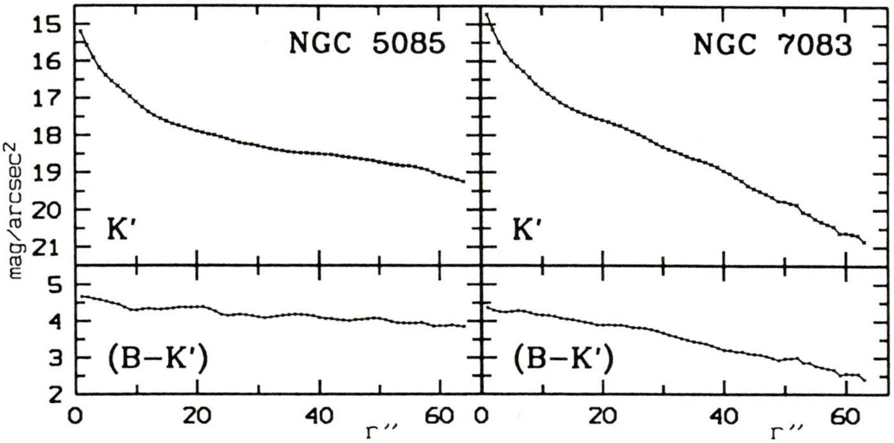

Fig. 2. Radial K′ intensity and (B-K′) color profiles for NGC 5085 and NGC 7083

4.1 Oval distortions

All the galaxies are classified as normal spirals and have not sign of bars on the visual images. The radial phase variation of the $m = 2$ harmonics (see Fig. 3) can be used to identify even weak oval distortions which appear as regions with constant phase. The innermost part of the phase variation has $\theta_2 = 90°$ due to a projection artifact of the bulge which was not removed. Disregarding this part, weak oval distortions may be present in NGC 5085 and NGC 5861 with relative amplitudes around 0.1 in both cases.

4.2 Pitch angle of spiral pattern

The spiral arms are well represented by logarithmic spirals as seen on Fig. 3 where they appear as straight line with a slope defining their pitch angle. The pitch angles i derived from both K′ and V maps are given in Table 3. For all five galaxies, the blue spiral arms are tighter wound than the K′ spirals (see Fig. 3). This suggests that the spiral perturbation is caused by a density wave (Lin and Lau 1979; Bertin *et al.* 1989a, 1989b) rather than by material arms. Stars and dust formed after a galactic shock in the gas associated to a spiral density wave would drift away due to differential rotation and may create this effect while a stochastic star formation process would not. The different in pitch angle i_t between two epochs of stars can be approximated by:

$$\cot(i_{t_1}) - \cot(i_{t_0}) = \frac{d\Omega}{d\ln(r)}(t_1 - t_0) \tag{1}$$

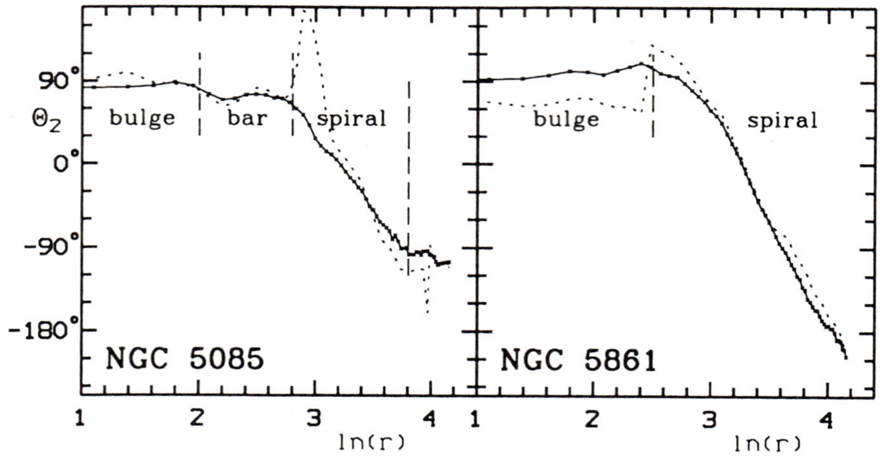

Fig. 3. Radial phase variation of the $m = 2$ harmonics for NGC 5085 and NGC 5861 in K′ (—) and V (···). Bulge, bar and spiral regions are indicated

where Ω is the angular velocity of the stars in the disk. It is unfortunately not possible to use the phase offset between the red and blue spirals to determine the location of co-rotation since the absolute phase difference is unknown. A significant part of the phase offset may be due to attenuation of dust rather than to an age difference in the stars (Yuan and Grosbøl 1981).

Table 3. Extend, amplitude and pitch angles of the spiral patterns

Galaxy	i_2^K	i_2^V	$r_{-2:1}$	$r_{-4:1}$	r_{co}	a_2
NGC 3223	-11°.5	-10°.6	20″	37″	45″	0.11
NGC 5085	-19°.1	-13°.4	18″	42″	55″	0.20
NGC 5247	-35°.1	-32°.8	16″	-	90″	-
NGC 5861	-14°.4	-13°.6	20″	44″	55″	0.15
NGC 7083	-29°.4	-	12″	23″	40″	0.18

4.3 Amplitudes

The radial variation of the relative amplitudes a_2 and a_4 of the spirals is shown in Fig. 4 for four of the galaxies. The relative amplitude for NGC 5247 cannot be obtained reliably due to the very low inter-arm K′ surface brightness. Numerous small knots along the arms are seen on the K′ images (see Fig. 1). Their size suggest that they are associated to a very young population since they otherwise would have dissolved due to intrinsic velocity dispersion and differential rotation. This will cause some overestimate of the amplitudes a_2 and a_4 although high frequencies will be more effected.

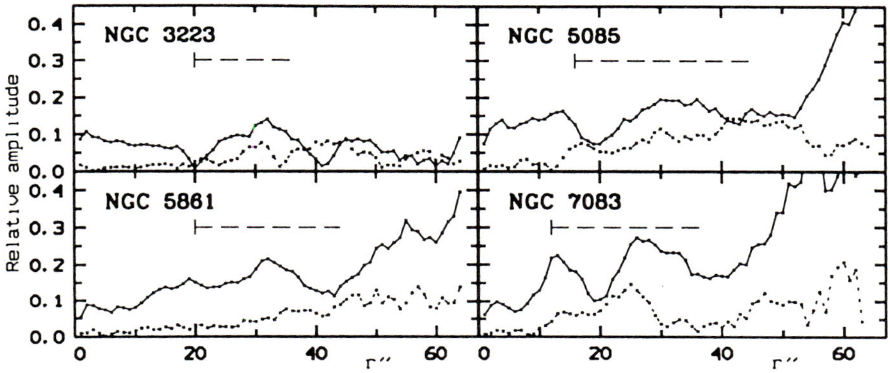

Fig. 4. Relative amplitudes of the spiral a_2 (—) and a_4 ($\cdots$) as function of radius for NGC 3223, NGC 5085, NGC 5961 and NGC 7083. The radial interval shown for each galaxy indicates the region of the main two armed spiral pattern

The main logarithmic spirals have relative amplitudes a_2 in the range of 0.1-0.2 with the tightest pattern having the smallest amplitude (*i.e.* NGC 3223). In the outer part, larger amplitudes may be found either due to warps or very low inter-arm surface brightness. The a_4/a_2 ratio for the spiral arms has typical values in the interval 0.3-0.5 which suggests that the spiral perturbation is marginally non-linear (Grosbøl 1993).

4.4 Resonances

The pattern speed of the underlying density wave is an important parameter for the disk dynamics but cannot be determined directly. It can be derived if the location of major resonances (*e.g.* 2/1, 4/1 or co-rotation) can be identified. The Inner Lindblad Resonance (ILR) is expected to be located just inside the radius where the two armed spiral in the old stellar disk starts since a stellar density wave would be absorbed there (Lin and Lau 1979). Two of the galaxies, NGC 3223 and NGC 5085, show significant peaks in a_4/a_2 at both the start and end of their logarithmic spiral while NGC 5861 and NGC 7983 only have peaks at its outer termination. The a_4/a_2 peaks at the end of the strong symmetric part of the pattern may be associated to the 4/1 resonance (Contopoulos and Grosbøl 1988). A possible location of the co-rotation is the radius where the axisymmetric K′ profile shows a break as this coincides with the end of the spiral structure. The locations for resonances as suggested by these consideration are given in Table 3 and may be used to determine the pattern speed of the density wave when combined with an accurate rotation curve.

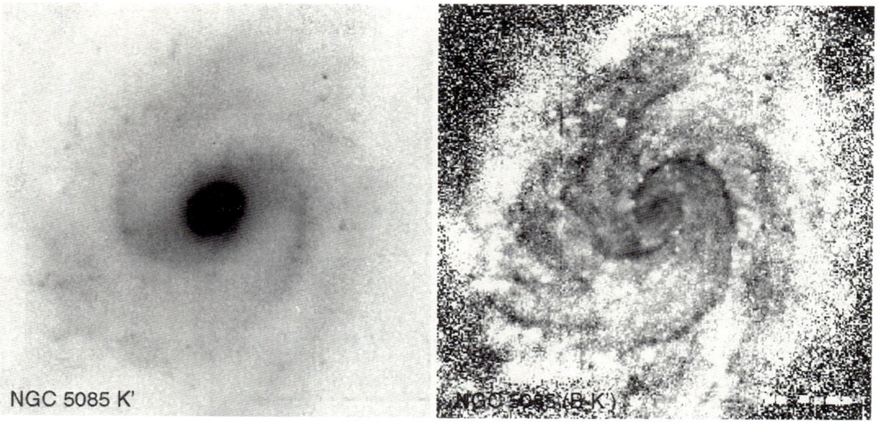

Fig. 5. K′ and (B-K′) maps of NGC 5085 (foreground stars removed). Black on the (B-K′) image corresponds to red colors (*e.g.* dust lanes) while white represent bluer regions

The location of dust lanes relative to the spiral perturbation can also be used to locate resonances. They are seen clearly on (B-K′) color maps as shown in Fig. 5 for NGC 5085 where also a direct K′ image is given for reference. For this galaxy, the dust spiral almost reaches the center whereas the stellar spiral determined from the K′ map first starts at a radius of $\approx 16''$. Further, the phase difference between dust/gas and stellar pattern becomes large around this radius in agreement with the predictions of the density wave theory (Lin and Lau 1979) near ILR.

5 Conclusions

The small size of the knots observed along the spiral arms on the K′ images suggests that they are associated to Population I objects. The intensity of these knots imply that the contribution of light in K′ from young objects must be considered and the amplitude of spiral arms derived from K′ photometry is an upper limit. Further, the study of the 5 spirals suggests that:

1. The dust/gas distribution is decoupled for that of the old stellar disk population.

2. The maximum amplitude $a_2 \lesssim 0.2$ and the ratio $a_4/a_2 \approx 0.3-0.5$ suggest that the spiral perturbations are marginal non-linear.
3. The radial amplitude and phase variations of the $m = 2, 4$ spiral modes indicate the location of resonances.
4. The dust/gas spiral may continue further in towards the center of the galaxies that the stellar spiral.
5. The variation of the pitch angle of spirals as function of color suggests that the spiral perturbation is due to a density wave.

Although K band photometry still should be correction for population effects, it is still much better, even uncorrected, that other band for the study of the mass distribution in the disk of galaxies. The position and shape of spiral perturbation can be determined with high accuracy which makes a detailed comparison of between observations and dynamic models possible.

References

Bertin, G., Lin, C.C., Lower, S.A., Thurstans, R.P., (1989a): ApJ **338**, 78
Bertin, G., Lin, C.C., Lower, S.A., Thurstans, R.P., (1989b): ApJ **338**, 104
Block, D.L., Bertin, G., Stockton, A., Grosbøl, P., Moorwood, A.F.M., Peletier, R.F. (1994): A& A **288**, 365
Block, D.L., Wainscoat, R.J. (1991): Nat **353**, 48
Contoupolos, G., Grosbøl, P. (1988): A& A **197**, 83
Danver, C.-G. (1942): Ann. Obs. Lund No. **10**
de Jong, R.S. (1995): Ph.D Thesis, Groningen
de Vaucouleurs, G., de Vaucouleurs, A., Corwin, Jr.,H.G., Buta, R.J., Paturel, G., Fouque, P. (1991): Third reference catalogue of bright galaxies (RC3), Springer, New York
de Vaucouleurs, A., Longo, G. (1988): Monograhs in Astronomy No. **5**, Austin, Texas
Elmegreen, D.M., Elmegreen, B.G. (1982): MNRAS **201**, 1021
Grosbøl, P. (1993): PASP **105**, 651
Lin, C.C., Lau, Y.Y. (1979): Stud. Appl. Math. **60**, 97
Rix, H.-W., Rieke, M.J. (1993): ApJ **418**, 123
Sandage, A., Tammann, G.A. (1981): A Revised Shapley-Ames Catalog of Bright Galaxies (RSA), Carnegie Inst. Wash. Pub. No. **635**, Washington, D.C.
Wainscoat, R.J, Cowie, L.L (1992): AJ **103**, 332
Witt, A.N., Thronson, Jr.,H.A., Capuano, Jr.,J.M. (1992): ApJ **393**, 611
Yuan, C., Grosbøl, P. (1981): ApJ **243**, 432

Measuring Spiral Arm Torques: M100 in K

Oleg Y. Gnedin, Jeremy Goodman, and James E. Rhoads

Princeton University Observatory, Princeton, NJ 08544, USA

Spiral arms exert gravitational torques on stars that lead to redistribution of angular momentum within galactic disks: in the inner part stars lose angular momentum and fall towards the center of galaxy and in the outer part stars gain angular momentum and move farther away from the center. These torques depend only on the non-axisymmetric mass distribution and can be measured directly from photometry, if light traces mass. The method to calculate the gravitational torques is described in Gnedin, Goodman, and Frei 1995 (hereafter GGF).

We used the GRIM II camera on the Apache Point Observatory 3.5 meter telescope to obtain a K' band (2.1μ) image of M100 (NGC 4321) for comparison with GGF's optical I,R, & V band CCD images. Due to the small field of view of the camera ($123''$) we used a mosaic of 187 15-second exposures to cover the inner $4' - 5'$. Each exposure was sky-subtracted, flat-fielded, and corrected for a fluctuating bias level in the camera electronics (see also Rhoads 1995). Foreground stars were removed using Z. Frei's semi-automatic procedure (Frei 1995). The resulting image is shown in Fig. 1.

The image was deprojected according to $i \sim 27°$, $PA \sim 158°$ (GGF). We decompose the surface light distribution into polar Fourier harmonics. Corresponding Fourier harmonics are in good agreement among the three optical bands. Generally, the harmonics in K follow the main optical features but have more noise. However, the $m = 1$ component is unusually bright in K. Computation of a rotation curve from the axisymmetric light and comparison with kinematic observations yields $M/L_K \approx 1.16$ in solar units.

To use photometry in the study of dynamical evolution of the stellar disks of galaxies we need a good tracer of the underlying mass (presumably in old stars). Studies by Rix (1993) and Rix & Zaritsky (1995) suggested K band light is the best indicator of the mass. However, Rhoads (1995) showed for NGC 1309 that young red supergiants may contribute a significant fraction of the galactic emission in K.

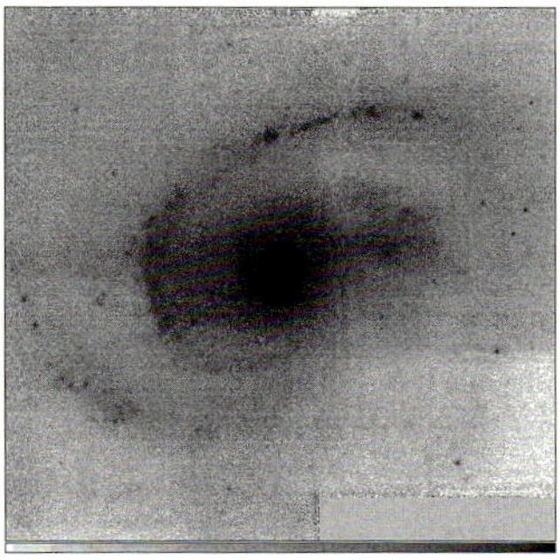

Fig. 1. M100 in K.

We have compared azimuthal brightness variations for five selected radii. We found that the variations of **even** Fourier harmonics have same amplitude and shape for all colors, from 0.5μ to 2.1μ. The maximum brightness contrast between arm and interarm regions is about 1.2 mag, or a factor of 3, at intermediate radii in M100. However, examine carefully fig. 1. In K band, the upper spiral arm appears much stronger than its lower counterpart. Bright knots along the arm are presumably young red supergiants in OB associations. They correlate with similar knots seen in optical V in GGF. Thus we observe that young stars are prominent in K. Since the Fourier harmonics represent global (azimuthally integrated) non-axisymmetric features, the observed $m = 1$ enhancement should have a serious impact on photometrically derived properties of spirals (such as lopsidedness of stellar disks).

Thus our first result is: we report that **young red supergiants in M100 contribute a significant fraction of the K light**, as proposed recently by Rhoads (1995). Therefore, K light may not be the best tracer of the old stellar population. We suggest that any conclusions concerning *dynamics* of spiral galaxies should rest on multi-band photometry, and not on K alone.

Second, we confirm our measurement of the torque reported in GGF, since it is dominated by the $m = 2$ component. Consistency among all bands in the $m = 2$ component indicates the robustness of the procedure. If the present epoch is representative, the timescale for angular momentum transport in M100 is about 6 Gyr (GGF). It may thus seriously affect the axisymmetric disk structure over a Hubble time. This work was supported in part by NASA grant NAGW-2491.

References

Frei, Z. 1995, PASP, in press

Gnedin, O. Y., Goodman, J., & Frei, Z. 1995, AJ, 110, in press [GGF]

Rhoads, J. E. 1995, this volume

Rix, H.-W. 1993, PASP, 105, 999

Rix, H.-W. & Zaritsky, D. 1995, this volume

Galaxy as Dynamical System with Accreting Halo

Peter Berczik and Sergei G. Kravchuk

Main Astronomical Observatory, Ukrainian National Academy of Sciences, Golosiiv, Kiev-022, 252650, Ukraine

Abstract. The dynamical evolution of the Galaxy system was analysed under the assumption that its Halo containes beside nonbaryonic matter some fraction of baryons in the form of "invisible" small hydrogen molecular clouds as was proposed by Pfenniger, Combes & Martinet (1994). The behaviour of the Galactic gaseous Disk surface density as a function of time and galactocentric distance strongly depends on baryonic Halo parameters. It was found that only model with additional baryonic Halo having initial mass $2 \cdot 10^{11} M_{\odot}$ and flat rotation curve, which mimics the Disk rotation curve, fits well modern observational data concerning the Galaxy evolution in the Solar vicinity.

1 The model

Now it has become apparent that there exists a strong link between structural properties of present-day galaxies and physical processes of their formation. The crucial role in the process of galaxy formation seems to play dark matter (White & Rees (1978)). Numerous assumptions have been made on its nature. But it is to be noted that a considerable amount of the dark baryonic matter could be tied up in hidden, low surface brighness galaxies (Davies (1993)). Recent observational data (e.g. Lequeux (1994), Bajaja, Huchtmeier & Klein (1994), Lequeux, Allen & Guilloteau (1993)) confirm the existence of large amount of molecular hydrogen in the outer regions of galaxies.

Pfenniger, Combes and Martinet (1994) supposed that at the level of individual disc galaxies the essential part of the dark matter could be in the form of cold gas cloudlets, essentially in molecular form. Such objects are really "invisible" (Wilson & Mauersberger (1994)). This hypothesis provides a reasonable explanation of a flat rotation curve, constant ratio of dark matter to HI mass in the outer spiral discs, the larger amount of visible gas in interacting galaxies with respect to the isolated ones, the gas consumption problem of spiral galaxies (e.g. Pfenniger, Combes & Martinet (1994)), the morphology of nearby star formation regions (Lepine & Duvert (1994)) etc. On the other hand galaxy chemical evolution simulations also require the existence of the permanent inflow of heavy - element deficient material into the galactic disc (Tossi (1983)).

In this paper the dynamical evolution of the Galaxy system, having the Halo which containe beside nonbaryonic matter some fraction of baryons in the form of "invisible" cold molecular cloudlets, is studied from the stage when the "visible Galaxy" had already been formed. Its gravitational potential is formed by the Bulge, Disk and Dark non-baryonic and baryonic Halo and changes slowly due to the accretion of baryonic matter from the Halo. For the Bulge and Disk this potential has the form (Li & Ikeuchi (1992)):

$$\Phi_{1,2}(x,y,z) = -G \cdot M_{1,2}/\sqrt{x^2+y^2+(a_{1,2}+\sqrt{b_{1,2}^2+z^2})^2},$$

where indices 1, 2 are referred to Bulge and Disc components respectively, and $M_1 = 2.05 \cdot 10^{10} M_\odot$, $a_1 = 0.0\ kpc$, $b_1 = 0.495\ kpc$, $M_2 = 2.547 \cdot 10^{11} M_\odot$, $a_2 = 7.258\ kpc$, $b_2 = 0.52\ kpc$. For non-baryonic Halo:

$$\Phi_{nbH}(x,y,z) = -G \cdot M_{nbH}/r_{nbH} \cdot (\ln(1+q) + 1/(1+q)) - \Phi_0,$$

where $q = \sqrt{x^2+y^2+z^2}/r_{nbH}$ and $M_{nbH} = 1.35 \cdot 10^{11} M_\odot$, $r_{nbH} = 13.0\ kpc$, $\Phi_0 = 1.4 \cdot 10^{11}\ m^2/s^2$. The baryonic Halo term was defined by numerical simulation.

According to Pfenniger & Combes (1994) the Halo baryonic dark matter was supposed to exist in the form of cold, gravitationally bound molecular cloudlets having radii of about 30 AU and masses of the order of Jupiter. Their total mass M_{HALO} is model parameters and to be determined via model fitting with observational data. For $t = 0$ baryonic Halo was treated as spherical structure of radius $A_{HALO} = 100\ kpc$. The total number of initially homogenuously distributed "particles were chosen to be $N = 2109$ that has provided relevant description of system's dynamics. The "particle" chaotic velocities were taken to be isotropic and about $\Delta V = 7.7\ km/s$. They were were assumed to be involved also into the rotational motion.

Owing to Halo low density the effective process of slowing-down of Halo cloudlets (gravitational and ram pressure retardation) is confined only to the region of the galactic Disk. In order to estimate the role of the process of "particle" capturing via ram pressure retardation, into the numerical model a Capturing Disk was introduced as homogenuous ellipsoid with semiaxies $a_{accr} = 20.0\ kpc$, $b_{accr} = 19.9\ kpc$, $c_{accr} = 1.0\ kpc$. For initial moment its mass was set equal zero. At fixed moment t its total mass included also gravitationally captured and temporarely ocupying "particles". When the Halo "particle" drop into this ellipsoid, with probability P it was added to the Disk. P was taken to be 0.01 what is a typical value for capturing by Giant Molecular Clouds (GMC). The influence of P is rather weak because the dominant process is the gravitational capturing. The gravitational potential formed by homogenuous Capturing Disk was calculated according to Chandrasekhar (1973) (see also Berczik & Kolesnik (1993b)). The modified Smoothed Particle Hydrodynamics (SPH) algorithm of individual time step (Berczik & Kolesnik (1993a)) was used for carring out integration of equation of motion for each "particle".

2 Results

All other model parameters were determined by fitting results of model calculations for different initial conditions with modern data on age, mass inflow rate and Disc surface density in the Solar vicinity $\sigma(t = 13.0 \cdot 10^9 year) \approx 50\ M_\odot/pc^2$ (e.g. Pagel (1994)).

It was found that only initial model having additional baryonic Halo with mass $2 \cdot 10^{11} M_\odot$ and flat rotation curve,

$$\Omega_{0z} = \sqrt{G \cdot M_{HALO}/A_{HALO}} / \sqrt{A_{HALO}^2/100 + x^2 + y^2}.$$

which mimics the Disk rotation curve, fits modern observational data. The resulting Capturing Disk surface density $\sigma(t)$ variations with time for different regions can be approximated by following expressions (σ in $M_\odot/pc^2$, t in $10^9 year$):

for the central regions ($0 \div 2\ kpc$) $\sigma(t) \approx 75 \cdot (1 - \exp(-t/2))$,

for the Solar vicinity ($9 \div 11\ kpc$) $\sigma(t) \approx 55 \cdot (1 - \exp(-t/5))$,

for the edge of the Capturing Disk ($18 \div 20\ kpc$) $\sigma(t) \approx 50 \cdot (1 - \exp(-t/8))$.

Tight resemblance between Halo and Disc rotation curves allows to assume that baryonic Halo formation could be tied up to the stage of the galactic Disc formation, next global, over the whole Disc, powerful star formation burst, significant mass expelling driven by the collective effect of supernovae and winds from massive stars and subsequent process of cloud formation. The accretion of Halo cloudlets onto Galactic disc can essentially modify the Galaxy chemical evolution and can serve as efficient mechanism of supporting supersonic turbulent motions observed in GMC. Such selfgravitating cloudlets captured by GMC could be a seeds triggering a formation of stars having initially heavy-element depleted cores what can seriously affect their observational properties.

References

Bajaja E., Huchtmeier W.K. & Klein U. 1994, A&A, **285**, 385

Berczik P. & Kolesnik I.G. 1993a, Kinem. i Fiz. Nebes. Tel, **9**, No 2, 3

Berczik P. & Kolesnik I.G. 1993b, Kinem. i Fiz. Nebes. Tel, **9**, No 5, 23

Chandrasekhar S. 1973, "Ellipsoidal Figures of Equilibrium", (Russian translation), Mir, Moscow

Davies J.I. 1993, in "Environment and Evolution of Galaxies", eds. Shull J.M. & Thronson H.A. Jr., Kluwer Acad. Publ.

Lepine J.R.D. & Duvert G. 1994, Ap. Sp. Sci., **217**, 195

Lequeux J. 1994, A&A, **287**, 368

Lequeux J., Allen R.J. & Guilloteau S. 1993, A&A, **280**, L23

Li F. & Ikeuchi S. 1992, ApJ, **390**, 405

Pagel B.E.J. 1994, in "The Formation and Evolution of Galaxies", eds. Munoz-Tunon C. & Sanchez F., Cambridge U.P.

Pfenniger D., Combes F. & Martinet L. 1994, A&A, **285**, 79

Pfenniger D. & Combes F. 1994, A&A, **285**, 94

Tossi M. 1983, A&A, **197**, 47

White S.D.M. & Rees M.J. 1978, MNRAS, **183**, 341

Wilson T.L. & Mauersberger R. 1994, A&A, **282**, L41

Near-Infrared Measurements of Kinematics in Luminous Galaxies

Niall I. Gaffney[1,2], Dan F. Lester[1,2], and Greg Doppmann[1]

[1]RLM 15.308, University of Texas at Austin, Austin, TX 78751, USA
[2]Visiting Astronomer at the Infrared Telescope Facility which is operated by the University of Hawaii under contract to the National Aeronautics and Space Administration

Abstract. We have examined the starburst galaxies M82 and NGC 6240. Studying the stellar kinematics derived using the 2.3 μm $(2\text{–}0)^{12}$CO stellar absorption bandhead and the 2.16 μm Brackett γ emission line, we found that the old stellar populations, that predate the starbursts, are significant contributor to the total K luminosity of both sources in both of these systems. We find evidence that as much as 1/4 of the K luminosity of M82 is not related to the starburst. Futher in NGC 6240 we find that the K luminosty is not related to the star formation taking place in its nuclear region. We therefore speculate these old stellar populations can be important contributors to the net luminosities of starbursting galaxies and must be accounted for when modeling these systems.

1 Kinematics in obscured systems

Starforming systems are regions rich in dust and gas. Therefore, they are regions often associated with high optical extinction. We are using the low extinction properties of the near–infrared, in particular the 2.3 micron stellar CO absorption bandhead and other 2 μm emission lines, to measure the stellar and gas kinematics in heavily obscured regions that are thought to be bright at 2 μm because of heavy star formation.

NGC 6240 — The details of our analysis of NGC 6240 can be found in Lester & Gaffney (1994). The mass–to–K luminosity we derive for this system is $\approx 1 M_{\odot}/L_{\odot}$, a ratio consistent with that seen in quiecent non–starbursting systems. We therefor conclude for this system that whatever starformation is taking place in this merging system does not currently dominate the K–luminosity of this system.

M82 — We have spatially resolved the stellar and gas kinematics in M82 in the near–infrared. At the nucleus, the stellar velocity dispersion in M82 is 75 km/s. Using the relation of Whitmore & Kirshner (1981) and colors from Frogel*et al.* (1983), we derive a bulge K–luminosity in M82 of –21.3. This implies that the old (pre–starburst) stellar population has a luminosity is 1/4 that of the starburst region. This meens that this population, which is not related to the current star formatoin, makes up a sizeable fraction of the total K light from M82.

2 M/L_K and the importance of the old pre–starburst stellar population

Another presentations at this conference (Moorwood) demonstrated that the M/L_K in other systems are also found to be 1 to 0.2. Other studies in this workshop (Quillen, Heraudeau) have claimed a ratio of order 1 to be consistend with a normal quescent region in which the light is entirely from red giants. Rieke (this workshop) has shown that more extreem systems such as NGC 253 with $M/L_K = 0.05$ clearly do exist. What removal of this bulge will do may be more complex than what has been explored here, as it may alter both the total luminosity and mass of the starforming regions being studied. However, determining the contribution of the underlying old stellar populations in these systems is important in understanding the nature of starbursts. Without this information, the luminosities of NGC 6240 and M82 might be compared as if their continuum were dominated by the same relative number of old and young stars, which at 2.2 μm is not the case.

References

Gaffney, N. I., Lester, D. F. 1994, ApJ, 352, 544
Whitmore, B. C., Kirshner, R. P. 1981, ApJ, 250, 43
Frogel, J. A., Persson, S. E., Aaronson, M. Matthews, K. 1978, ApJ, 220, 75
Shen, J., Lo, K. Y. 1995, ApJ, in press

IV

IR OBSERVATIONS OF DISK GALAXIES

Galaxies with the DENIS 2 Micron Survey: A Preliminary Report

Gary A. Mamon[1,2]

[1] Institut d'Astrophysique, 98 bis Blvd Arago, F–75014 Paris, France
[2] DAEC, Observatoire de Paris, F–92195 Meudon, France

Abstract. The DENIS *IJK* survey of the entire southern sky is beginning observations this year. It will return one million images, representing 4 Terabytes, among which nearly 10^8 stars and up to one million galaxies. The main extragalactic applications of such a multi-band, digital, near-IR 2D survey are listed, and the importance and feasibility of a spectroscopic followup are discussed. First estimates of the capabilities for galaxy extraction are given, and illustrated with preliminary images of three very bright galaxies.

1 Introduction

The DENIS (DEep Near-Infrared Survey of the Southern Sky) survey is a European effort to map the entire southern sky ($-88° \leq \delta \leq +2°$) in the I ($0.8\,\mu$m), J ($1.25\,\mu$m), and K_s ($2.15\,\mu$m) bands (see Epchtein et al. 1994; Ruphy, in these proceedings). The survey uses the ESO 1m telescope, full-time, and is expected to take 3 or 4 years. DENIS should produce one million images in each of the 3 bands (observed concurrently), yielding nearly one hundred million stars, and up to one million galaxies (see Sect. 3 and Table 1 below). The 4 Terabytes of images will be analyzed partly on site in Chile, partly in Paris and partly in Leiden. An analogous effort is being taken by the American 2MASS project (see Lonsdale, in these proceedings), which will map *both* hemispheres, with dedicated telescopes, in the J, H, and K_s bands. DENIS is scheduled to commence in the fall of 1995 (the I camera has only been on-line since July 1995), while 2MASS is expected to begin the Northern survey in early 1997. The reader is referred to Ruphy (in these proceedings) for additional details and results from preliminary star counts.

2 Extragalactic Applications

As emphasized throughout this meeting, the Near-Infrared (NIR) has two important advantages for extragalactic and cosmological science:

1) The NIR bands lie represent the optimum region of the electromagnetic spectrum between *low dust extinction* and *low dust emissivity*.
2) The NIR light is expected to be better correlated with the *mass of the stellar component* of galaxies than the optical light (enhanced by recent star formation) or Mid- to Far-Infrared light (tracing warm dust, associated with recent star formation).

The first advantage allows one to probe galaxies behind the plane of the Milky Way (e.g. Mamon 1994), which is all the more important that the most important concentration of mass in the local Universe, the *Great Attractor*, seems to be centered at $b = 0°$ (Kolatt, Dekel and Lahav 1995). In comparison, optical surveys (e.g. CfA2, SSRS2) are typically limited to $|b| > 20°$, while IRAS is limited at $|b| \leq 12°$, mainly by stars and cirrus (Meurs and Harmon 1988). Moreover, the near transparency of dust to NIR light allows one to have a full view of external galaxies.

It seems reasonable to extrapolate the second advantage to the fact that NIR light should trace best the underlying total mass distribution of the Universe, hence the importance of NIR surveys for cosmological analyses of the large-scale structure of the Universe.

With these advantages in mind, the applications of such 2D surveys as DENIS and 2MASS are

1) A large (few thousand) set of bright galaxies from which one will obtain statistics of their morphological features, such as profiles in surface brightness, color, axis-ratio, and position angle. This will be especially true with the I-band images, for which the spatial resolution is superior, and hence a large enough database of galaxies with enough details.
2) The 2D distribution of galaxies in each of the three bands. In particular,
   Establishment of catalogs of binaries, loose groups, compact groups, and clusters
   a) Quantification of large-scale structure (2-point correlation functions, counts in cells ...)
   b) Mapping of large-scale structure through the Galactic Plane, in particular to observe the large but only slightly overdense Great Attractor.
3) The relation between color and environment, i.e. *color-segregation*, which itself is related to morphology and bulge/disk ratio.
4) The verification of galaxy counts at the bright end (where current surveys usually suffer from low statistics, see Gardner, Cowie, and Wainscoat 1993).

In addition, 2MASS, with its full-sky coverage, should probe the cosmic dipole to estimate the acceleration of the Local Group.

A spectroscopic followup will be necessary to obtain

1) A 3D view of the local Universe
2) Internal kinematics of groups and clusters

Unfortunately, it is a costly proposition to measure tens to hundreds of thousands of galaxy redshifts. A comparison of the efficiency of different instruments available in the southern hemisphere (Mamon 1995) indicates that the 2dF 400-fiber spectrograph on the 4m AAT is by far the most efficient instrument, only requiring 200 ($B < 17$) to 400 ($B < 19$) clear nights to cover a hemisphere of galaxies (without regard of the extinction near the Galactic Plane). However, in view of the pressure on the 2dF and the shallowness of the DENIS limiting magnitudes (the targets are so sparse that the 2dF would only use a fraction of its fibers) it is not realistic to expect much observing time for a spectroscopic followup of DENIS. The alternative would be to use an upgrade of the FLAIR-II 100-fiber spectrograph mounted on the UKST Schmidt telescope (see Parker and Watson 1995), which, with additional fibers and automated fiber-positioning, could be faster than the 2dF for a shallow ($B_{\rm lim} = 17$) spectroscopic followup. It is subject to lower user-pressure than the 2dF and a 300-fiber version of FLAIR would use all of its fibers at $B = 17$, which is the rough equivalent blue limit for a K-catalog obtained with cooled optics. which thus matches well the (see Table 1 below).

3 DENIS Capabilities for Galaxy Extraction

The capabilities of DENIS for detecting galaxies with decent photometry and separating them from stars have been estimated first with simulated images, and will later be checked by comparing with sets of 6-times coadded images (see Mamon *et al.* 1996 for more details on the strategy for optimizing the algorithms and estimating the magnitude limits). We will effectively be limited by the accuracy of the photometry for spiral galaxies and by star/galaxy separation for ellipticals. The preliminary magnitude limits and the catalog sizes we expect are listed in Table 1 below, where completeness and reliability are required at the 95% level *for each galaxy type* not explicit in the last column. The K limits are expected to improve with the projected cooling of the optics (only the detectors currently sit within dewars), which could lead to catalogs 3 times larger in K.

Table 1. DENIS galaxy catalogs with 95% completeness/reliability

Magnitude limit	Size (2π sr)	Incompleteness ($> 5\%$)
$K < 12.2$	30 000	$\|b\| < 2°$, Face-on Sc→Ir
$J < 14.4$	110 000	$\|b\| < 5°$
$J < 15.6$	300 000	$\|b\| < 5°$, Ellipticals
$I < 17.0$	1 100 000	$\|b\| < 10°$

4 Bright Galaxies

In Figure 1 are shown the 3 brightest galaxies (NGC 1512: SBa, $B_T = 11.1$; NGC 1515: Sbc, $B_T = 12.0$; NGC 1527: S0, $B_T = 11.7$) of the RC3 catalog (de Vaucouleurs *et al.* 1991) that were in the regions available on computer-disk at the DENIS Paris Data Analysis Center in Sept. 1994. The figure illustrates galaxies in the DENIS J and K_s bands (9×1 s exposures), and for comparison, in 70 min (NGC 1512 and NGC 1515) and 75 min (NGC 1527) exposures in blue (b_J) on the UKST Schmidt telescope (obtained from the Digitized Sky Survey).

In the DENIS J image, the bar of NGC 1512 is clearly seen with hints of the spiral arms starting at the edge of the bar. In the corresponding K_s image, both the bar and the beginning of the spiral pattern are only very marginally seen, whereas the nucleus is clearly visible. The disk of NGC 1512, seen nearly face-on, is invisible in both J and K_s.

The J image of the nearly edge-on spiral NGC 1515 traces the inner disk with hints of the inner spiral structure. In K_s, the galaxy is much smaller, with only the smallest hints of the inner spiral structure.

Finally, the nearly edge-on lenticular NGC 1527, being nearly featureless in the optical is of course even more so in the DENIS images, only the inner nucleus and the beginning of the disk are seen in K_s.

Altogether, these 3 images illustrate 1) the greater sensitivity of J relative to K_s in DENIS, which is mainly caused by the much higher atmospheric/instrumental background in the latter band; 2) very bright galaxies $B_T \leq 12$ can show bars and spiral arms in the DENIS J and K_s bands.

References

de Vaucouleurs, G., de Vaucouleurs, A., Corwin, H.G., Buta, R.J., Paturel, G., Fouqué, P. (1991): Third Reference Catalog of Galaxies (Springer Verlag, Berlin)

Epchtein, N. et al. (1994): DENIS: A deep near infrared survey of the southern sky, in Science with Astronomical Near-Infrared Sky Surveys, Epchtein, N., Omont, A., Burton, B., Persi, P., Eds. (Kluwer, Dordrecht) [also in Ap&SS **217**], 3–9

Gardner, J.P., Cowie, L.L., Wainscoat, R.J. (1993): Galaxy number counts from $K = 10$ to $K = 23$, ApJ **415**, L9–L12

Kolatt, T., Dekel, A., Lahav, O. (1995): Large-scale mass distribution behind the galactic plane, MNRAS **275**, 797–811

Mamon, G.A. (1994): How well should the DENIS survey probe through the Galactic Plane? in Unveiling Large-Scale Structures behind the Milky Way, Balkowski, C., Kraan-Korteweg, R.C., Eds. (ASP, San Francisco) 53–61

Mamon, G.A. (1995): The DENIS 2 micron survey and its cosmological applications, in Wide-Field Spectroscopy and the Distant Universe, Maddox, S.J., Aragón-Salamanca, A., Eds. (World Scientific, Singapore) 73–80

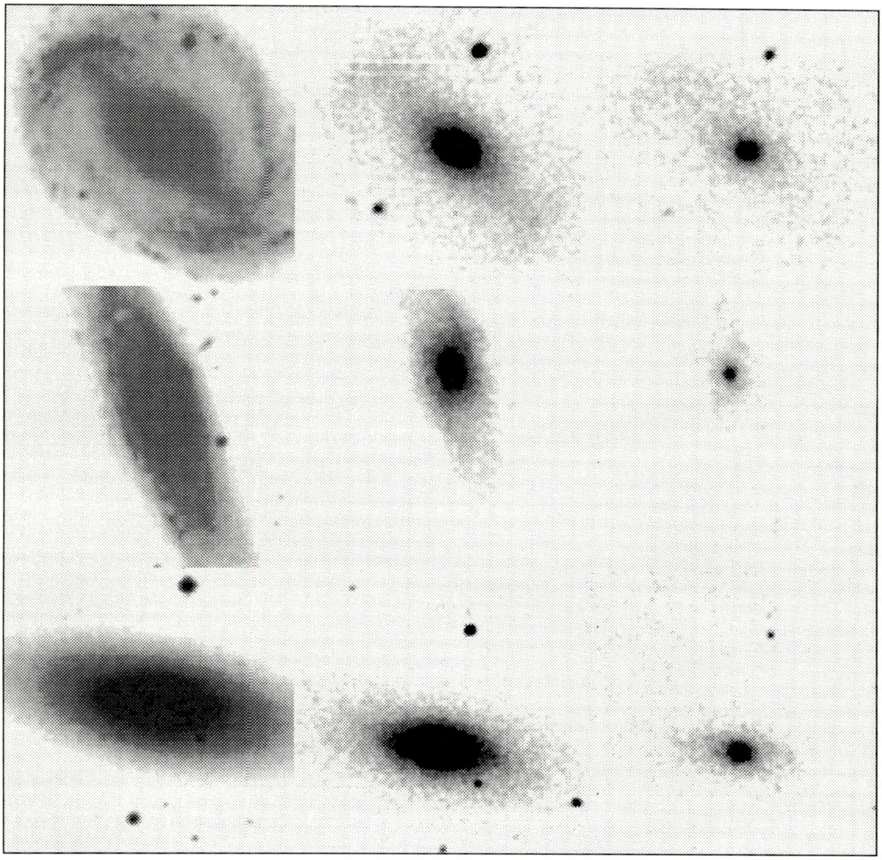

Fig. 1. Three bright galaxies seen in b_J (*left* from Digitized Sky Survey) and the DENIS J (*middle*) and K_s (*right*) bands; *top:* NGC 1512, *middle:* NGC 1515, and *bottom:* NGC 1527. Fields are 6′ wide. NIR images are 3 × 3 boxcar smoothed. Greyscales (white to black) are $9 \rightarrow 60\sigma$ (b_J) and $2 \rightarrow 20\sigma$ (J and K_s, where σ is after smoothing). Because NGC 1515 and 1527 are situated near the edges of the DENIS images, there is a small vertical misalignment with the optical images.

Mamon, G.A., Banchet, V., Boisson, C., Cayatte, V., Engelmann, F. (1996): A first look at galaxies in the DENIS survey, in Euroconference on Near Infrared Sky Surveys, Persi, P. Ed., in press

Meurs, E.J.A., Harmon, R.T. (1988): The extragalactic sky viewed with IRAS, A&A **206**, 53–62

Parker, Q.A., Watson, F.G. (1995): A FLAIR for wide-field spectroscopy, in Wide-Field Spectroscopy and the Distant Universe, Maddox, S.J., Aragón-Salamanca, A., Eds. (World Scientific, Singapore) 33–39

How Good Is the Near-Infrared Tully-Fisher Relation?

Gary M. Bernstein[1], Puragra Guhathakurta[2], and Somak Raychaudhury[3]

[1]Department of Astronomy, 830 Dennison Bldg., University of Michigan, Ann Arbor, MI 48109, USA
[2]University of California Lick Observatory, Santa Cruz, CA 95064, USA
[3]Inter University Centre for Astronomy & Astrophysics, Post Bag 4, Ganeshkhind, Pune 411 007, India

Abstract. A study of the Tully-Fisher relation (TFR) in H, I, and B filter bands using high-quality data for 25 Coma-region spirals confirms the common suspicion that the infrared TFRs have lower scatter than the blue TFR. We find, however, that the I-band TFR has equivalent or slightly lower scatter (0.10 mag RMS!) than the H-band TFR, and that internal extinction corrections are important even in H band, so that infrared array detectors offer little advantage over I-band CCD data for TFR surveys. Extensions to the 25-galaxy sample, and comparisons with other authors, show that there are spirals which depart by 1 mag or more from the original infrared TFR, and that the slope of the TFR may vary. There are hints that $I - H$ colors may be useful in detecting and diagnosing the differences between spirals which have common rotation speeds yet have up to threefold discrepancies in infrared luminosity.

1 Motivations and Expectations for the IR Tully-Fisher Relation

Tully and Fisher (1977) noted a correlation between 21-cm width W and absolute photographic magnitude for spiral galaxies, and proposed its use as an extragalactic distance indicator. We expect some sort of luminosity–linewidth correlation, since both the total luminosity and the linewidth should be determined primarily by the amount of mass within the first few scale lengths of the disk. The exact form of the TFR is not, however, known a priori, because at present we are little able to understand or predict the various physical processes linking the rotation speed and the luminosity. For example, the relative distributions of stellar and total mass, the mass-to-light ratio of the stellar population, the influence of dust on the observed luminosities, and the influence of non-circular motions on the observed linewidth are all ill understood at this time. Thus while the TFR in principle is physically motivated, our knowledge of its exact form and scatter is entirely empirical. Detailed studies of the TFR give us strong insights to the formation and structure of spiral galaxies.

The TFR is also of great utility as an extragalactic distance indicator. The TFR has been applied to thousands of spiral galaxies for use in peculiar velocity surveys—it is probably fair to say that more spirals have been observed in the infrared for this purpose than for any other. The TFR is most useful, of course, if the scatter in the relation is minimized. Early TFR studies used blue or visual magnitudes, but it was quickly realized (Aaronson, Huchra, & Mould 1979) that the TFR scatter might be lower using near-infrared magnitudes. The reasons are familiar to anyone here: H-band flux is much less strongly affected by the age of a stellar population and the presence of dust than is B-band flux.

The questions we ask, therefore, are: (1) does going to the infrared indeed decrease the scatter in the TFR—and how far to the red must one go to gain these benefits? In particular, would peculiar velocity surveys benefit from use of near-IR array cameras instead of the now-predominant I- or r-band CCD data? (2) How does the effect of dust on total luminosity change with wavelength? (3) What does the scatter in the TFR tell us about the formation of spiral galaxies? Is all of the variation in global parameters of the family of spiral galaxies attributable to a single parameter (linewidth), or do we find secondary global parameters which are important in determining the nature of a particular spiral galaxy? To answer these questions, we have taken high-quality images of distant spirals in B, I, and H bands. The pioneering H-band observations of Aaronson *et al.* were done using aperture photometers. The advent of near-IR arrays allows us to put the H band on an equal footing with the shorter wavelengths. Since we are interested in the intrinsic scatter in the TFR, we use distant spirals, for which the uncertainties in distance modulus are minimized. A similar study of the $BVIH$ TFR was conducted by Pierce & Tully (1988) using Virgo and Fornax cluster galaxies, but it is likely that the line-of-sight depths of these clusters contributes to their measured scatter. An introduction to our program is in Guhathakurta *et al.* (1993), and results from the I-band are in Bernstein *et al.* (1995, Paper I), which contains many details omitted here. Future papers will contain the further results discussed at this conference.

2 Data Acquisition and Reduction

Fukugita *et al.* (1991) compile a list of all spirals within 4° of the Coma cluster core which have a 21-cm linewidth in the literature, and select those in the proper redshift range as cluster members. We observed all 30 of these galaxies having inclination angles $i > 45°$. As detailed in Paper I, the TFR data rule out at high confidence the possibility that these galaxies are located at a common Coma cluster distance. In this paper we will assume that these spirals are in fact in the field, moving with the Hubble flow. An additional 10 galaxies in the cosmographic vicinity of the Coma cluster (but definitely not cluster members) were observed from a list kindly provided by G. Gavazzi. Deep images of each galaxy were obtained in B, I, and H bands. Typical exposure times and depths are shown in Table 1. The S/N of the galaxy images drops towards the red for comparable exposure time; it takes several times longer exposure in H than in I, and longer yet compared to B, to produce an image reaching a fixed number of scale lengths into the disk—a well-known disadvantage to near-IR work. Surface photometry on each image yields an ellipticity e of the outer isophotes of each galaxy, and a total magnitude is caculated by extrapolating an aperture magnitude to infinity using an exponential surface profile. The images are deep so that the extrapolation is almost always < 0.1 mag.

Table 1. Multiband Tully-Fisher Relations

Band	Exposure (min)	Depth (mag arcsec^{-2})	m	a	RMS scatter (mag)
H	5–10	21.7–22	-6.74 ± 0.27	0.66 ± 0.19	0.14
I	10–15	24.5–25	-5.61 ± 0.18	1.42 ± 0.13	0.10
B	5	27	-3.73 ± 0.38	2.25 ± 0.28	0.20

A population of spirals will have higher TFR scatter if the outer disks are not circular. The assumption of circularity is implicit, first, in the use of the apparent ellipticity e to derive the inclination i. Second, non-circular motions will ruin the correlation between W and the mass (hence luminosity) of the galaxy (Franx & de Zeeuw 1992). We exclude galaxies having tidal tails, varying e's, or other morphological signs of non-circularity. One might think that H-band images would be of most use here, since features due to dust and star formation are muted, giving us a better view of the true disk shape. Figure 1a shows such a case: the galaxy appears normal in B, but the I and H images clearly show a superposed background spiral! Figure 1b, to the contrary, shows a more common situation: the B image, because it is deeper, clearly shows tidal tails which are undetected on the I or H images (the galaxy is indeed an outlier from the TFR). We find that B images are *more* useful than I or H for measuring disk inclinations, because one gets better

S/N on the outer disk, where dust and star formation are minimal, and where the most valuable information on disk shape is to be found.

The exclusion of disturbed and $i < 45^\circ$ galaxies leaves 25 in the sample. Inclination angles i are determined from e by the usual projection formula. Rotation widths W are determined from high-quality Arecibo 21-cm profiles, and corrected to a face-on value W_0 by the factor $\sin i$. Only galaxies with steep-sided 21-cm profiles are acceptable to us. For two galaxies, the 21-cm S/N is low, and we substitute our own rotation widths obtained from long-slit Hα spectroscopy (we have rotation curves for all 25 galaxies and will discuss their use in a future publication). The total magnitude of each galaxy is corrected to the value it would have if the galaxy were at a standard distance of $7000\,\mathrm{km\,s^{-1}}$.

3 Fits to the TFR

3.1 Inclination Corrections

A preliminary TFR is derived by fitting the data to the equation

$$M = m \log W_0 + b,$$

where M is the distance-corrected H, I, or B magnitude. The residuals to this preliminary fit are shown in Figure 2, plotted against disk ellipticity e. As we expect, absorption and scattering by dust makes the edge-on galaxies appear too faint. *The dust signature is clearly detected even in H band.* A better fit to the TFR is obtained when we allow an additional term to describe the dust-related inclination correction:

$$M_x = m_x \log W_0 + b_x + a_x e,$$

where $x \in \{H, I, B\}$ and a_x we call the inclination coefficient. We claim no physical motivation for our assumption that the magnitude correction is linear in e. While others have used logarithmic corrections, these are based on a slab model which is clearly not applicable to spiral galaxies. In Table 1 we list the best-fit a_x values, and in Figure 3 these are plotted against wavelength. The infrared inclination correction is much larger than a naive Whitford-law scaling from B band would suggest, emphasizing the oft-made point that the inclination correction is not due to a foreground slab of dust, but rather a complicated radiative transfer problem involving intermingled stars and dust. We offer our a_x values as input to those modelling this process, and refer the reader to other contributions in this Volume for a more detailed examination.

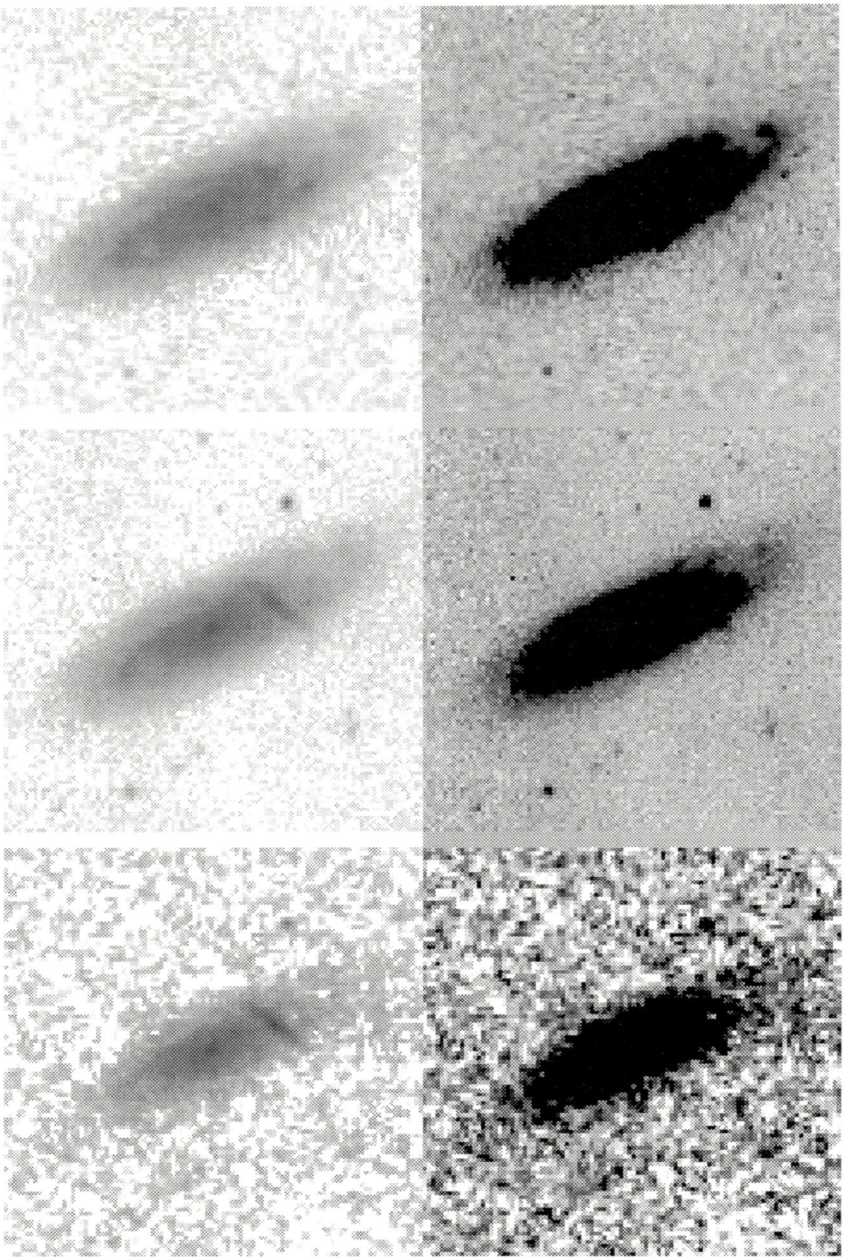

Fig. 1a. Images of UGC5041 in (from top) B, I, and H bands. Left-hand column shows interior structure in logarithmic stretch, and right-hand column shows outer structure in linear stretch. Note that the superposed background galaxy is not apparent in the B image.

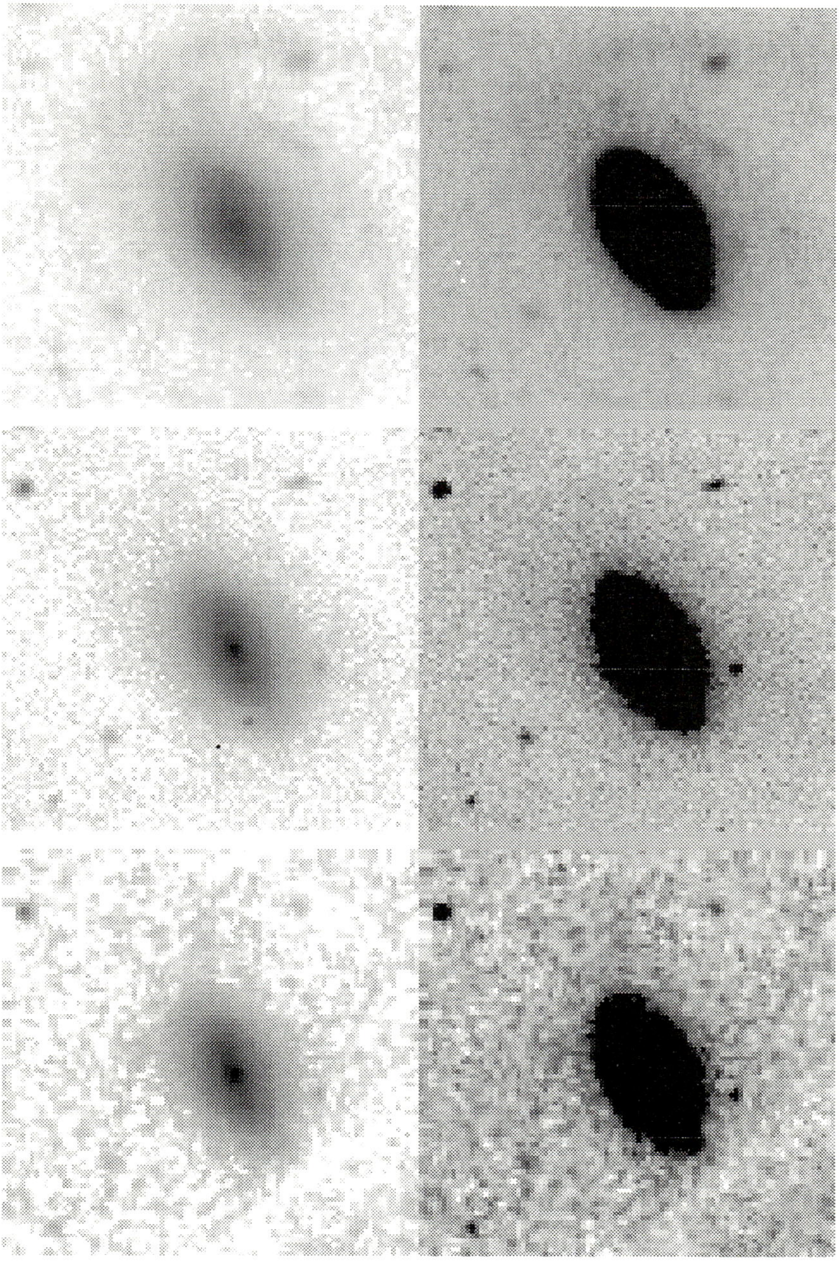

Fig. 1b. Images of Zw159-075 in (from top) B, I, and H bands, as in Figure 1a. The tidal tails, which make this galaxy unsuitable for TFR study, are noticable only on the deeper B image.

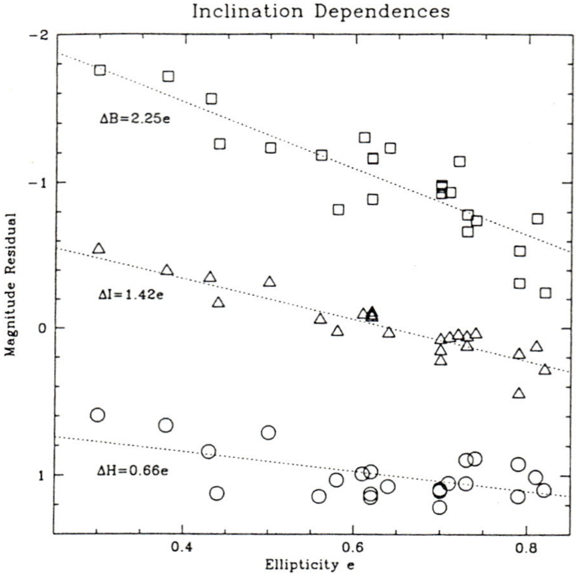

Fig. 2. Residuals to the preliminary TFR fit are plotted versus disk ellipticity e. Edge-on galaxies appear too faint, the expected signature of dust. The effect is apparent even in H band.

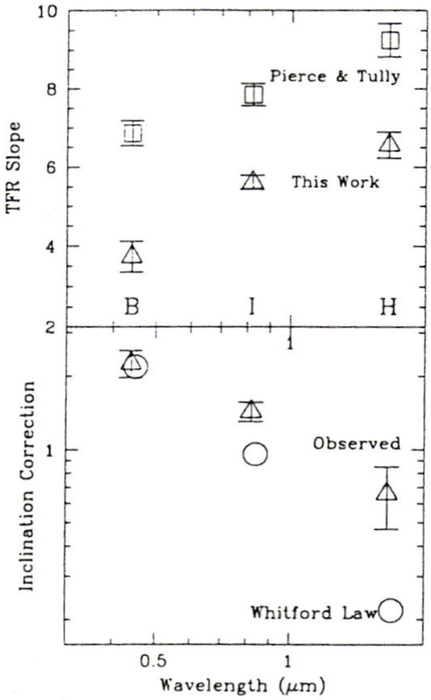

Fig. 3. TFR slope (upper panel) and inclination coefficient (lower panel) are plotted vs wavelength. Triangles are from the Coma region sample described here; squares in the upper panel are Pierce & Tully Virgo/Fornax data; circles in the lower panel show the results of a naive extrapolation of the B data to the IR using the Whitford law.

3.2 Slope and Scatter

Table 1 shows the best-fit slopes m_x and RMS scatter σ for the inclination-corrected TFR in each band. In Figure 4, we plot inclination- and distance-corrected magnitudes versus $\log W_0$ for our 25-galaxy sample in the H, I, and B bands. Several results are immediately apparent. First, in Figure 3, we plot TFR slope vs wavelength for our data and the Pierce & Tully data. Both datasets show slope increasing toward the red, but our Coma-region slopes are significantly shallower at all wavelengths than are the Virgo-Fornax results—we are facing either a severe statistical fluctuation, or evidence that the form of the TFR is somehow influenced by environment.

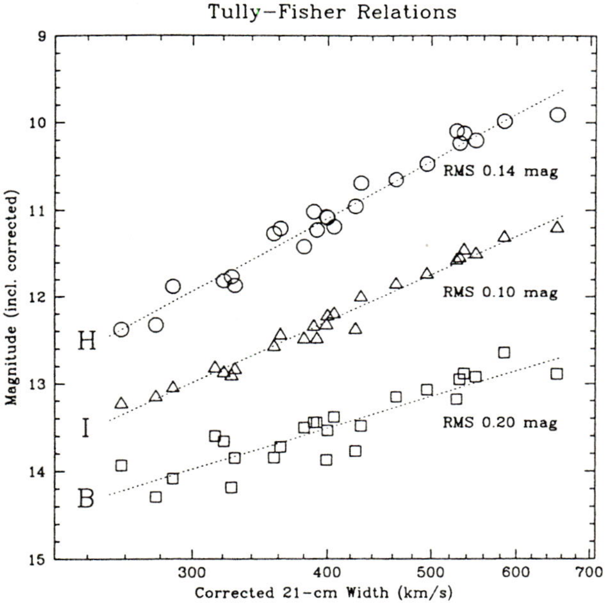

Fig. 4. TFR relations in H, I, and B bands. Magnitudes have been corrected for inclination.

Second, we notice that the TFR scatter is stunningly low in the I and H bands at 0.10 and 0.14 mag RMS, respectively. Most other estimates of the TFR scatter are $\approx$ 0.30 mag. Some of the diminution in scatter is certainly attributable to our exclusion of obviously non-circular galaxies. Even so, the scatter is much lower than previous work would suggest, and we will examine this further in the next section. If errors in W_0 dominate the TFR scatter, the steeper slope of the H-band TFR will lead to slightly larger RMS magnitude error. Our observation of slightly higher scatter in H band than in I is attributable to this and to statistical fluctuation. Thus we conclude that *the more difficult H-band observations offer no increase in TFR accuracy over I-band data.* An H-band TFR survey might be useful, though, in regions of higher Galactic extinction.

The B-band scatter of 0.20 mag is significantly larger than in the infrared bands. The B-band TFR residual does not correlate with any other galaxy parameter we have measured. Zaritsky (1995) proposes that the B-band TFR residuals of nearby galaxies correlate with evidence of recent accretion events, such as abundance gradients.

4 Varying Slopes, Outliers, and Hidden Parameters

Our initial 25-galaxy Coma-region sample yields a TFR with half or one-third the scatter found by other authors, and a slope markedly shallower than generally found. This could be due to an extremely rare statistical fluctuation, or could be indicative of an underlying difference in the populations. We have collected data on many other galaxies around Coma and other distant clusters in order to seek an explanation—statistical or physical—for this discrepancy. As we expand our sample further into the Great Wall area, we indeed find galaxies which are up to 1 mag too faint for the I-band TFR derived above (far too large to be due to peculiar velocities). These galaxies are better fit by the Pierce & Tully I-band TFR. Is there a second parameter for spirals which, along with W_0, determines total luminosity? In our expanded TFR sample, we have tested for correlations between TFR residual and many parameters: galaxy size, surface brightness, 21-cm or IRAS fluxes, rotation curve shape, $B - I$ color. No significant correlations have been found. The only distance-independent parameter in which the extreme TFR outliers seem to distinguish themselves is infrared color. In Figure 5, the filled triangles show the inclination-corrected colors for the original sample vs W_0. Four new galaxies with large I-band TFR residuals are plotted as open symbols, and are seen to have relatively red IR colors. We hope to obtain H data for more of the expanded sample, to see if this trend continues. If so, the $I - H$ color could be used as a second parameter or as a "trigger" for deciding whether spirals should fit a steep or shallow TFR.

5 Summary

Our multiband TFR study confirms some preconceptions and confounds others. Indeed the infrared TFR seems more accurate than B-band TFR, but the B images are in fact most efficient for diagnosing the structure of the outer disks of spirals. Though the internal extinction correction in H band is half that in I, the overall TFR scatter in H is no lower, and there is little reason to conduct peculiar velocity surveys in H band, given the lower efficiency dictated by the brighter sky.

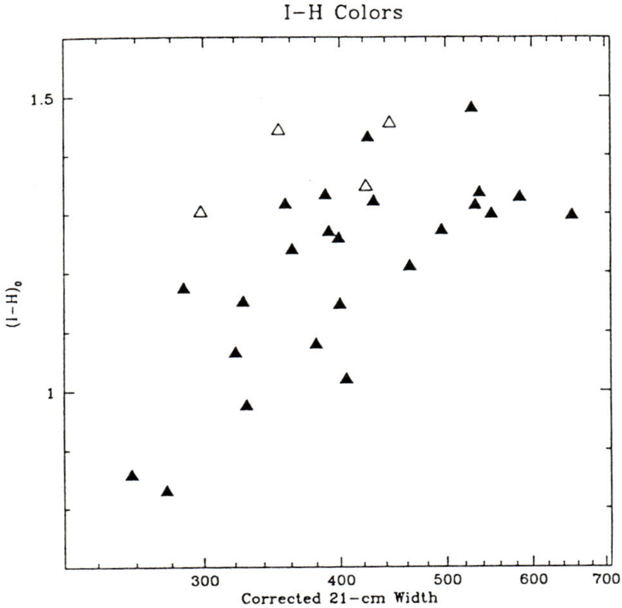

Fig. 5. Inclination-corrected IR color vs 21-cm width for the original sample (filled symbols) and for four newly observed galaxies (open symbols) which are outliers from the original *I*-band TFR.

Our initial sample hews to the TFR with frighteningly small scatter. The *H*-band data may prove useful in diagnosing the difference between our initial sample and the larger population of spirals, which seems to have steeper TFR slope with larger scatter. Infrared colors are valuable indicators of population or extinction differences which could explain why galaxies with similar rotation speeds can have such dissimilar luminosities.

References

Aaronson, M., Huchra, J., & Mould, J. 1979, *ApJ* **229** 1

Bernstein, G., Guhathakurta, P., Raychaudhury, S., Giovanelli, R., Haynes, M. P., Herter, T., & Vogt, N. P. 1994, *AJ* **107** 1962 (Paper I)

Franx, M. & de Zeeuw, T. 1992, *ApJ* **392** L47

Fukugita, M., Okamura, S., Tarusawa, K., Rood, H. J., & Williams, B. A. 1991, *ApJ* **376** 8.

Guhathakurta, P., Bernstein, G., Raychaudhury, S., Haynes, M., Giovanelli, R., Herter, T., & Vogt, N. 1993, *PASP* **105** 1022

Pierce, M. J. & Tully, R. B. 1988, *ApJ* **330** 579

Tully, R. B. & Fisher, J. R. 1977, *A&A* **54** 661

Zaritsky, D. 1995, *ApJ* **448** L17

Near Infrared Imaging of Late Type Virgo Cluster Galaxies

Alessandro Boselli[1], Giuseppe Gavazzi[5], Hans Hippelein[2], James Lequeux[4], Daniele Pierini[3], Richard Tuffs[1], Heinrich Völk[1], and Cong Xu[1]

[1] MPI für Kernphysik, D-69117 Heidelberg, Postfach 103980, Germany
[2] MPI für Astronomie, Königstuhl 17, D-69117 Heidelberg, Germany
[3] Osservatorio Astronomico di Brera, Via Brera 28, 20121 Milano, Italy
[4] DEMIRM, Observatoire de Paris, 61 Av. de l'Observatoire, 75014 Paris, France
[5] Universita degli Studi di Milano, Dipartimento di Fisica, Via Celonia 16, 20133 Milano, Italy

Abstract. We present near infrared (NIR) images of spiral, irregular and blue compact dwarf galaxies belonging to the Virgo cluster. Using the MAGIC camera on the Calar Alto 2.2-m telescope, 89 galaxies have been measured in the K' band, and 17 in the H band. These images are compared with optical images (H_α and R band) to assess the distribution of star-formation within the galaxian disks.

1 Introduction and Observations

The Virgo cluster constitutes a unique laboratory for the statistical study of a complete volume-limited sample of late-type galaxies, including a full representation of dwarf galaxies. We are measuring a sample of 118 cluster member galaxies later than S0 from the Virgo Cluster Catalogue (VCC) of Binggeli et al. (1985, 1993) in key spectral regions to the limiting capability of available facilities. This sample spans the range $11 < B_T < 18$, is complete to $B_T = 18$, and is approximately equally divided between giant spirals on the one hand and dwarf and irregular galaxies on the other. It will be observed with the Infrared Space Observatory (using ISOCAM, ISOPHOT and LWS).

To allow the effect of the cluster environment on observed properties to be statistically investigated, the sample is divided into contrasting cluster-core and cluster-periphery subsamples. The cluster-core subsample (45 galaxies) is bounded by a 2.0 degree radius circle centred on M87, while the inner boundary of the cluster-periphery subsample (73 galaxies) is at a radius of 4 degrees from the position of maximum projected galaxy density.

The 89 brightest galaxies in the sample have been imaged in the NIR using the MAGIC camera (Herbst et al. 1993) on the Calar Alto 2.2-m telescope in K'. 17 of these were also measured in H band. In addition, 56 galaxies were imaged in H_α using the Calar Alto 3.5-m or 2.2-m telescopes. All but one galaxy brighter than $B_T = 15.5$ were measured in the NIR. Good detections were achieved for all the spiral and most of the dwarf galaxies, thus allowing a morphological comparison between H_α and the NIR.

2 Results for selected objects

K' observations of giant spirals like the Milky Way map the distribution of the older stars comprising the bulk of the mass of a galaxy, rather than just the younger stars which contribute more to the visible. There may however be a larger contribution to the integrated NIR emission from red supergiants in star forming regions in Blue Compact Dwarf (BCD) and dwarf irregular galaxies. This can be investigated by comparing NIR images with H_α images. To illustrate the data obtained we present such a comparison here for three of the observed spiral and dwarf galaxies.

2.1 NGC 4579 (SAB(rs)b)

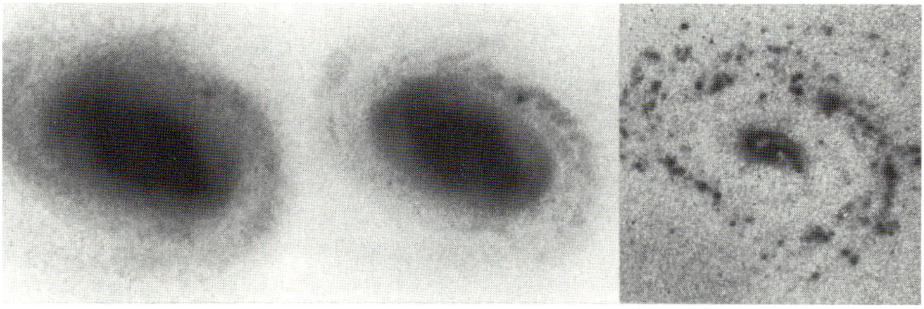

Fig. 2.1. NGC 4579 ($B_T = 10.56$) in (from left to right) K', R and H_α. The square area depicted on each image has side 153 arcsec (or 11 kpc for a distance of 15 Mpc).

This is an anemic (van den Bergh 1976) and HI-deficient (Cayatte et al. 1991) galaxy close to the centre of the cluster. As in the red image, a prominent nucleus, a bar, an outer ring and the spiral arms are visible in the K' band. The nuclear ring, which can be clearly seen in the H_α image, is not visible in the NIR image. No young stars are visible along the bar, while some HII regions are present along the outer ring and the spiral arms.

2.2 IC 3617 (SBm/BCD)

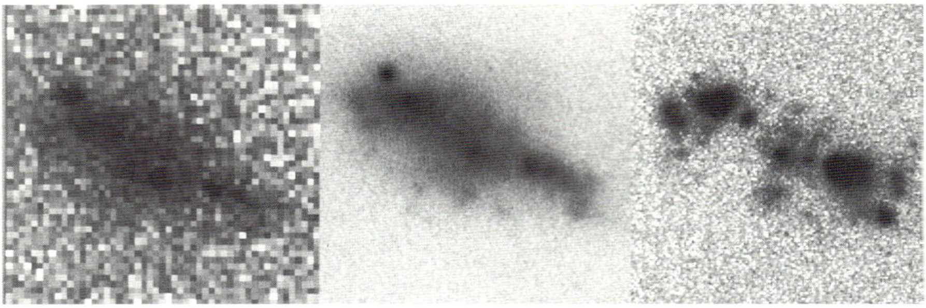

Fig. 2.2. IC 3617 (B_T = 14.67) in (from left to right) K', R and H_α. The square area depicted on each image has side 80 arcsec (or 5.8 kpc for a distance of 15 Mpc).

IC 3617 is a hybrid low surface brightness blue compact galaxy located in the periphery of the cluster. The clumpy structure clearly visible in the red and H_α images can also be seen in the K' band. The HII regions in the NE and SW regions are clearly visible in the NIR image.

2.3 IC 3355 (SBm)

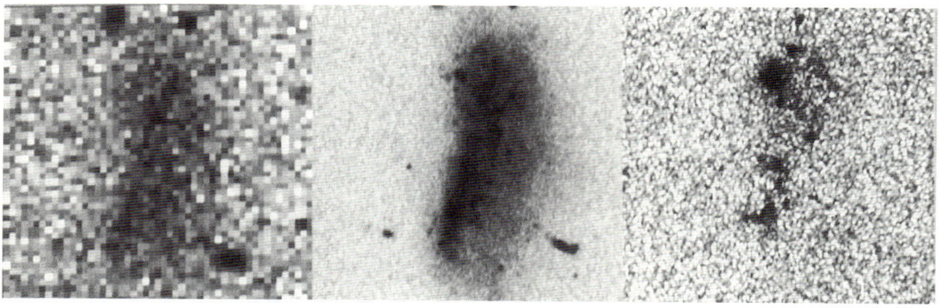

Fig. 2.3. IC 3355 (B_T = 14.82) in (from left to right) K', R and H_α. The square area depicted on each image has side 80 arcsec (or 5.8 kpc for a distance of 15 Mpc).

IC 3355 is located close to the centre of the Virgo cluster. In contrast to the morphology in H_α the galaxy has a smooth, diffuse appearance in the K' band, without clumps. The HII regions seen in H_α in IC3355 are probably younger than those seen in IC3617.

3 Discussion

The spiral structure, rings and bars of the observed galaxies, which are evident in the optical bands, are generally also observable in the NIR bands, albeit more diffuse and less clumpy. In some spiral galaxies, however, some giant HII regions can be seen in the NIR images. At these wavelengths, the emission of the HII regions is probably dominated by relatively young ($\sim 10^7$ yr) red supergiant stars (Mas-Hesse & Kunth, 1991), while their H_α emission is due to massive, young (age $< 10^6$ yr) OB stars (Kennicutt 1990).

Low surface brightness galaxies are faint and diffuse in the NIR. Compact structures in the NIR are more common in BCD galaxies and contribute significantly to the spatially integrated emission. As for the giant spirals, those HII regions seen in the NIR are probably dominated by relatively young, red supergiant stars. On the other hand, HII regions visible in the H_α images but not seen in the NIR are probably younger still, not yet having produced significant numbers of red supergiants. The comparison of NIR with H_α images to evaluate the age of HII regions within the galaxies will be a useful tool, when applied in a statistical way throughout the sample, to unravel the recent star-formation history of these objects and investigate dependencies on morphological type and environment.

References

Binggeli, B., Sandage, A. & Tammann, G.A. (1985): AJ **90**, 1681.

Binggeli, B., Popescu, C.C. & Tammann, G.A. (1993): A&AS **98**, 275.

Boselli, A., Gavazzi, G., Combes, F., Lequeux, J. & Casoli, F. (1994): A&A **285**, 69.

Cayette, V., van Gorkom, J., Balkowski, C. & Kotani, C. (1991): AJ **100**, 604.

Combes, F., Dupraz, C., Casoli, F., & Pagani, L. (1988): A&A **203**, L9.

Herbst, T.M., Beckwith, S.V.W., Birk, Ch., Hippler, S., McCaughrean, M.J., Mannucci, F. & Wolf, J. (1993): in *Infrared Detectors and Instrumentation* p. 605, Fowler, A.M. ed. (SPIE Conference 1946)

Kennicutt, R. (1990): in *The interstellar medium in galaxies* p405, Thronson, A., Shull, J., eds. (Dordrecht Kluwer)

Mas-Hesse, J. & Knuth, D. (1991): A&AS **88**, 399.

Peletier, R.F. & Willner, S.P. 1991: ApJ **382**, 382.

van den Bergh, S. (1976): ApJ, **206**, 883.

Near-IR Surface Photometry of 400 Late-type Galaxies

Daniele Pierini[1,2] and Giuseppe Gavazzi[1,2]

[1]Universitá di Milano, Dipartimento di Astrofisica, I-20133 Milano, Italy
[2]Osservatorio Astronomico di Brera, I-20121 Milano, Italy

Abstract. NICMOS3 H band (1.6μ) images of 400 spiral galaxies in clusters and in the isolated regions of the Coma Supercluster were obtained with the TIRGO[3] 1.5 m and the Calar Alto[4] 2.2 m telescopes.

1 Results

As part of the multifrequency survey of galaxies in clusters and of isolated galaxies in the Coma Supercluster (Gavazzi & Boselli, 1995), during the 1994-1995 observing campaigns we obtained H band NICMOS3 images of 400 late-type galaxies using the 256x256 pixel arrays:
1) ARNICA (Lisi et al., 1993; Hunt et al. 1994) attached to the Italian 1.5 m TIRGO telescope (field of view of 4.3x4.3 arcmin^2, pixel size of 1 arcsec) (in collaboration with C.Baffa, A.Boselli, I.Randone, L.Hunt, F.Lisi, and G.Trinchieri);
2) MAGIC (Herbst et al. 1993a,b) attached to the 2.2 m Calar Alto telescope of MPI (field of view of 6.8x6.8 arcmin^2, pixel size of 1.6 arcsec) (in collaboration with A.Boselli, R.Tuffs, H.Hippellein, H.Voelk, R.Lenzen, and J.Lequeux).

We observed 29 galaxies in the cluster A262 (Perseus-Pisces), 36 in the Cancer cluster and 334 the Coma Supercluster.

In the region of the Coma Supercluster ($11^h30^m \leq \alpha \leq 13^h30^m$; $18^\circ \leq \delta \leq 32^\circ$) the CGCG catalogue (Zwicky et al., 1961-68) lists 290 spiral galaxies with $m_p \leq 15.7$. Among these we observed 262 objects. The subsample of galaxies with $m_p \leq 15.6$ (229 objects) has been completed and is suitable for statistical analyses.

Thanks to the large field of view of the MAGIC camera (in the present optical set-up) we were also able to obtain for the first time H and K$'$ mosaics of the three

[3] TIRGO is operated by CAISMI-CNR, Arcetri Observatory, Florence, Italy
[4] Calar Alto is operated by MPIA, Heidelberg, Germany jointly with the Spanish National Commission for Astronomy

nearby galaxies NGC2366, NGC2403 and NGC4236, which are commonly used as calibrators of the Tully-Fisher relation (Pierini et al, 1995). Galaxies belonging to the Virgo cluster were observed as well, as reported in Boselli et al. (this conference).

2 Future Developements

Image reduction and photometric calibration of the frames have been completed. Extraction of ligth profiles along elliptical annuli up to the 21.5 mag arcsec^{-2} isophote is under way. For all galaxies we will derive total H magnitudes and photometric diameters ($a_{21.5}$). With the data in our possession we will:
1) derive the NIR light distributions for the various morphological classes;
2) compare these and the typical NIR photometric diameters with those obtained in the visible (V or/and B band);
3) use the subsample of galaxies with proper inclination and whose 21 cm line profiles are available for determining galaxy distances using the Tully-Fisher relation (Tully and Fisher, 1977);
4) re-determine the H and B band luminosity functions of Coma + A1367 and compare them with those obtained for isolated galaxies in the Coma Supercluster (see Gavazzi, Randone & Branchini, 1995).
5) determine the extinction properties of galaxies by comparing the H with the V and/or B surface brightness distributions.

References

Gavazzi, G., Randone, I. and Branchini, E., 1995 ApJ, 438, 590
Gavazzi, G. and Boselli, A., 1995, A&A, in press
Hunt, L.K., Maiolino, R., Moriondo, G., 1994a, Arcetri Technical Report, 2/1994
Herbst, T.M., Beckwith, S.V.W., Birk, Ch., Hippler, S., McCaughrean, M.J., Mannucci, F., Wolf, J., 1993a, SPIE, 1946
Herbst, T.M., Rayner, J.T., 1993b, proc. Conf. " IR Arrays"
Lisi, F., Baffa, C., Hunt, L.K., 1993, SPIE, 1495, 594
Pierini, D., Gavazzi, G., Boselli, A., Tuffs, R., 1995, submitted to A&A
Tully, B. and Fisher, J., 1977, A&A, 54, 661
Zwicky, F., et al., 1961-68, "Catalogue of Galaxies and Clusters of Galaxies", California Institute of Technology Press, Pasadena

Nonaxisymmetric Structures in Stellar Disks

Dennis Zaritsky[1] and Hans-Walter Rix[2]

[1]UCO/Lick Observatory, Univ. of Calif. Santa Cruz, Santa Cruz, CA, 95064, USA
[2]Max-Planck-Institut fur Astrophysik, Karl-Schwarzschild-Strasse 1, D-85748, Garching bei München, Germany

Abstract. We review our recent study of low-order non-axisymmetric features in K′ (2.2μm) images 18 nearby face-on disk galaxies. We find that about one third of these galaxies are substantially lopsided ($A_1/A_0 \gtrsim 0.20$) at 2.5 disk exponential scale length, and that their disks lie in a potential of characteristic ellipticity $0.045^{+0.03}_{-0.02}$ in the disk plane. However, the spiral pattern couples significantly to the estimate of the intrinsic ellipticity, and our measurement may represent an upper limit on the "true" potential triaxiality. We also discuss some kinematic implications of these observations.

1 Introduction

Basic questions regarding the shapes of stellar disks, in particular regarding deviations from axisymmetry in the disk plane, remain unanswered. Do disks have non-zero intrinsic ellipticity in the disk plane? Are spiral arms small ($\ll 1$) or large (> 1) fractional mass density enhancements within the disk? Does the mass distribution have mirror symmetry about the galaxy's nucleus? These questions are important because deviations from axisymmetry create radial streaming motions and thereby alter the dynamics and evolution of galaxies.

Knowledge the shape of stellar disks in detail can help address at least three different issues. First, an understanding of the morphology of galaxies is key in understanding galaxy formation. For example, the presence of thin disks already implies that those systems had inefficient early star formation and that their gas had sufficient time to settle into a disk. The degree to which the disk is symmetric contains information regarding the influence of the gas (and subsequent stars) in the collapse dynamics and constrains the dark matter halo ellipticity. Second, radial flow velocities within disks can be estimated from measurements of the magnitude of the disk asymmetries. This knowledge has repercussions on the study of chemical and stellar population gradients in galaxies (*e.g.* strong flows could fuel central engines or starbursts). Third, this work can lead to a better understanding of galaxy properties because some measurements are based on the assumption of disk symmetry. For example, Franx and de Zeeuw (1992) demonstrated that even a relatively small amount of disk ellipticity ($\epsilon < 0.1$) introduces significant scatter into the Tully-Fisher (TF; 1977) relation and argued in turn that the observed TF scatter provides an upper limit on the disk ellipticity. In this paper, we will concentrate on two specific distortions of the stellar disks of nearby galaxies, lopsidedness and non-zero ellipticity.

2 The Sample

The ideal sample for addressing these questions would have the following characteristics: (1) the images would be taken in a wavelength where the effects of dust are minimized; this directs us toward the near-IR ($\sim 2\mu$m), where dust absorption is much less than in the optical and dust emission is much less than in the mid-IR, (2) the images would be taken in a wavelength where the light comes primarily from the phase mixed stars in the galaxy, rather than from the young, luminous stars that have had insufficient time to disperse from their birth place (which again directs us toward the near-IR, where the most common types of stars emit most of their flux), (3) the images would cover a large angular area on the sky, to enable us to study the outer, well-resolved, isophotes of nearby galaxies, (4) the selection criteria would be chosen to simplify the analysis (the ambiguity between intrinsic ellipticity and projection effects is the most serious obstacle in obtaining a precise determination of the intrinsic ellipticity of disk galaxies), and (5) the sample would be large, because this project will ultimately rely on a statistical analysis in order to account for projection effects.

To satisfy requirements (1), (2), and (3) as much possible, we chose to observe in the K′ -band (2.2μm) with a NICMOS3 (256 ×256) array. To obtain a sample of nearly face-on galaxies we inverted the Tully-Fisher relationship,

$$i = \sin^{-1}\left(\frac{W'_{20}}{2v^{*} \cdot \mathrm{M}_B}\right), \tag{1}$$

where W'_{20} represents the width of the HI emission profile at 20% of the maximum emission level, corrected for the internal velocity dispersion of the gas, and $v^{*} = 158$ km sec^{-1} (Pierce and Tully 1992). The magnitude, M_B, is calculated by assuming pure Hubble flow and $\mathrm{H}_0 = 75$ km s^{-1} Mpc^{-1}. This determination of the inclination angle is somewhat uncertain because even galaxies with perfectly face-on stellar disks may have a linewidth larger than the vertical gas velocity dispersion (Lewis 1984), because many gas disks are warped at large radii. We searched the Huchtmeier-Richter catalog (Huchtmeier and Richter 1989) for HI linewidths and the CfA Redshift database (Huchra 1993) for apparent galaxy magnitudes and redshifts, in which most entries are unrelated to the question at hand (therefore avoiding any morphological bias). We selected spirals with $i < 25°$ and examined images on POSS or ESO sky survey images to restrict the sample to isolated field spirals.

Name	Type	M_B	v_{HI}	W'_{20}	$i(°)$	$\tilde{A}_1(2.5R_{exp})$	ϵ_ϕ
NGC 600	SBd	−19.0	1843	80	17	0.11 ± 0.08	0.09 ± 0.06
NGC 991	Sc	−19.2	1535	82	17	0.09 ± 0.05	0.10 ± 0.03
NGC1015[a]	SBa	−19.7	2630	87	15	0.23 ± 0.08	0.04 ± 0.03
NGC1302[b]	Sa	−20.2	1703	105	16	0.03 ± 0.02	0.18 ± 0.03
NGC1309[b]	Sbc	−20.3	2135	156	23	0.37 ± 0.07	0.09 ± 0.01
NGC1325A	Sb	−17.9	1333	46	14	0.58 ± 0.25	0.06 ± 0.07
NGC1376[b]	Scd	−21.0	4162	179	21	0.05 ± 0.08	0.04 ± 0.03
NGC1642[b]	Sc	−20.6	4621	143	19	0.01 ± 0.02	0.04 ± 0.01
NGC1703[b]	SBb	−19.6	1526	65	11	0.17 ± 0.08	0.06 ± 0.04
NGC2466[b]	Sc	−20.7	5364	175	23	0.17 ± 0.04	0.10 ± 0.02
NGC2485[c]	Sa	−20.8	4612	200	25	0.27 ± 0.10	0.11 ± 0.05
NGC2718	Sab	−20.8	3843	130	16	0.00 ± 0.07	0.27 ± 0.07
NGC6814	Sbc	−19.5	1565	112	21	0.19 ± 0.04	0.09 ± 0.02
NGC7156	Scd	−20.5	3984	125	17	0.08 ± 0.04	0.06 ± 0.02
NGC7309	Sbc	−20.6	4000	136	18	0.19 ± 0.05	0.12 ± 0.02
NGC7742	S0	−19.4	1649	85	17	0.05 ± 0.01	0.02 ± 0.01
IC 2627	Sbc	−19.6	2082	42	8	0.17 ± 0.04	0.07 ± 0.03
ESO-436	Sc	−20.2	4079	85	13	0.16 ± 0.14	0.10 ± 0.07

[a] HI profile remeasured by Tully (see Huchtmeier and Richter 1989).
[b] I-band image was obtained.
[c] HI profile has broad wings.

We list in Table 1 our final sample of 18 galaxies. We present the galaxy's name in column (1), its type in column (2) from the Revised-Shapley Ames catalog (Sandage and Tammann 1981), M_B in column (3), its recessional velocity and linewidth from the Huchtmeier-Richter HI catalog corrected for an internal velocity dispersion of 10 km sec^{-1} (Tully 1988) in columns (4) and (5) respectively, its inclination from Eq. 1 in column (6), and two measures of asymmetry that will be discussed later. The median luminosity of galaxies in this sample is $1.1L^*$ and their median distance is 31 Mpc (images of the sample galaxies are shown in Figure 1).

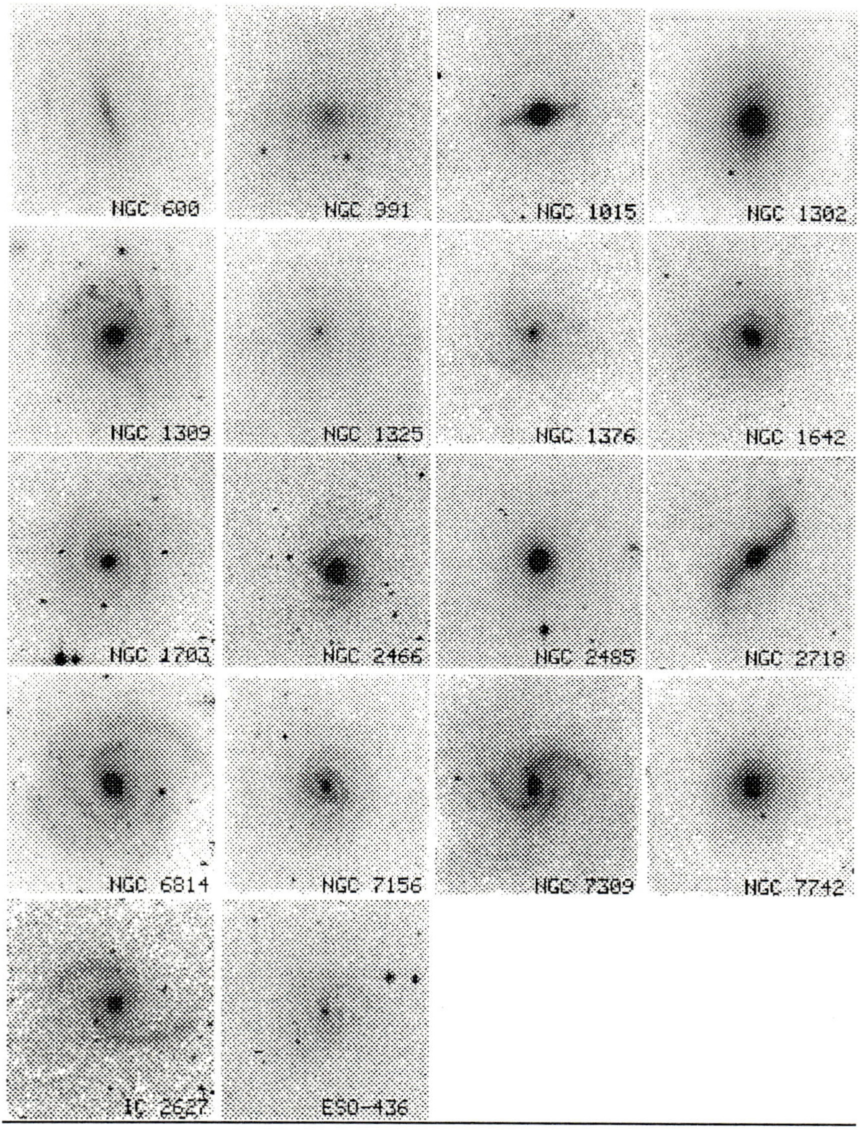

Figure 1: K' images of the sample galaxies. Each panel covers about 100 arcsec on a side.

3 The Analysis

a) Lopsidedness or $m = 1$

Fourier analysis is a natural way to analyze the non-axisymmetric component of the light in a face-on galaxy. It has the advantage over model fitting that no prior assumption is made about the intrinsic light distribution. Fourier analysis techniques have been used frequently in the past (cf. Grosbøl 1987, Elmegreen *et al.* 1989). Once we fix the center of a polar coordinate system at the brightest point of the galaxy in K′ , the surface brightness distribution distribution $\mu(R, \varphi)$ can be expressed as a Fourier series:

$$\mu(R,\varphi)/\langle\mu(R)\rangle = \sum_{m=1}^{\infty} A_m(R) e^{im[\varphi - \varphi_m(R)]}, \tag{2}$$

where φ denotes the azimuthal angle, m the azimuthal wavenumber, and A_m and φ_m are the associated Fourier amplitude and phase, respectively.

At large radii a sizable fraction of our galaxies exhibit lopsidedness with $A_1/A_0 > 0.2$. The most prominent examples are IC 2627, NGC 1309, NGC 1325, NGC 1642, NGC 2485, and NGC 6814. Only one of these galaxies, NGC 1309, has an apparent companion visible in our I-band CCD images or on the sky survey plates.

Because the amplitude of the "true" lopsidedness, $\tilde{A}_1(2.5R_{exp})$, is a positive definite quantity, the expectation value of its measurement, $A_1(2.5R_{exp})$, is the quadrature sum of the error and the true value. Therefore, we plot $\tilde{A}_1(2.5R_{exp})$, the error corrected measure of the asymmetry at 2.5 disk scale-lengths, in Figure 2. We chose $R = 2.5R_{exp}$ as the fiducial radius because this is the largest radius for which estimates of $\tilde{A}_1$ exist for most of our sample.

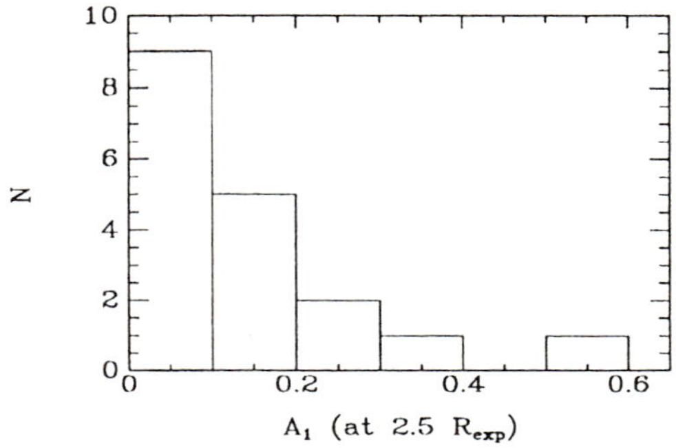

Figure 2: Histogram of $\tilde{A}_1(2.5R_{exp})$.

It is important to note that the lopsidedness measurement is not affected by an improper choice for the polar coordinate origin. If the galaxies were concentric, but we had chosen the wrong center, then $A_1(R)/A_0(R)$ would diverge as $1/R$ for $R \longrightarrow 0$, as long as there is a finite intensity gradient. We do not observe any such effect.

As a further check, we obtained large field, $20' \times 20'$, CCD images in I for six sample galaxies: NGC 1302, NGC 1309, NGC 1376, NGC 1642, NGC 1703 and NGC 2466. We performed a Fourier decomposition using the same grid as that used for K$'$ data. The agreement of the $m = 1$ distortion between I and K$'$ shows that these asymmetries are neither to gradients in the sky background nor due to asymmetric dust extinction.

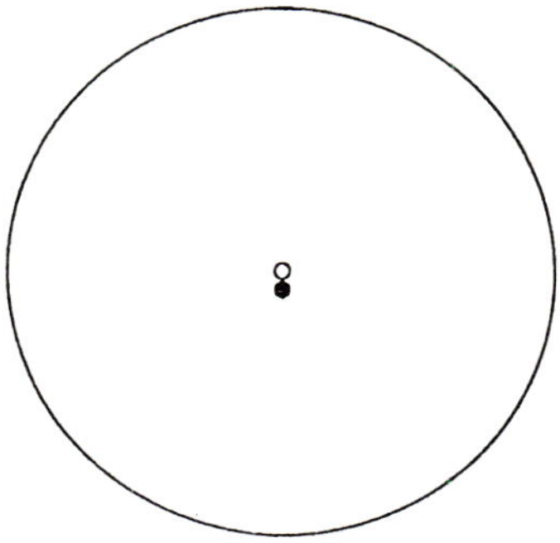

Figure 3: Illustration of the central offset necessary to account for $\tilde{A}_1(2.5R_{exp}) = 0.2$. The outer circle represents isophote at $R = 2.5R_{exp}$, the solid circle represents origin of outer center, and the open circle represents origin of galaxy with chosen lopsidedness.

b) Disk Ellipticity

The $m = 2$ amplitudes reflect inclination effects, bulge and disk ellipticity, and spiral arms. Therefore, they cannot be used to place unambiguous constraints on the shapes of disks. Instead, we concentrate on the mean ellipticity at a radius-independent, but arbitrary, position angle. By fitting a specific image model, we average radially over the distortions created by the spiral arms. As a specific mode, we assume that the disks reside in a non-rotating, logarithmic potential with ϵ_Φ. For radii $R > R_{exp}$ the isophotes are then given by $\epsilon_{iso}(R) = \epsilon_\Phi(1 + R_{exp}/R)$ (Franx and de Zeeuw, 1992). To fit our model, we first derive the radial range over which the disk dominates the total light ($\mu_{disk} > 2\mu_{bulge}$) by fitting an axisymmetric bulge-disk model to the whole image. Then we estimate ϵ_Φ by fitting a disk model with an ellipticity profile given by $\epsilon_{iso}(R)$ over a restricted radial range. Our fits provide an estimate of the scale length, R_{exp}, the ellipticity of the potential, ϵ_Φ, the position angle, PA, and their associated uncertainties for each galaxy. The results for ϵ_Φ are presented in Table 1. With a few exceptions (e.g. NGC 1302 and NGC 2718), the values for ϵ_Φ are found to be < 0.1. There are reasons to suspect that NGC 1302 and NGC 2718 should be excluded from this analysis (we present our final results for samples with and without these two galaxies).

To study the coupling between the arms and the disk ellipticity we used simulated data, constructed from a perfectly axisymmetric disk and a superimposed spiral with properties similar to NGC 1703. We add noise and sky errors to this image matching the noise properties of our data. Finally, we determine the disk's ellipticity using the same fitting procedure as before and find $\epsilon = 0.03$. Therefore, we treat the interplay between disk ellipticity and spiral arms by quadratically adding an extra error term of $\Delta\epsilon_{Sp} = 0.03$ to our ellipticity measurements.

We estimate the probability distribution of intrinsic potential ellipticities, $P(\epsilon_\Phi^I)$ from our sample in both a parametric and a non-parametric fashion. In the non-parametric approach, we do not need to adopt a functional form for $P(\epsilon_\Phi^I)$. This is an important asset because we have no justification for any particular choice. However, such a technique will not recover the "true" distribution even for an infinite number of measurements. In contrast, we obtain an unbiased estimate of $\epsilon^I\Phi$ if we adopt a parametric functional form for the $P(\epsilon_\Phi^I)$. Our philosophy is to estimate $P(\epsilon_\Phi^I)$ from the results of the non-parametric approach and then to use this result to choose a functional form for the parametric estimate.

As shown in Rix and Zaritsky (1995), one can determine the probability distribution of intrinsic ellipticities, $P_j(\epsilon_\Phi^I)$, from the observed parameters for each galaxy j. This estimate is based on the assumption that the angle specifying the disk orientation at given inclination is distributed uniformly. The probability distribution for the entire sample is then the combination of the probabilities of the individual distributions. We compare this distribution to that we would have obtained if we observed a set of perfectly round disks with inclinations and uncertainties corresponding to the disks in our sample. The latter distribution can be thought of as a "point-spread function", because it reflects the estimate of $P(\epsilon_\Phi^I)$ for an input δ-function. The median of the distribution of ϵ_Φ^I for the axisymmetric disks is 0.025, and is 0.07 for our sample galaxies. We concluded that the underlying stellar disks in these galaxies are slightly elliptical because the observed median value of ϵ_Φ^I is significantly larger for the observed disks than for the model axisymmetric disks. However, these non-parametric estimates do not provide a good measure for a "characteristic" ellipticity, ϵ_0 and its uncertainty.

The nonparametric distribution prompted us to chose an exponential as the parametric trial function for P;

$$P(\epsilon_\Phi^I, \epsilon_0) = \frac{1}{\epsilon_0} \exp\left[-\frac{\epsilon_\Phi^I}{\epsilon_0}\right] \qquad (3).$$

We determine the most likely value of ϵ_0 using the maximum likelihood technique detailed by Rix and Zaritsky (1995). The results are illustrated in Figure 4, where we show a histogram of the measured ϵ_Φ along with the expected distributions of measurements for $\epsilon_0 = 0.025, 0.045$, and 0.075. Maximizing $\mathcal{L}(\epsilon_0)$ yields the following results: (1) if we assume a spurious ellipticity due to spiral arms of 0.03, which is added in quadrature to the measurement error, we find $\epsilon_0 = 0.045^{+0.03}_{-0.02}$, (2) if we do not include this source of error, then more of the observed ellipticity becomes attributable to the underlying disk ellipticity and $\epsilon_0 = 0.055^{+0.03}_{-0.02}$, (3) if we assume that our estimate of $\cos(i)$ from the HI linewidths is an upper limit, and that the galaxies are distributed uniformly between $\cos(i)_{HI}$ and $\cos(i) = 1$, then less of the observed distortions is due to inclination effects and ϵ_0 increases to $0.055^{+0.04}_{-0.02}$, and (4) if we assume the same conditions as in (1), but eliminate the two most strongly distorted galaxies (NGC 1302 and NGC 2718), we find $\epsilon_0 = 0.030^{+0.03}_{-0.015}$. Therefore, we conclude that the galaxy disks in our sample have small but finite ellipticity. The characteristic ellipticity of the potential, within the disk plane, is $\epsilon_0 = 0.045^{+0.03}_{-0.02}$, and this result is quite insensitive to the details of the statistical treatment. The potential ellipticity, $\epsilon_\Phi = 0.1$, inferred for the Milky Way by Kuijken and Tremaine (1994) is somewhat higher than the mean found here, but still consistent with our distribution for ϵ_Φ determined from external galaxies.

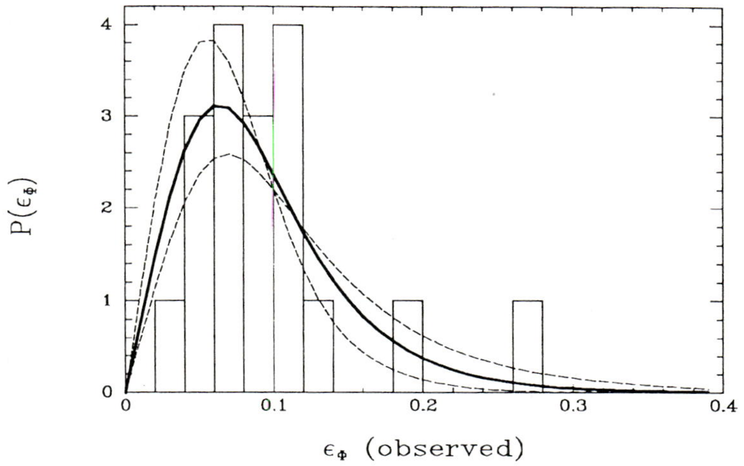

Figure 4: Parametric Determination of disk ellipticity. The figure illustrates the parametric determination by comparing the observed distribution (the histogram) to three distributions with different parameter choices. The sharpest peaked curve corresponds to $\epsilon_0 = 0.025$. *The other dashed line corresponds to* $\epsilon_0 = 0.075$, *and the solid line corresponds to* $\epsilon_0 = 0.045$

The largest source of uncertainty in this analysis is the cross-talk between two-arm spirals and disk ellipticity. Even though we made an attempt to account for this through simulations, it is conceivable that spiral arm effects are bigger than we estimated. In this case, our ϵ_0 estimates can be considered as a conservative upper limit on the triaxiality of the potential.

c) Kinematic Implications

Both lopsidedness and non-zero ellipticity have kinematic implications for the stellar disks. Although it is common lore that some galaxies appear lopsided, the study of Baldwin, Lynden-Bell, and Sancisi (1980) appears to be the only systematic study of such asymmetries in the literature. Regardless of the cause or longevity of such $m = 1$ distortions, they, like disk ellipticity, will lead to a radial velocity of the LSR in these galaxies, which we calculate to be

$$\langle v_R \rangle = 7.4 \text{ km s}^{-1} \left[\frac{v_c}{200 \text{ km s}^{-1}}\right] \left[\frac{\tilde{A}_1}{0.11}\right] \left[\frac{2.5 R_{exp}}{R}\right]. \qquad (4)$$

The average galaxy in our sample will have a comparable $\langle v_R \rangle$ to the value inferred for the solar neighborhood (14km sec^{-1} ; Blitz and Spergel 1990).

This dynamical inference is valid regardless of the origin and the longevity of the $m = 1$ distortions, if the distortion persists longer than one orbital time. However, the question of the longevity of lopsidedness is both important and puzzling. As first discussed by Baldwin *et al.* (1980), if lopsidedness is due to elliptic orbits in an *axisymmetric* logarithmic potential, any initial azimuthal alignment of the orbit apocenters winds quickly. Using values characteristic for our sample, $v_c = 200$ km sec^{-1} , $R_{outer} = 1.5 R_{inner}$ and $R_{outer} = 8$kpc, the "wind-up time" is only about 10^9 yrs. This estimate of the lifetime in conjunction with our observation of significant lopsidedness in about 1/3 of our sample suggests that the generation of lopsidedness has a typical time-scale of 3×10^9 yrs. Alternatively, we could suppose that disk stars merely respond to a potential distortion of the form $\Phi(R, \varphi) = \Phi(R) + (\epsilon_{lop}/2) \cos \varphi$. Then all closed orbits are aligned eccentric ellipses and there is no winding problem. However, the stellar orbits "oppose" the potential they reside in and cannot create such distortions self-consistently. Therefore the winding problem is merely replaced by the question of how to create a lopsided potential.

Using the non-zero ellipticity estimates, we can also estimate the magnitude of the velocities produced from disk ellipticity. Elliptical streamlines arising in triaxial disks will result in nonzero radial velocity components for the local standard of rest (LSR) at all points along the orbit except at apocenter and pericenter. As shown by Rix and Zaritsky, the radial velocity of the LSR, averaged over all azimuthal angles is

$$\langle v_R \rangle = 6.4 \text{ km s}^{-1} \left[\frac{v_c}{200 \text{ km s}^{-1}} \right] \left[\frac{\epsilon_\Phi}{0.045} \right]. \tag{5}$$

Using characteristic values for our sample, $v_c = 200$ km sec^{-1} and $\epsilon_0 = 0.045$, we find that along a streamline the radial velocity with respect to the galaxy center varies from ~ -9 km sec^{-1} to ~ 9 km sec^{-1} with $\langle v_R \rangle = 6.4$ km sec^{-1} . This value is again consistent with the Blitz and Spergel value for the the solar neighborhood, although it is equally consistent with the non-detection ($v_R = -1 \pm 9$km sec^{-1}) found by Kuijken and Tremaine (1994).

Lastly, elliptical streamlines also lead to variations of the azimuthal velocity along the orbit. As pointed out by Franx and de Zeeuw (1992), this variation contributes to the scatter in the TF relation – even if the true inclinations can be determined perfectly. Because the measured H I linewidth is presumed to be proportional to $2v_{tangential}(\varphi)/\sin i$, it will vary among identical galaxies seen at the same inclination but at different angles φ. This effect leads to a scatter (in magnitudes) of at least $\Delta M = 2.4\epsilon_\Phi$, for the TF relationship given by Pierce and Tully (1992), which for $\epsilon_0 = 0.045$ implies a scatter of 0.11 magnitudes. However, an immediate comparison to the scatter in the TF relation from observed samples of galaxies is complex. The K$'$ photometry and H I linewidths presumably probe triaxiality at different radii. While our data probe typically $3-9$ kpc, the H I flux arises typically from radii larger by a factor of two. (S. Rao, *priv. comm.*). In addition, most TF observations are intended for distance estimates, therefore the samples are rarely statistically complete and may be biased against morphological or kinematic peculiarities.

References

Baldwin, J., Lynden-Bell, D., & Sancisi, R. (1980): MN, **193**, 313
Blitz, L., & Spergel, D. (1990): ApJ, **370**, 205
Elmegreen, B.G., Elmegreen, D.M., & Seiden, P. (1989): ApJ, **288**, 438
Franx, M., & de Zeeuw, T., (1992): ApJL, **392**, L47
Grosbol, P. (1987): in *Selected Topics on Data Analysis in Astronomy,* ed L. Scarsi, V. DiGesu, & P. Crane (Singapore: World Scientific), p. 57
Huchra, J. (1993): CfA Redshift Catalogue
Huchtmeier, W.K., & Richter, O.-G. (1989): *HI Observations of Galaxies*, (New York: Springer)
Kuijken, K. & Tremaine, S. (1994): ApJ, **421**, 178
Lewis, B. M. (1984): ApJ, **285**, 453
Pierce, M., & Tully, R.B. (1992): ApJ, **387**, 47
Tully, R.B. (1988): *Nearby Galaxies Catalog*, (Cambridge: Cambridge University Press)
Tully, R.B., & Fisher, J.R. (1977): AA, **54**, 661
Rix, H.-W., & Zaritsky, D. (1995): ApJ, , 447, 82
Sandage, A., & Tammann, G.A. (1981): *A Revised Shapley-Ames Catalogue of Bright Galaxies*, (Washington: Carnegie Institute of Washington)

Azimuthal Color Gradients in M99

Rosa A. González and James R. Graham

Astronomy Department, University of California, Berkeley CA 94720-3411,USA

Abstract. We present optical and IR surface photometry of M99 (NGC 4254) at g, r_S, i, J and K'. A central motivation for our research is the fundamental idea of density wave theory that the passage of a spiral density wave triggers star formation as it shocks the molecular gas in the disk. If this is the case color gradients, due to stellar population age gradients, should occur across the spiral arms. We have found a gradient consistent with this scenario in a reddening-free Q-like parameter at 6 kpc galactocentric distance across one of the arms of M99; this photometric parameter, $Q_{r_S Jgi}$, is very sensitive to red supergiants. We rule out that the change in $Q_{r_S Jgi}$ across the arm is mainly due to dust. We also see a difference in $Q_{r_S Jgi}$ going from the interarm regions to the arms that indicates that arms cannot be due exclusively to crowding of stellar orbits. We have measured the angular speed Ω_p of the spiral pattern and the location of the corotation radius, for the first time from the drift velocity of the young stars away from their birth site. The measured $Q_{r_S Jgi}$ implies a star formation rate for M99 within the range of 10–20 $\mathrm{M_\odot yr^{-1}}$, a disk stellar mass surface density of $\sim$ 80 $\mathrm{M_\odot pc^{-2}}$ and a maximum contribution of $\sim$ 20 percent from red supergiants to the K' light in the region under consideration.

1 Introduction

Past studies of spirals have used optical surface photometry to probe the dynamics of stellar disks (*e.g.* Schweizer 1976; Talbot *et al.* 1979; Iye *et al.* 1982; the work of the Elmegreens and colaborators). However, even at I the observed morphology of disk galaxies is strongly influenced by extinction. Combined optical and near infrared data extending to a wavelength of 2 μm offer a better opportunity to correct for extinction and reddening. This is crucial, since such effects can lead to incorrect estimates of the contribution of young stars and the true arm–interarm density contrast. Also, given that the effect of dust extinction is much less severe in the IR relative to the optical ($\mathrm{A_K} = 0.08\mathrm{A_B}$), the IR surface brightness distribution is a more direct measure of the stellar disk mass surface density distribution (Rix and Rieke 1993).

2 Observations

The optical images of M99 –type ScI; distance = 16 Mpc (Jacoby *et al.* 1992), for 78 pc/″; R_{25} = 2.′69 (Elmegreen *et al.* 1992)– were obtained in 1994 April 15 and 16 UT with the 1-m Nickel telescope at Lick Observatory. The pixel scale is 0.″37/pixel, for a field of view of the Ford CCD 2048^2 (binned 2 × 2) of 6.′3. The photometric calibration in the optical bands was done in the Thuan-Gunn system (Thuan and Gunn 1976, Wade *et al.* 1979, Kent 1985). The infrared images were obtained in 1994 March 14 UT with the 1.3-m telescope at Kitt Peak; the Infrared Imager (IRIM) uses a 256x256 NICMOS3 detector and has a field of view of 8.′5 (Probst 1993, Heim *et al.* 1994). The photometric calibration of the IR images was done with faint UKIRT standard stars (Casali and Hawarden 1992, Wainscoat and Cowie 1992). Table 1 gives a summary of the observations.

Table 1. Observation Log

Object	Filter	$\lambda_{\rm eff}$	FWHM	Exposure	Telescope
M99	g	5000Å	830Å	40 min	Lick 1-m
	r_S	6800Å	1330Å	35 min	
	i	7800Å	1420Å	40 min	
	J	1.25μm	0.29μm	56 min	Kitt Peak 1-m
	K'	2.16μm	0.33μm	56 min	

3 Azimuthal Color Gradients

3.1 A Red-Supergiant-Sensitive Q-Parameter

With our data, we have constructed the photometric parameter $Q_{r_S Jgi} = (r_S - J) - \frac{E(r_S-J)}{E(g-i)}(g-i) = (r_S - J) - 0.82(g-i)$. $Q_{r_S Jgi}$ is reddening-insensitive for screen absorption (Mihalas and Binney 1981) assuming the extinction curves of Schneider *et al.* (1983), and Rieke and Lebofsky (1985). However, extinction and reddening properties are highly dependent on geometry. In general, for a mixture of dust and stars the value of $Q_{r_S Jgi}$ increases with optical depth relative to the case of screen absorption. For example, $Q_{r_S Jgi}$ is negligibly affected by a mix of dust and stars with plane-parallel geometry in a galaxy with $\cos(\omega) = 0.87$ (Bruzual *et al.* 1988) and in the "cloudy galaxy" model of Witt *et al.* (1992), but is no longer reddening-insensitive for the Witt *et al.* "starburst galaxy" model with $\tau_V > 2$ or their "dusty galaxy" model with $\tau_V > 3$ (the Witt *et al.* models have spherical symmetry, and consist of mixtures of stars and dust with different relative spatial distributions). It is unlikely that τ_V is > 2 for a face-on galaxy (Bruzual *et al.* 1988, Peletier and Willner 1992), so $Q_{r_S Jgi}$ should be reddening-insensitive for the disk of M99.

Being very degenerate with respect to spectral type –*i.e.* having almost the same value for all except the reddest stars– (Johnson 1966, Lee 1970, Koornneef

1983), it turns out that $Q_{r_S\,Jgi}$ is, precisely for that reason, an excellent tracer of star formation. Writing $Q_{r_S\,Jgi}$ in a slightly different way helps to appreciate this fact:

$$Q_{r_S\,Jgi} = log_{10}\frac{I_g^{2.05}\,I_J^{2.50}}{I_{r_S}^{2.50}\,I_i^{2.05}}$$

A combination of small numerator and large denominator in the quotient will yield a small and even negative $Q_{r_S\,Jgi}$; a large numerator plus small denominator will result in a high $Q_{r_S\,Jgi}$. It is easy to achieve a low value of $Q_{r_S\,Jgi}$ with a single Planckian peaking around 7000Å; however, to obtain a high value of $Q_{r_S\,Jgi}$ it is necessary to have a double-humped spectral energy distribution with a valley around 7000Å, and increasing intensity both redward and blueward of that wavelength. Therefore, $Q_{r_S\,Jgi}$ will have a higher value for a mixture of blue and red stars than for just about any single star. Extreme values of $Q_{r_S\,Jgi}$ occur when the most massive stars of a young population become red supergiants.

3.2 The Stellar Drift and the Pattern Speed

If the passage of a spiral density wave and the consequent shocking of molecular clouds triggers the formation of stars, then the mean age of the young stellar population is related to the distance from the arm, at a fixed radius, through the difference between the angular rotation speed, $\Omega(\mathrm{R})$, and the pattern speed, Ω_{p}, *i.e.* $\mathrm{d} = [\Omega(\mathrm{R}) - \Omega_{\mathrm{p}}]\mathrm{R}t_{\mathrm{age}}$. In this case, stellar population synthesis models (Charlot and Bruzual 1991, Bruzual and Charlot 1993) predict that the evolution of $Q_{r_S\,Jgi}$ across an arm is dramatic enough that age gradients can be spatially resolved in a nearby spiral even if the ratio of young to old (age 5×10^9 yr) stars is as small as 1–2 percent by mass.

We have focused our attention on a patch in the N1 arm (following the nomenclature of Iye *et al.* 1982), at $\sim$ 5.9 kpc from the center of the galaxy, that from its $J - K'$ color appears to have little contamination by dust ($J - K'$ is a good dust tracer and suffers less contamination from young stars than, say, $g - K'$). Even in annuli having width equal to the spatial resolution of the data, this spot exhibits an azimuthally asymmetric profile in $Q_{r_S\,Jgi}$ that is consistent with coherent star formation induced by the passage of the spiral density wave; to increase the signal-to-noise, though, we have collapsed the patch from r $\sim 60''$ to $\sim 95''$.

Figure 1 shows, superimposed, the $Q_{r_S\,Jgi}$ profile and the $J - K'$ profile of this patch. In this region, $Q_{r_S\,Jgi}$ increases even as the amount of dust, inferred from $J - K'$, starts to decrease away from the dust lane. This is the opposite of what the Bruzual *et al.* and Witt *et al.* models of radiative transfer in mixed stars and dust predict for a diminishing optical depth, and therefore we can rule out that the change in $Q_{r_S\,Jgi}$ is due mainly to dust. Figure 2 shows a full azimuthal cut of $Q_{r_S\,Jgi}$ at R $\sim 75''$, as well as the K' azimuthal light profile for the same radii. $Q_{r_S\,Jgi}$ changes at the location of every arm, and the arm Q-profiles are more similar to each other than the arm light profiles among themselves; a similar process seems to be at work in the three arms that is compatible with a change of age of the stars across them. If arms were due to crowding of stellar orbits exclusively we should not see any change in the Q-parameter going from the interarm regions to the arms.

Figure 3 shows the $Q_{r_S\,Jgi}$ profile *vs.* distance from the arm for the region of interest and two model $Q_{r_S\,Jgi}$ curves with 1 and 2 percent by mass of young stars, respectively. Both models have been calculated with a Salpeter IMF with $M_{lower} = 0.1M_\odot$ and $M_{upper} = 10M_\odot$, and a duration of the burst of 2×10^7 yr (Gruendl 1995); in both models the young burst is superimposed on a background of stars 5×10^9 yr old. The choice of 1–2 percent of young population seems reasonable compared both to the number of times that the region under consideration has encountered the spiral pattern (once every $\sim 10^8$ yr) in the lifetime of the disk ($\sim 10^{10}$ yr) and to the amount of light in the arms contributed by old stars in the B band as estimated by Schweizer (1976). We have chosen the origin of the burst, d=0, at the location of the dust lane in the data, *i.e.* the location of the shock (Roberts, Huntley and van Albada 1979). To transform the time in the models into distance from the site of star formation we have adopted the flat rotation curve with $V_{rot} = 140$ km s^{-1} derived by Phookun *et al.* (1993) from HI data, and then chosen the difference between Ω(5.9 kpc) and Ω_p that allows a good fit of the model $Q_{r_S Jgi}$ to the data. Taking $\Omega(5.9\ \text{kpc}) - \Omega_p = 6.5$–$8.0$ km s^{-1} kpc^{-1} yields $\Omega_p = 17.2$–15.7 km s^{-1} kpc^{-1}, and places R_{CR} at 0.6–0.7 R_{25}. This is slightly larger than the 0.54 R_{25} found by Elmegreen *et al.* (1992), but in better agreement with modal theory of spiral structure (Bertin *et al.* 1989), which predicts that the corotation zone is at ~3 radial scale lengths, *i.e.* at a location where the gaseous mass is plentiful enough to play a significant dynamical role. (By fitting the radial light profile of M99 between 59″ and 111″, we have measured a scale length that varies between 0.24 R_{25} at g and 0.20 R_{25} at K'.)

3.3 The Star Formation Rate

By attributing the K' flux in the interarm region downstream of the shock to a mixture of 1–2 percent young population ($\sim 2 \times 10^8$ yr old, estimated from the distance from the dust lane) and 99–98 percent old (age 5×10^9 yr) stars, we find a disk stellar mass surface density of 80–75 $M_\odot pc^{-2}$. With 1 percent of young stars formed in 2×10^7 yr these numbers imply a star formation rate of 10 $M_\odot yr^{-1}$, if integrated over the area inside 0.7 R_{25}, and 20 $M_\odot yr^{-1}$, if all the area inside R_{25} is included. This result is in the range of 0–20 $M_\odot yr^{-1}$ found by Kennicutt (1983) for Sc galaxies. It is important to note, though, that he derived his result from $H\alpha$ flux and UBV colors, *i.e.* mainly from the output of stars more massive than 10 $M_\odot$, which make up $\sim$ 6 percent of the mass of all stars; we are sampling stars less massive that 10 $M_\odot$ and therefore a much bigger fraction of the newly formed stars. Combined with a detected color gradient that is consistent with the evolution in time of a stellar population whose birth was triggered at the location of the dust lane, this general agreement suggests that at least some of the star formation in M99 is associated with the spiral density wave.

The fit to the observed $Q_{r_S Jgi}$ with 1–2 percent young stars by mass and an upper mass limit of the burst of 10 $M_\odot$ implies that less than 0.1 percent by mass of red supergiants should be present. If the mass fraction of young stars is just a couple of a percent, their peak contribution to the K' light is 15–25 percent between the ages of 4.5×10^7 and 10^8 yr. This is consistent with recent results (Rix and Rieke 1993, Rhoads 1995) and implies that K' is in general a good tracer of disk mass.

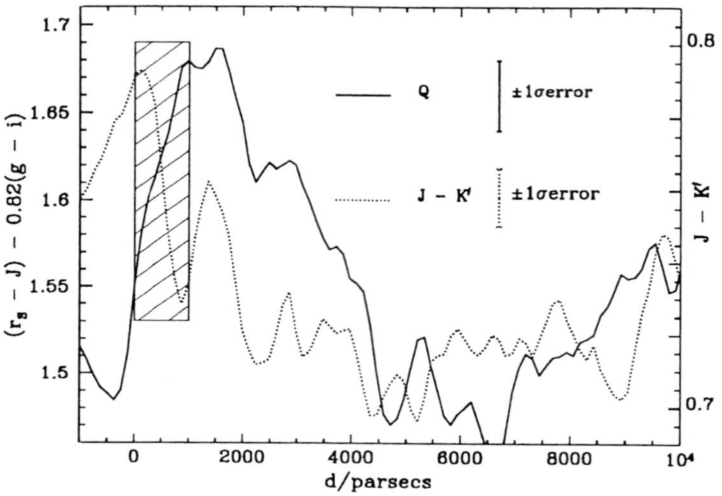

Fig. 1. $Q_{r_S Jgi}$ (*solid line*, left-hand Y-axis) and $J - K'$ magnitude (*dotted line*, right-hand Y-axis) for M99 at R ~ 5.9 kpc. $Q_{r_S Jgi}$ increases in the shaded region even as the amount of dust starts to decrease; the X-origin is at the location of the dust lane.

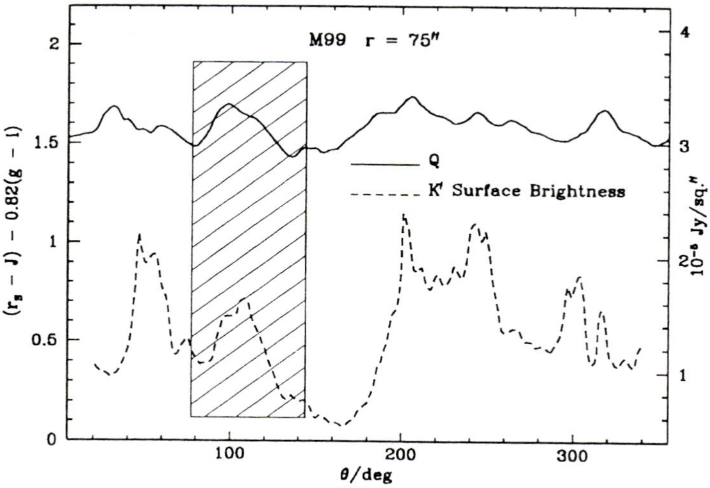

Fig. 2. $Q_{r_S Jgi}$ (*solid line*, left-hand Y-axis) and K' surface brightness (*dashed line*, right-hand Y-axis) for M99 at R ~ 75″. The asymmetric arm $Q_{r_S Jgi}$ profiles are more similar to each other than the arm light profiles among themselves; we have shaded the region under consideration.

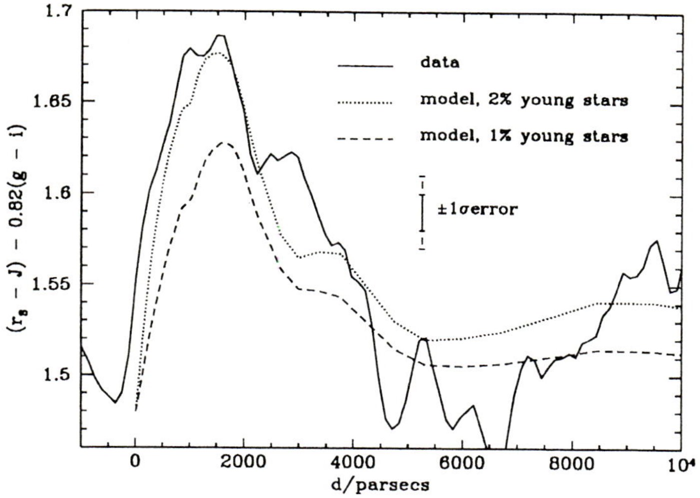

Fig. 3. Observed $Q_{r_S Jgi}$ profile (*solid line*) *vs*. distance from the dust lane in pc for arm N1 of M99 at R $\sim$ 5.9 kpc, and two model $Q_{r_S Jgi}$ curves with 2 percent (*short-dashed line*) and 1 percent (*long-dashed line*) by mass of young stars. Models have a Salpeter IMF with $M_{lower} = 0.1M_{\odot}$ and $M_{upper} = 10M_{\odot}$, and a duration of the burst of 2×10^7 yr; the young stars are superimposed on a background population 5×10^9 yr old. We have used $\Omega(5.9 \text{ kpc}) - \Omega_p = 7.0$ km s^{-1} kpc^{-1} to scale the models to the data (see text). Inner error bars show data statistical error only.

4 Conclusions

We have identified a Q-like photometric parameter that is very sensitive to red supergiants. Thanks to very deep optical and near infrared images, we have a positive detection of an azimuthal gradient in $Q_{r_S Jgi}$ due to a stellar population *age* gradient across a spiral arm of M99. ¿From the drift velocity of the young stars away from their birth site, we have measured the angular speed Ω_p of the spiral pattern and the location of the corotation radius. According to the models that better resemble our data, spiral structure is both the product of a density wave that involves the whole stellar disk (hence the close to 100 percent by mass of old stars in the arm) and the cause of organized star formation as the wave encounters and shocks molecular gas.

References

Bertin, G., C. C. Lin, S. A. Lowe and R. P. Thurstans (1989), *Ap. J.* **338** 78
Bruzual, G. and S. Charlot (1993), *Ap. J.* **405** 538
Bruzual, G., G. Magris-C. and N. Calvet (1988), *Ap. J.* **333** 673
Casali, M. M., and T. G. Hawarden (1992), *JCMT–UKIRT Newsletter* **4** 33
Charlot, S. and G. Bruzual (1991), *Ap. J.* **367** 126
Elmegreen, B. G., and D. M. Elmegreen (1985), *Ap. J.* **288** 438
Elmegreen, B. G., D. M. Elmegreen and L. Montenegro (1992), *Ap. J. S. S.* **79** 37
Elmegreen, B. G., D. M. Elmegreen and L. Montenegro (1993), *P. A. S. P.* **105** 644
Elmegreen, B. G., D. M. Elmegreen and P. E. Seiden (1989), *Ap. J.* **355** 52
Elmegreen, D. M. (1981), *Ap. J. S. S.* **47** 229
Elmegreen, D. M., and B. G. Elmegreen (1984), *Ap. J.* **54** 127
Elmegreen, D. M., and B. G. Elmegreen (1990), *Ap. J.* **355** 52
Gruendl, R. A. (1995) Ph.D. Thesis, in preparation.
Heim, G. B., N. C. Buchholz and R. W. Luce (1994), in *Instrumentation in Astronomy VIII*, eds. D. L. Crawford and E. R. Craine (Bellingham, Washington, USA:SPIE) p.1024.
Iye, M., S. Okamura, M. Hamabe and M. Watanabe (1982), *Ap. J.* **256** 103
Johnson, H. L. (1966), *Ann. Rev. Astron. Astrophys.* **4** 193
Kennicutt, R. C., Jr. (1983), *Ap. J.* **272** 54
Kent, S. M. (1985), *P. A. S. P.* **97** 165
Koornneef, J. (1983), *Astr. Ap.* **128** 84
Lee, T. A. (1970), *Ap. J.* **162** 217
Mihalas, D., and J. Binney (1981), *Galactic Astronomy*, 2nd ed. (San Francisco:Freeman).
Peletier, R. F., and S. P. Willner (1992), *A. J.* **103** 1761
Phookun, B., S. N. Vogel and L. G. Mundy (1993), *Ap. J.* **418** 113
Probst, R. G. (1993), *Instrument Manual for the IRIM Infrared Camera System*, NOAO.
Rieke, G. H., and M. J. Lebofsky (1985), *Ap. J.* **288** 618
Rix, H. W., and M. J. Rieke (1993), *Ap. J.* **418** 123
Roberts, W. W., Jr., J. M. Huntley and G. D. van Albada (1979), *Ap. J.* **233** 67
Schulman, E., J. N. Bregman and M. S. Roberts (1994), *Ap. J.* **423** 180
Schweizer, F. (1976), *Ap. J. S. S.* **31** 313
Talbot, R. J., Jr., E. B. Jensen and R. J. Dufour (1979), *Ap. J* **229** 91
Thuan, T. X., and J. E. Gunn (1976), *P. A. S. P.* **88** 543
Wade, R. A., J. G. Hoessel and J. H. Elias (1979), *P. A. S. P.* **91** 35
Wainscoat, R. J., and L. L. Cowie (1992), *A. J.* **103** 332
Witt, A. N., H. A. Thronson, Jr., and J. M. Capuano, Jr. (1992), *Ap. J.* **393** 611

Spiral Disk Asymmetries and Other Evidence of Accretion in Nearby Galaxies

Dennis Zaritsky[1]

[1] UCO/Lick Observatory, Univ. of Calif. Santa Cruz, Santa Cruz, CA, 95064, USA

Abstract. Correlations between the stellar disk asymmetries measured in the K-band magnitude, $B-V$ color, residuals from the B-band Tully-Fisher relationship, chemical abundance gradients of nearby spiral galaxies are discussed within the context of a model in which small satellites or H I clouds are accreted onto the outer disks of spiral galaxies. The data suggest that accretion events at the current time are common.

Although the frequency of galaxy interactions is poorly constrained, such events do occur (cf. Arp 1966) and they affect galaxy evolution (Lonsdale, Persson, & Matthews 1984). Although major mergers are spectacular, the more common interaction is presumably the accretion of a small companion galaxy or gas cloud by a larger galaxy.

Which observables might attest to recent accretion in nearby spirals? By analogy to major mergers (Lonsdale, Persson, & Matthews 1984), accretion events might elevate the star formation rate (SFR). If so, it may be possible to identify galaxies that have recently accreted by their Hα fluxes, colors, or absolute blue magnitudes, M_B. A second possible result of accretion, one which has often been invoked in models of Galactic chemical evolution (cf. Sommer-Larsen 1991 and references therein), is the dilution of existing metal enriched gas by infalling low-metallicity gas. Because satellite orbits decay faster into the disk plane than they do in radius (Quinn & Goodman 1986), gas-phase chemical abundance anomalies in the *outer disk* may be a further sign of recent accretion. Lastly, one might expect even minor interactions to affect a galaxy's morphology (cf. Schweizer *et al.* 1990).

The data used in this study are from published studies and discussed by Zaritsky (1995). The presence of young (age $<$ 1 Gyr) stars can be ascertained using either a luminosity normalized color, such as $B-V$, or a mass normalized color, such as that obtained by comparing M_B for galaxies of similar mass. I use the Tully-Fisher relationship given by Pierce & Tully (1992) to calculate a galaxy's mass-normalized color. The difference between the observed M_B and the predicted M_B is the B-band TF residual magnitude (hereafter referred to as ΔB). The abundance distributions are characterized by O/H gradients and the morphology by measurements of disk asymmetries in the K-band (discussed by Rix & Zaritsky 1995). The data are shown in Figure 1.

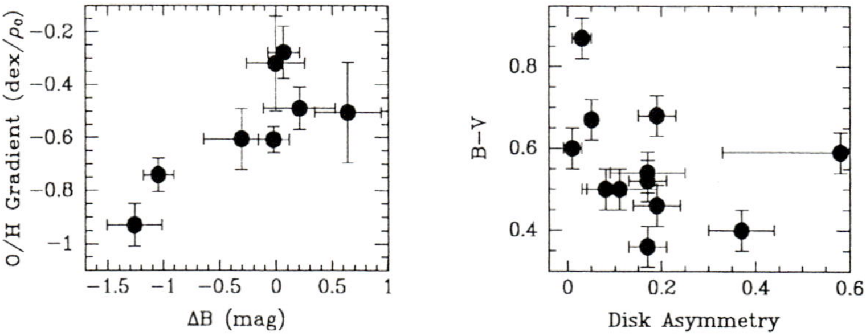

Fig. 1. The right panel shows the correlation between ΔB and the slope of the O/H abundance gradient (in units of dex/isophotal radius) for the sample of galaxies for which Cepheid distances are available. The left panel shows the correlation between disk asymmetry (m=1 amplitude measured at 2.5 scale lengths) and $B - V$ for a sample of galaxies from Rix and Zaritsky (1995).

A quantitative discussion of the correlations is given by Zaritsky (1995) but here I conclude that the slope of the abundance gradient correlates with ΔB and that there is a trend of that galaxies with more asymmetric disks have elevated SFR. Both of these results are consistent with a model in which there has been recent infall.

References

Arp, H. (1966): Atlas of Peculiar Galaxies, (California Institute of Technology, Pasadena)

Lonsdale, C.J., Persson, S.E., & Matthews, K. (1984): ApJL **287**, 95L

Pierce, M., & Tully, R.B. (1992): ApJ, **387**, 47

Quinn, P.J., & Goodman, J. (1986): ApJ, **309**, 472

Rix, H.-W., & Zaritsky, D. (1995): ApJ, **447**, 82

Schweizer, F., Seitzer, P., Faber, S.M., Burstein, D., Dalleore, C.M., & Gonzales, J.J. (1990): ApJ **364**, L33

Sommer-Larsen, J. (1991): MNRAS, **249**, 368

Zaritsky, D. (1995): ApJL, 448, L17

Mass Models from Near-Infrared Surface Photometry

Ph. Héraudeau[1], F. Simien[1], and G.A. Mamon[2]

[1]Observatoire de Lyon, F-69561 Saint-Genis-Laval cedex, France
[2]Institut d'Astrophysique, 98 bis boulevard Arago, F-75014 Paris, France

Abstract. We analyze the surface-brightness distribution of six inclined Sa-Sc galaxies in terms of bulge, axisymmetric-disk, and spiral-arm components. The data are K'-band frames completed, in the outer parts of the galaxies, by I-band surface photometry. The derived mass models are used to calculate circular-rotation curves which, compared to observed gas kinematics, set constraints on the mass-to-light ratios.

1 Introduction

Most photometric decompositions of spiral galaxies were performed in the visible band; the derived 3D mass models, however, suffered from the extinction by dust and from an excessive contribution of the younger population. Recently, the steady increase in the dimensions of receptors sensitive to the near infrared (NIR) has allowed the separation into main stellar components for a growing sample of spiral galaxies (see, e.g., Peletier et al. 1994; de Jong 1995; Andredakis et al. 1995). The far better "light traces mass" properties of the NIR provide a decisive advantage for all applications requiring an accurate estimate of the global stellar mass; we wanted to test the relevance of a simple mass model in terms of global gas kinematics, and we present our preliminary results.

As part of a wider program, we have obtained K'-band images on a sample of 30 inclined Sa-Sc spirals with the IRAC2-A camera attached to the 2.2-m ESO/MPI telescope at La Silla; the observations, reduction procedures, and resulting photometric data are presented in Héraudeau, Simien, & Mamon (1996). For the present application, six highly-inclined objects have been selected, for which wider-field I-band surface photometry was available, together with good kinematical data.

2 Two-dimensional photometric decompositions

We have adopted a simple model for the light distribution: an oblate bulge and an infinitely thin, axisymmetric disk, on which the spiral arms are superposed (Simien & Héraudeau 1994, Héraudeau & Simien 1995). For the bulge, we fitted either the classical de Vaucouleurs $r^{1/4}$ law, or an exponential law (Andredakis & Sanders 1994; de Jong 1995). For the main part of the disk, we fitted an exponential to the interarm regions, supposed to represent the evolved-disk population. For the outermost regions, our K'-band light distribution was extrapolated by farther-reaching I-band data, assuming a constant $(I - K')$ color; for NGC 7541, we used a frame obtained by one of us (PhH) at the Haute-Provence Observatory, and for the other five galaxies we used data from Mathewson et al. (1992), kindly sent to us in computer-readable form; in these regions, where the spiral pattern is vanishing in most cases, the disk light is adequately represented by either a prolongation of the inner exponential, or a "convex" falloff. Despite its oversimplification, the model has provided an overall good fit.

The photometric profiles along the major axis are given in Figs. 1a and 2a, for two galaxies, NGC 7541 (an Sbc with a small, exponential bulge), and NGC 6788 (an Sab with a bright $r^{1/4}$ bulge). Table 1 presents the bulge-to-total (B/T) ratios. In our limited sample (these six galaxies together with a few additional ones from our observations), we are able to confirm a trend in the bulge luminosity law: intrinsically bright bulges are preferentially $r^{1/4}$, and faint ones exponential (Andredakis et al. 1995).

3 Mass models and rotation curves

After the separation of the bulge, disk and arm components, the inclination of the galaxy, as given by the axial ratio of its disk isophotes, allowed to calculate the spatial density of luminosity in the oblate bulge. The overall mass distribution (3D for the bulge, 2D for the disk and arms) was then derived by assigning an M/L ratio to each component. At this stage, we adopted the same M/L for the underlying-disk and spiral-arm components, although this may be a rough approximation (Rix & Rieke 1993; Gnedin et al. 1995), so we were left with two free parameters, $(M/L)_{\text{bulge}}$ and $(M/L)_{\text{disk}}$.

We have assumed a pure circular rotation of the gas, even if there are obvious radial streamings, especially in the vicinity of the spiral arms; and we have not separated the gas and disk-star masses. Following Kalnajs (1983), Buchhorn & Mathewson (cited by Freeman 1992), and Simien & Héraudeau (1994), we then calculated, at each point within the equatorial plane, the force (and the contribution to the square rotation velocity) of each pixel of the flat component; the contribution of the bulge was also added (Monnet & Simien 1977), and the M/L ratios adjusted to fit the observations. Table 1 presents the results, and Fig. 1b shows the example of NGC 7541. The case of NGC 6788 (Fig. 2b) deserves special attention: a free fit to the rotation curve leads to an unrealistically low $(M/L)_{\text{bulge}} = 0.25$; since the bulge dimensions and luminosity are clearly sufficient to set tight constraints on its spatial structure, the likely explanation is that the gas motion is not circular in the inner regions, a result already stressed by Kent (1988) for about half his sample of

Sa spirals. This is why we fitted NGC 6788 with an additional condition: the same value for the bulge and disk M/L ratios.

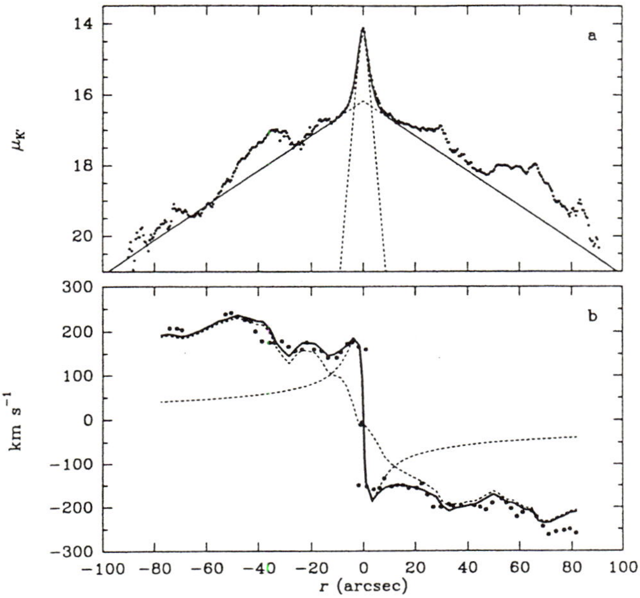

Fig. 1. NGC 7541. **a** Major-axis photometric profile; ***dots***: K'-band data extrapolated by I-band data; ***dashed curves***: PSF-convolved exponential bulge and disk; ***solid curve***: axisymmetric B+D model. **b** Circular rotation curve; ***dots***: data from Rubin et al. (1980); ***dashed curves***: calculated bulge and disk+arms contributions; ***solid curve***: combined model.

4 Discussion

Fitting the rotation curves leads to results which can be considered, at this early stage of our investigation, as globally satisfactory in most cases. Asymmetries and wiggles are accounted for reasonably well, but there is an obvious need for a less naive approach of the gas motion near spiral arms. And NGC 6788 is a reminder that the center of a massive bulge is another place where the gas kinematics may not be straightforward to use.

The M/L ratios derived are not inconsistent with other works (Gnedin et al. 1995; Quillen 1995). The similar values found for the bulge and the disk, as well as the tight correlation between the colors of these components (Terndrup et al. 1994), may eventually lead to the conclusion that only one free M/L parameter is needed, at least to a first approximation. Then, the main role of the photometric decomposition would be (merely) to separate the different geometries of the components in the overall potential.

Still, fine tuning a photometric decomposition can be rewarding in another respect: for galaxies with a contrasted spiral pattern, bringing out the underlying disk can lead to a better estimate of the inclination, a key parameter for, e.g., applications connected to the Tully-Fisher relation.

We also note (as Kent 1986), that the outer rotation needs only a moderate amount of dark matter inside D_{25}; a more accurate estimate of the "visible" mass would then lead to slightly tighter constraints on the unseen matter.

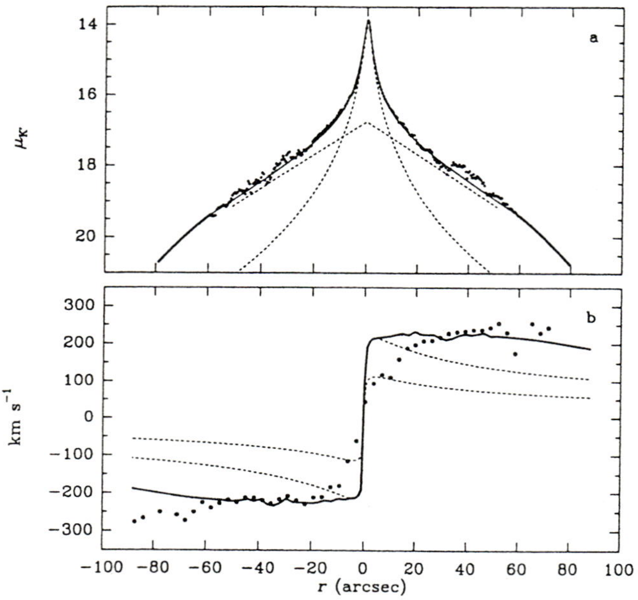

Fig. 2. NGC 6788. **a** Same as Fig. 1a, but $r^{1/4}$ bulge. **b** Circular rotation curve; *dots*: data from Mathewson et al. (1992); *dashed curves*: bulge contributions with $M/L = 0.9$ and 0.25; *solid curve*: adopted bulge+disk+arms model ($M/L = 0.9$).

Table 1. Preliminary results

Object	Type	Bulge	B/T	$(M/L)_{\rm disk}$	$(M/L)_{\rm bulge}$
NGC 6788	Sab	$r^{1/4}$	0.46	0.9*	0.9*
NGC 6810	Sab	$r^{1/4}$	0.59	1.0	1.0
ESO 320-26	Sb	exp	0.12	1.1	0.8
NGC 7606	Sb	$r^{1/4}$	0.07	1.0	1.3
NGC 7541	Sbc	exp	0.07	1.0	0.9
NGC 7083	Sc	$r^{1/4}$	0.10	0.8	0.7

Note: * assuming $(M/L)_{\rm bulge} = (M/L)_{\rm disk}$

References

Andredakis, Y.C., Peletier, R.F., Balcells, M. (1995): MNRAS, in press

Andredakis, Y.C., Sanders, R.H. (1994): MNRAS **267**, 283

de Jong, R.S. (1995): Thesis, University of Groningen

Freeman, K.C. (1992): in Thuan, T.X., Balkowski, Ch., Tran Than Van, J. (Eds.) Physics of nearby galaxies: Nature or nurture? (Frontières, Gif-sur-Yvette), p. 201

Gnedin, O.Y., Goodman, J., Rhoads, J.E. (1995): preprint

Héraudeau, Ph., Simien, F. (1995): Astro. Lett. & Communications **31**, 219

Héraudeau, Ph., Simien, F., Mamon, G.A. (1996): submitted to A&AS

Kalnajs, A. (1983): in Athanassoula, E. (Ed.) Internal kinematics and dynamics of galaxies (Reidel, Dordrecht), p. 87

Kent, S.M. (1986): AJ **91**, 1301

Kent, S.M. (1988): AJ **96**, 514

Mathewson, D.S., Ford, V.L., Buchhorn, M. (1992): ApJ **81**, 413

Monnet, G., Simien, F. (1977): A&A **56**, 173

Peletier, R.F, Willner, S.P. (1993): ApJ **418**, 626

Peletier, R.F, Valentijn, E.A., Moorwood, A.F.M., Freudling, W. (1994): A&AS **108**, 621

Quillen, A. (1995): this Workshop

Rix, H.-W., Rieke, M. (1993): ApJ **418**, 123

Rubin, V.C., Ford, W.K., Thonnard, N. (1980): ApJ **238**, 471

Simien, F., Héraudeau, Ph. (1994): *La Lettre de l'OHP* No. 13, p. 1

Terndrup, D.M., Davies, R.L., Frogel, J.A., DePoy, D.L., Wells, L.A. (1994): ApJ **432**, 518

NIR imaging and modeling of the core of M100

J.H. Knapen[1], R.F. Peletier[2,3], I. Shlosman[4], J.E. Beckman[3], C.H. Heller[5] & R.S. de Jong[6]

[1]Université de Montréal, Dép. de Physique, C.P. 6128, Succ. Centre Ville, Montréal, Québec, H3C 3J7 Canada; and Observatoire de Mont Mégantic
[2]Kapteyn Institute, Postbus 800, NL-9700 AV Groningen, the Netherlands
[3]Instituto de Astrofísica de Canarias, E-38200 La Laguna, Tenerife, Spain
[4]Dept. of Physics and Astronomy, Univ. of Kentucky, Lexington, KY 40506-0055
[5]Univ. Sternwarte, Geismarlandstraße 11, D-37083 Göttingen, Germany
[6]Univ. of Durham, Physics Dept., South Road, Durham DH1 3LE, UK

Abstract. High-resolution NIR and optical images are used to constrain a dynamical model of the circumnuclear star forming (SF) region in the barred galaxy M100 (=NGC 4321). Subarcsecond resolution allowed us to distinguish important morphological details which are easily misinterpreted when using images at lower resolution. Small leading arms observed in our K-band image of the nuclear region are reproduced in the gas flow in our model, and lead us to believe that part of the K light comes from young stars, which trace the gas flow.

In the optical, the central region of M100 shows a set of tightly wound star-forming spiral armlets. The arms can be traced outward through the bar of the galaxy and connect to the main spiral arm pair in the disk. They are accompanied by well-defined dust lanes (Knapen et al. 1995a). In the NIR however the central region looks markedly different. In the $0\farcs8$ 2.2μm K-image we obtained at UKIRT (Knapen et al. 1995a), one distinguishes, from inside out, a small bulge; an inner barlike region, with identical position angle and ellipticity to the large-scale bar; two leading armlets; two symmetric emission peaks; and an oval ring-like zone where the SF armlets are found in the optical. The K contours in this zone are smooth and show hardly any indication of spiral arms or dust lanes. This change in morphology from optical to NIR is caused by different distributions of the stellar populations, along with the much reduced absorption by dust at 2.2μ. The two symmetrically placed peaks of K emission are massive starburst regions of roughly the same age (Knapen et al. 1995a), where at least part of the K emission is likely to come from young stars (e.g. O stars and K supergiants). The fact that the K isophotes become progressively elongated and skewed towards the position angle of the bar both outside *and* inside the "ring" is a strong indicator in favor of a double inner Lindblad resonance in this galaxy.

The need for high ($< 1''$) resolution NIR imaging in this work becomes especially clear when comparing the results from our work with those from two recent papers, where the morphology of the inner $\sim 15''$ region is misinterpreted from K-band images with resolutions of $1\farcs9 \pm 0\farcs2$ (Shaw et al. 1995), and $2\farcs9$ (Sakamoto et al. 1995). Both authors claim the existence of a $\sim 10''$ radius nuclear bar (already reported by Pierce 1986), whereas we resolved this structure into an inner part of the bar of some $5''$, the leading arms, and two symmetric peaks of K emission (Knapen et al. 1995a). As a consequence, only our high-resolution K imaging shows that the inner and outer parts of the bar are aligned, and not at different position angles such as claimed by the other authors.

We have modeled the stellar and gas dynamical processes of the nuclear ring-like structure and associated features by means of 3D numerical simulations using a method described by Heller & Shlosman (1994). Non-linear orbit analysis was used to verify the positions of the ILRs by determining the spatial extent of the family of orbits oriented along the minor axis of the bar. Our modeling shows explicitly that the dominant morphology in the center of M100 can be explained by the gas response to the stellar bar potential (Knapen et al. 1995b). We find that a system of trailing and leading shocks in the gas in the vicinity of the ILRs shows a robust behavior. We have been able to identify and model the main regions of SF there, corresponding to four compression zones. Two zones of SF correspond to the so-called "twin peaks," and two additional ones are found where a pair of large-scale trailing shocks interacts with a pair of leading shocks. Young massive stars ($\sim 10^7$ yrs and less) are present in the resonance region in addition to the old population. Emission from young stars, which trace the gas flow, and from hot dust can explains why we see the leading arms in K emission, whereas they are theoretically expected to occur in the gas.

References

Heller, C.H. & Shlosman, I. 1994, ApJ 424, 84

Knapen, J.H., Beckman, J.E., Shlosman, I., Peletier, R.F., Heller, C.H. & de Jong, R.S. 1995a, ApJL 443, L73

Knapen, J.H., Beckman, J.E., Heller, C.H., Shlosman, I. & de Jong, R.S. 1995b, ApJ, in press (Dec. 1, 1995)

Pierce, M.J. 1986, AJ 92, 285

Sakamoto, K., Okumura, S., Minezaki, T., Kobayashi, Y., & Wada, K. 1995, AJ Nov. 1995

Shaw, M., Axon, D., Probst, R., & Gatley, I. 1995, MNRAS 274, 369

Edge-on Spiral Galaxies: From Optical to Near-Infrared

Richard de Grijs and Pieter C. van der Kruit

Kapteyn Astronomical Institute, P.O. Box 800, NL-9700 AV Groningen, The Netherlands

Abstract. The light distribution of a small sample of edge-on spiral galaxies is examined from optical to near-infrared wavelengths, in *B, V, R*, and *I*, by fitting model distributions to the light profiles. In general, the best fitting vertical model is more peaked than expected for an isothermal sheet distribution (sech^2), i.e. it is either an exponential light distribution or a sech(z)-model. It is found that the vertical scale parameters for both the thin and the thick disks are confined within narrow ranges. The mean ratio of the radial to the vertical scale parameter in *I*, h_R/z_0, is 5.9 $\pm$ 1.2.

1 Scale Parameter Analysis

We present the results of a study of the structural parameters in edge-on galaxies. Radially, we assume an exponential distribution; vertically, several models have been proposed and tested to account for the distributions observed. In general, the best fitting vertical model is more peaked than expected for an isothermal sheet distribution (sech^2), i.e. it is either an exponential light distribution or a sech(z)-model.

1.1 Constant Scaleheight

We have found that the vertical scale height in our sample of edge-on spiral galaxies remains constant as a function of radius, although it seems to loose strength in the outer parts. (Our ongoing 2-D modeling may change this picture.) The probable cause of this constancy is that in dynamically stable disks Toomre's (1964) Q-parameter is constant. Combined with an exponentially decreasing stellar velocity dispersion (as found in exponential disks), this leads to a constant scale height.

1.2 Ranges of Vertical Scale Parameters

It turns out the range of vertical scale parameters is quite narrow for both the thin and the thick disk components – the scaleheights of the thin disks vary from about 100 to 250 pc, and for the thick disks the range is from 470 to 620 pc – and the scale heights do not vary significantly as a function of passband.

1.3 Scale Parameter Ratios

Since we are looking at edge-on galaxies, we can easily measure the ratio of the radial to the vertical scale parameter, h_R/z_0. So far, this ratio is only known for a few galaxies. The mean I-band h_R/z_0 ratio of our sample is 5.9 ± 1.2.

2 Implications and Conclusions

1. The best fitting vertical model is sharper than expected for an isothermal sheet distribution, i.e. it is either an exponential light distribution or a sech(z)-model.
2. In the presence of a dust layer and different stellar populations, variations in the scale height as a funtion of passband would be expected. However, we do not find any such variation.
3. The mean (I-band) scale parameter ratio is 5.9 ± 1.2. A number of studies have dealt with the determination of the h_R/z_0 ratio for our Galaxy, including van der Kruit (1987) who finds $h_R/z_0 = 8.5 \pm 1.3$, closer to the theoretical value of about 10 needed for disks of Sc galaxies with maximum rotation (Bottema 1993), but Van der Kruit and Searle (1982) found a ratio of 4.6 ± 1.2 for another sample of edge-ons.

References

Bottema, R. (1993): A&A **275**, 16
Kruit, P.C. van der (1987): A&A **173**, 59
Kruit, P.C. van der, Searle, L. (1982): A&A **110**, 79
Toomre, A. (1964): ApJ **139**, 1217

Dynamical Evolution and Population Analysis of Galactic Disks from Optical and Infrared Colour Profiles in Edge-on Galaxies

Andreas Just[1], Cecilia Scorza[2], Roland Wielen[1], and Burkhard Fuchs[1]

[1] Astron. Rechen-Inst., Mönchhofstr. 12–14, D-69120 Heidelberg, Germany
[2] Landessternwarte, Königstuhl, D-69117 Heidelberg, Germany

Abstract. The stellar population of galactic disks is composed of subpopulations with increasing age and scale height. This is well known for the solar neighbourhood and can also be deduced in edge-on galaxies from the vertical luminosity profiles, which are essentially exponential after correction of dust extinction, and from the colour profiles showing a central blue part. We show that constraints on the heating rate, the SFR and the metal enrichment can be derived with physical models of a selfgravitating disk combined with the method of photometric evolutionary synthesis. The stellar composition of IC 2531 is very similar to the solar cylinder but with the young component more concentrated to the midplane. In NGC 891 the central disk of young stars is much brighter consistent with a constant SFR.

Galactic disks are heated by some gravitational scattering process. The velocity dispersion and scale height increases with the age of the stellar populations as can be observed directly in the solar neighbourhood. This evolutionary effect can also be derived from the vertical structure of spiral galaxies seen edge-on. Firstly the intrinsic luminosity profiles are not isothermal but more or less exponential requiring stellar populations with intermediate scale heights. Secondly there is a young blue population concentrated to the midplane of the disk. Since the dust extinction is strong in edge-on galaxies, a multicolour analysis including especially NIR-bands is necessary to extract the intrinsic properties of the stellar populations in the disk.

We use a physical model of a selfgravitating thin disk which is exponential in radial direction (Just et al. 1995). The stellar disk is composed of isothermal subpopulations with increasing age and velocity dispersion. The surface density of the stars is determined by the initial mass function and the star formation history, the velocity dispersion by the heating function. The metal enrichment is computed by star formation scenarios consistent with the adopted star formation rates. The dust extinction is modeled by an exponential dust distribution in vertical and radial direction. The emissivity of the stars is computed from the luminosities in the different bands of single age populations which are computed with the method of photometric evolutionary synthesis (Einsel et al. 1995). With these models we get information about the heating of the disk, the star formation history, the metal enrichment and also constraints on the initial mass function for the galactic disks form characteristics of the vertical luminosity and colour profiles.

We apply these models to published profiles of the edge-on galaxies IC 2531 in U,B,V,R,I,K bands (Wainscoat et al. 1989) and on CCD data of NGC 891 in V,I, and K. We find that the general features of the luminosity and colour profiles are well reproduced by the models. For IC 2531 the same SFR and metallicity enrichment as in the solar neighbourhood can be used. We find a heating function $\sigma \propto t^2$ instead of $\sigma \propto t^{1/2}$ and a ratio of 0.8 for the radial scale length of the dust to that of the stars. Synthetic and observed colour profiles are shown in Fig. 1. In NGC 891 the central blue disk is much stronger yielding a constant SFR which is confirmed by the central luminosity excess observed in the K band (Aoki et al. 1991).

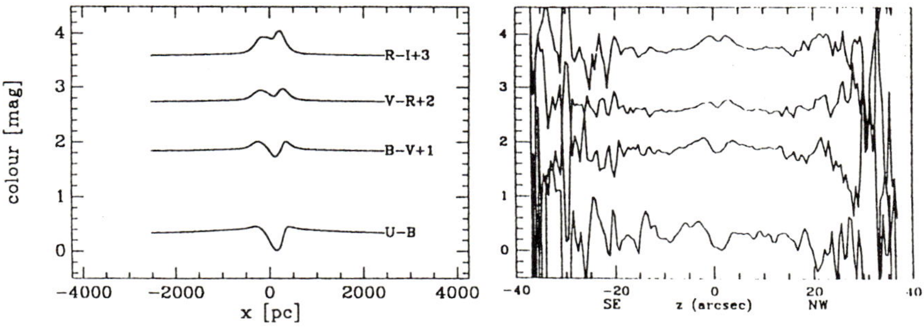

Fig. 1. Synthetic (left) and observed (right) colour profiles of IC 2531 in different bands of a vertical cut with an offset of 40″ to NE of the minor axis. The inclination of the disk is 89.4°.

References

Aoki, T.E., Hiromoto, N., Takami, H., Okamura, S. (1991): PASJ **43**, 755–780

Einsel, C., Fritze-v. Alvensleben, U., Krüger, H., Fricke, K.J. (1995): A&A **296**, 347–358

Just, A., Fuchs, B., Wielen, R. (1995): Submitted to A&A

Wainscoat, R.J., Freeman, K.C., Hyland, A.R. (1989): ApJ **337**, 163–190

Orbits in the Bar of NGC 4314

P.A. Patsis[12], E. Athanassoula[1] and A.C. Quillen[13]

[1] Observatoire de Marseille, 2 Place Le Verrier, F-13248 Marseille cedex 4, France
[2] Max-Planck Istitut für Astronomie, Königstuhl 17, D-69117 Heidelberg, Germany
[3] Ohio State University, Dept. of Astronomy, 174 West 18th Ave., Columbus, OH 43210, USA

Abstract. We examine the stability of all families of simple periodic orbits in the potential of the barred galaxy NGC 4314, as calculated by Quillen et al. (1994) using near-infrared photometry in the J, H, and K band. We underline the importance of the "sin" terms of the potential in explaining the observed twisting of the isophotes with respect to the bar major axis and compare the stable periodic orbits found with the structure of the bar of NGC 4314.

1 The Potential

Quillen et al. (1994, hearafter QFG) calculated the gravitational potential of the barred galaxy NGC 4314 using near-infrared observations in the J, H and K bands, and approximated it on the z=0 plane by the function:

$$\Phi(r,\theta) = \Phi_0(r) + \sum_{m>0} \Phi_{mc}(r)\,\cos(m\theta) + \Phi_{ms}(r)\,\sin(m\theta), \qquad (1)$$

where m=2,4,6. We calculated orbits in this model (hereafter model "T"), as well as in the potential including only the axisymmetric, cos(2θ) and cos(4θ) terms (hereafter "model C"). In both cases the calculations have been done in a frame rotating with Ω_p = 44.96km/s/kpc. This gives a corotation radius $r_c \approx 70''$, as proposed by QFG. For the adopted distance D=10 Mpc, this radius corresponds to 3.5 kpc.

2 Conclusions

The area of the **1/1 resonance** is very close to the center. The orbits we have found in this region do not play a major role in the dynamics of this galaxy. Nevertheless the interconnections of the several families in this area, as well as the evolution of the characteristic at the **2/1** region in model "C", have theoretical interest. A detailed description of these orbits will be given elsewhere (Patsis, Athanassoula and Quillen 1995, in preparation).

At the **3/1 region** we find, in both models, unstable 3/1 orbits with one side roughly parallel to the bar major axis, and stable 3/1 orbits with one side rougly parallel to the bar minor axis. The former develop loops along the bar minor axis and the later along the bar major axis. For every orbit one has its mirror image with respect to either the bar major or the bar minor axis and puting them together one gets rectangular-like figures with four loops.

At the region of the **4/1** resonance, the orbits of the x_1 family become diamond shaped and develop loops at their ends along the bar major axis. A second branch of this family can be found in lower x's in the (x,E) diagrams in both cases. They have "rectangular" shapes. These orbits have been found also in other barred potentials (Contopoulos and Grosbøl 1989; Athanassoula 1992). In the "T" model, however, the orbits of this branch as the orbits of every family, are asymmetric. The stability curve at this region for model "C" does not level off, but continues upwards, intersects the α=1 axis and bifurcates a new asymmetric family. In model "T" the stability index levels off at the 4/1 resonance. The branch with the asymmetric "rectangulars" starts coexisting with x_1 for the same energies as in model "C", but has only a small stable part. If we follow the x_1 family on the corresponding (E,x) diagram, the values of x start decreasing with increasing E values at about the same energy at which the asymmetric family bifurcates in model "C". We note that the stable orbits of the x_1 family in the "decreasing" part do not support the bar all the way to corotation. For increasing energies, the size of their loops, as well as their extension at the left and right size of the bar increases, while the extent of their projection on the y-axis decreases. Thus, if they are able to support some structure in NGC 4314, this has to be sought at the sides and not along the bar of the galaxy.

References

Athanassoula E., (1992). MNRAS **259**, 328-344
Contopoulos G., Grosbøl P.; 1989, Astron. Astroph. Rev **1**,26
Quillen A.C., Frogel J.A., Gonzalez R.A.; 1994, ApJ **437**,162 (QFG)

Multiband analysis of central-region features

Ph. Héraudeau[1], F. Simien[1], and G.A. Mamon[2]

1 Observatoire de Lyon, F-69561 Saint-Genis-Laval cedex, France
2 Institut d'Astrophysique, 98 bis boulevard Arago, F-75014 Paris, France

Abstract. Surface photometry in the J, H, and K' bands for four inclined Sa-Sb galaxies is used to analyze the morphology and/or absorbing properties of several features located in the inner regions.

1 Introduction

As part of a wider surface-photometry program on nearby galaxies (Hérau-deau, Simien, & Mamon 1996), we have used the IRAC2-A camera attached to the 2.2-m ESO/MPI telescope to get J, H, and K' images on four spirals with a large angular diameter and a high inclination (≈ 70 degrees). These data are completed by B and R images from the ESO-LV database (Lauberts & Valentijn 1989), for a wider frequency coverage. Here we present our preliminary analysis of dust lanes, dust patches, and other features appearing in the innermost regions.

2 Preliminary results

1 ESO 320-26

In this Sb spiral, we find: a) a bulge with a marginal evidence for triaxiality; b) a bright, highly flattened inner structure parallel to the major axis (a bulge + disk decomposition shows that this structure is unlikely to be the inward prolongation of the disk, being brighter by more than one magnitude); and, c) a small ring (axial ratio 0.5, diameter $\approx 8''$ or 1.4 kpc) in the region of this inner structure; a $(J - K')$ map shows that this ring is redder than the surroundings regions by ≈ 0.2 mag, but has colors similar to those of the nucleus.

2 ESO 267-37 (NGC 4219)

This Sbc galaxy is the reddest of the four objects. It exhibits two well-defined spiral arms with two very absorbing patterns which follow the inner ridge of the arms, at $r \approx 10''$. Color-color diagrams suggest an extinction corresponding to $A_V \approx 1$ mag. The $(J - K')$ color image shows two red spiral arms which are more closely related to the absorbing regions than to the stellar spiral arms.

3 ESO 184-67 (NGC 6788)

The overall light and mass distribution of this Sab galaxy has been studied by Héraudeau, Simien, & Mamon (1995). It has embryonic, low-contrast spiral arms deep into the central regions ($r \approx 10''$) of its bright bulge. ¿From $(B - K')$ and $(J - K')$ data, there is no evidence that these arms have colors different from the neighborhood.

4 ESO 142-35 (NGC 6810)

This object (Sab) exhibits a conspicuous asymmetry of the inner structure. What appears on visible-band images like a bright crescent-shaped structure is actually the unabsorbed half of the bulge. Due to the presence of strong extinction patterns, this bulge appears twisted, even in K', without being intrinsically barred or triaxial. Regions of the bulge slightly beyond the nucleus have "normal" colors; in comparison, regions situated 5-10$''$ ahead of the nucleus, within the dust lanes, appear reddened by $A_V \approx 1.6$ to 2 magnitudes. The reddest point (2.5 mag) does not belong to a very contrasted dust lane. In terms of Galactic-law extinction, $(J - H)$ vs. $(H - K')$ diagrams of the strong lanes are not in good agreement with similar color-color diagrams involving B; this may reveal that the model is irrelevant, but the lower accuracy of the (photographic-plate) B data could also play a role. The nucleus, classified Seyfert 2, is only marginally reddened in $(H - K')$.

A more detailed account of this work is in preparation.

References

Héraudeau, Ph., Simien, F., Mamon, G.A. (1996): submitted to A&AS

Héraudeau, Ph., Simien, F., Mamon, G.A. (1995): this Workshop

Lauberts, A., Valentijn, E.A. (1989): The surface photometry catalogue of the ESO-Uppsala galaxies. ESO, Garching

V

MODELING THE DUST

The history and future of extinction models

J. Edwin Huizinga

European Southern Observatory, Karl-Schwarzschildstraße 2,
D-85748 Garching bei München, Germany

Abstract. Extinction models play a crucial role in our understanding of spiral galaxies. But, although they have gradually improved over the last four decades, the models are still very far from the complex reality actually observed in spiral galaxies. Here, first an overview of the history of extinction models is presented. Next, the current models are compared with spiral galaxies, and the implications of the lack of realistic structure in the models are discussed. Finally, some alternative routes towards more realistic extinction models are suggested.

1 Introduction

The amount of internal extinction in spiral galaxies cannot be measured directly. The classic approach to study extinction in spiral galaxies has therefore been based on the inclination dependence of total magnitudes, isophotal diameters and surface brightnesses. The reasoning here is as follows. If galaxies are completely transparent, we can see all the star-light from all directions, and total magnitudes are independent of inclination. With increasing inclination (from face-on to edge-on), the path length along the line-of-sight through the disks increases, resulting in higher surface brightnesses and larger isophotal diameters. In contrast, if galaxies are completely opaque, we can only see the outer 'crust' of stars, and surface brightnesses and isophotal diameters are independent of inclination. With increasing inclination, the projected surface-area of the galaxies decreases, resulting in fainter total magnitudes.

So, by observing a large enough sample of galaxies seen under different inclinations one could in principle determine whether they are optically thick or transparent. Though in theory this seems relatively straightforward, the lack of consensus in this field illustrates that in practice it is not (e.g. Holmberg 1958, Heidmann, Heidmann and de Vaucouleurs 1972, Tully 1972, Valentijn 1990, Choloniewski 1991, Burstein, Haynes and Faber 1991, Huizinga and van Albada 1992, Cunow 1992, Han 1992, Giovanelli et al. 1994, Valentijn 1995, Peletier et al. 1995). The first problem is to define a representative sample of galaxies, for which the intrinsic properties are independent of inclination. This topic is extensively discussed elsewhere (e.g. Chołoniewski 1991, Burstein, Haynes and Faber 1991, Huizinga and van Albada 1992), and is of no concern to us here. In principle, it is possible to define a representative sample, and to determine how apparent galaxy properties depend on inclination. With the resulting empirical relations the observations of inclined galaxies can be corrected to face-on. But, to interpret the observations quantitatively in terms of face-on extinctions, or in terms of face-on optical depths, we need extinction models.

Although extinction models play a crucial role in our understanding of spiral galaxies, they have improved only slowly. Current models are still far from the complex reality actually observed in spiral galaxies. The following section reviews the history of extinction models for spiral galaxies, and discusses some of the adopted modelling techniques. In Sect. 3, the current models are compared with the actual light and dust distributions observed in spiral galaxies, and the limitations of the extinction models are discussed. Finally, in Sect. 4, arguments for more realistic models are presented, together with the possible ways in which these can be achieved.

An excellent overview of present status of the study of internal extinction in spiral galaxies can be found in the proceedings of the NATO workshop *The Opacity of Spiral Disks* (Davies and Burstein 1995).

2 The evolution of extinction models

Figure 1 illustrates the evolution of global extinction models over the past four decades. This section presents a condensed overview of the history of these extinction models, and discusses some of the new insights each model provided.

2.1 Analytical models

The history of extinction models begins with the first statistical extinction study presented by Holmberg (1958). For a sample of 116 spiral galaxies, he fitted the inclination dependence of surface brightness to a cosec(i) law, which corresponds to a geometry in which all the dust is located in front of all the stars; the Screen model. Holmberg realized that the Screen model is unrealistic, and therefore doubled the optical depths derived from it. He did, however, not correct the amounts of face-on extinctions derived from the Screen model. This then led to the widespread believe that spiral galaxies are largely transparent, with face-on extinctions of only 0.28 mag for Sc and 0.43 mag for Sa and Sb spirals. In the same paper, Holmberg also studies the inclination dependence of galaxy colours. There, he introduces both the

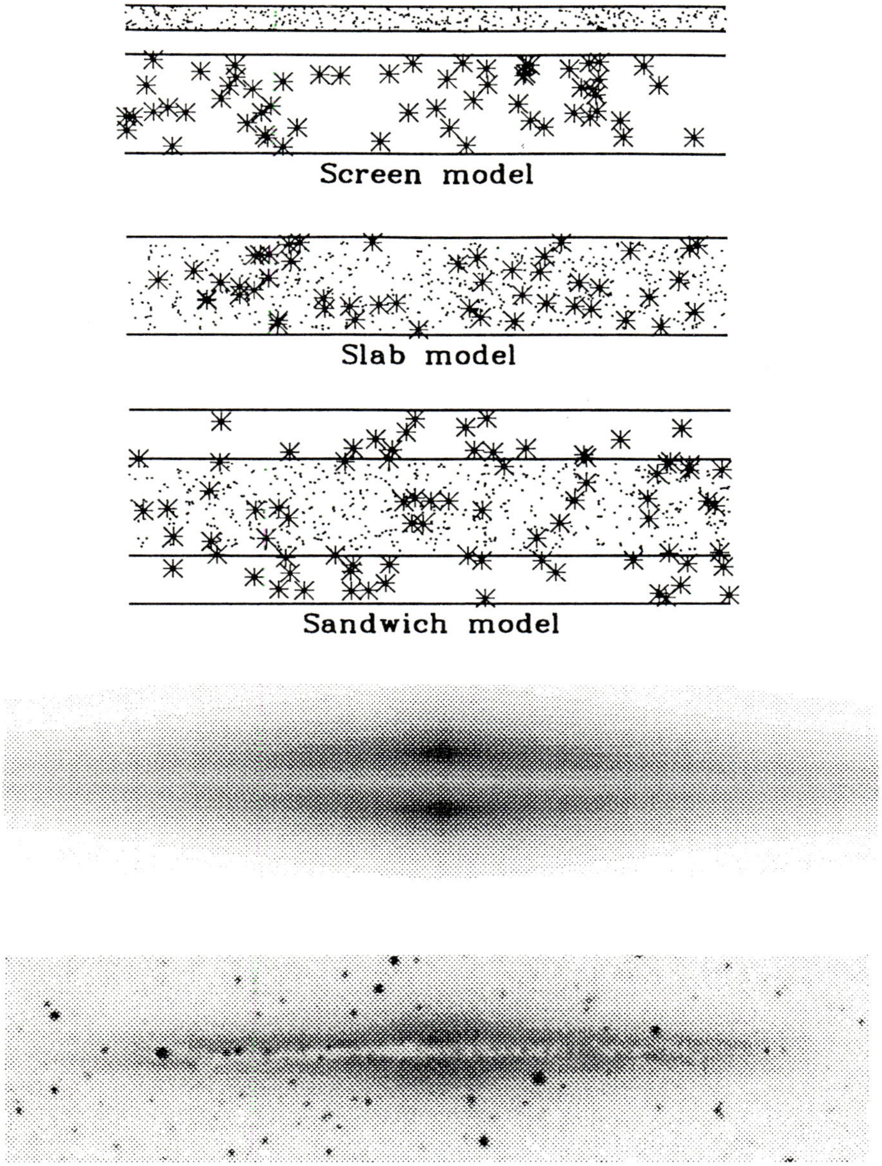

Fig. 1. The evolution of extinction models (screen ⇒ slab ⇒ sandwich ⇒ triplex) compared to the complex reality of NGC 891

homogeneous slab and the sandwich model, which are discussed in more detail below. Unfortunately, these models went unnoticed, and nearly all subsequent extinction studies adopted the naive Screen model.

In the second model, dust and stars are uniformly mixed; the (homogeneous) Slab model. This model introduced the concept of saturation. Once the model is optically thick, one cannot tell the difference between an optical depth of τ=5 or τ=10. Also, reddening is no longer proportional to the relative optical depths in the different bands.

In the third model, a dust layer is embedded in a thicker slab of stars; the Sandwich model (e.g. Tully and Fouqué 1985). Heidmann, Heidmann and De Vaucouleurs (1972), had already considered a model in which the dust was concentrated in the mid-plane of a slab of stars. In the Sandwich model, a fraction of the stars will always be unobscured. And, as Disney, Davies and Phillipps (DDP, 1989) clearly pointed out, if the dust layer is completely optically thick, the model will behave as if completely transparent.

DDP also introduced the so called Triplex model, in which the dust and light have exponential radial and vertical profiles, with independent scale-lengths and scale-heights. This model introduced extinction gradients, with optically thick central regions, and transparent outer parts. So, the inclination dependence of total magnitudes, which are dominated by the bright central regions can behave optically thick, whereas the fainter isophotal diameters can behave optically thin.

It must be pointed out that the main motivation behind these early models was not to create *the* extinction model for spiral galaxies. The main motivation was to derive functional forms to describe the inclination dependence of total magnitudes, surface brightnesses and isophotal diameters. It was realized that the the final correction from a face-on total magnitude to the dust-free, or extinction-free, total magnitude was highly uncertain and very model dependent.

2.2 Numerical models

Christensen (1990) presented numerical models which included exponential disks for both the dust and light, and a stellar bulge component. The next step towards more realistic extinction models was the introduction of scattering. Although scattering had of course been included in many astrophysical radiative transfer problems, Kylafis and Bahcall (1987) were the first to use it in a global model for a spiral galaxy, NGC 891 (Fig. 1). They studied the vertical light profiles in order to determine the optical depth of the prominent dust-lane. Their modeling technique was generalized by Byun et al. (1994). This paper presents extinction models, including first order scattering (see Sect. 2.4), for spiral galaxies with varying Bulge-to-Disk ratios. These models are used to study the inclination dependence of total magnitudes, radial scale-lengths, (B-I) colours, and asymmetries in highly inclined spirals.

For several astronomers the first results from extinction models that included the effects of scattering came apparently as a surprise. In the *Third Reference Catalogue of Bright Galaxies*, RC3, de Vaucouleurs et al. (1991) predicted that by including the effects of scattering, the models would need larger optical depths to account for the observed inclination dependence of e.g. total magnitudes. The opposite was true. What actually happens is the following. Photons that travel in the plane of the disk have a higher chance of being scattered 'upwards' or 'downwards' out of the disk, than vice versa. For small optical depths, this effect actually results in a brightening of the face-on model! This re-distribution of light increases the difference between edge-on and face-on total magnitudes, without actually absorbing any photons. Of course, this is only valid for a smooth distribution of dust. What happens in more realistic, clumpy models remains to be seen.

More general models are those of Bruzual, Magris and Calvet (1988), who study the effects of extinction, including scattering, on galaxy spectra in a Slab model. Witt, Thronson and Capuano (1992) study the effects of extinction, including scattering in spherical models. They use different radial profiles for the dust and light, to simulate different star-dust geometries, such as an embedded nuclear source and a cloudy galaxy. Their models again show that large amounts of dust can be hidden in a galaxy, without resulting in large amounts of extinction or reddening. Finally Hobson and Padman (1993) study models in which the dust is concentrated in spherical clouds, which themselves are randomly distributed in a homogeneous stellar layer.

2.3 Discrete absorption features

Besides the attempts to create global extinction models, several studies have concentrated on specific absorption features in spiral galaxies. Well-known is the work by Elmegreen (1980), who models strong dust-lanes and other distinct absorption features in several nearby face-on spiral galaxies. She finds optical depths corresponding to τ_V=7-10 in the densest dust-lanes.

Similar results are obtained by Rix and Rieke (1993, see also Rix 1995) who use optical and NIR data to study the dustlanes in M51. They also use the observed contrast between the strong dustlanes and their surroundings to put limits on the amount of diffuse extinction.

Regan and Vogel (1995) use optical and NIR data to model dust features in the barred spiral galaxy NGC 1530. Using an extinction model, they can constrain the optical depth and vertical extend of the dust-lane. By comparing the derived optical depths along the dust-lane to the observed CO intensities, they show that the CO emission is probably enhanced in the central regions due to heating by a central source.

Witt et al. (1994, 1995) use the unique geometry of the dust layer in NGC 4826 ('The Evil Eye' galaxy) to measure the K'-band dust albedo for the very first time. Even for Galactic dust this is still a highly uncertain number.

Block et al. (1994, 1995), use (*B*-*K*') and (*V*-*K*') colour maps of face-on spiral galaxies to study the detailed distribution of dust. Besides a diffuse dust component they find widespread filamentary dust features associated with the dust-lanes along the spiral arms, but also dust associated with cold interarm molecular clouds, undetected in CO observations. Using the models of Witt et al. (1992), they determine the optical depths in both the diffuse and filamentary dust components, and derive total dust masses. The dust masses derived in this way are upto 10 times the dust masses derived from the IRAS FIR data, which is only sensitive to the warm dust component.

These last studies show that observations and modelling of dust extinction in the optical and NIR can not only be used to study extinction as such, but also as a more general diagnostic tool. The next logical step would be to use the derived dust and light distributions as input to fully 3-dimensional extinction models, and study the effects of the clumpy distributions of the light and dust on the model properties.

2.4 Modelling techniques

All the models illustrated in Fig. 1 can be treated analytically (see e.g. Disney, Davies and Phillips 1989). But the more advanced models, those that include either scattering, structure or both, have to be treated numerically. This may partly explain the hesitance in adopting more realistic models, and the tendency to use the spherical models of Witt et al. (1992) to study extinction in plane parallel geometries.

Until recently, there were two popular techniques to include non-isotropic scattering in non-uniform radiative transfer models; the Iteration and the Monte Carlo method. A new, cellular technique is presented by Madore, elsewhere in these proceedings.

The Iteration technique

The Iteration method was introduced by Kylafis and Bahcall (1987), and generalized by Byun et al. (1994). It briefly works as follows. The model first calculates the integrated light along the line-of-sight, taking only pure absorption into account. Next, at each point along the line-of-sight, it determines the amount of light received from all other points in the model, again only including pure absorption, and determines the amount of light that is scattered into the line-of-sight. This gives the amount of direct and once scattered light that escapes from the model. The total light is then approximated using a series expansion. The disadvantages of this technique are that it is not exact and rather time consuming.

The Monte Carlo technique

Monte Carlo techniques as used in extinction models (e.g. Witt, Thronson and Capuano 1994, De Jong 1995) are based on Witt's (1977) models for reflection nebulae. The basic principle is as follows. From a random point in the model, a 'photon' is emmited in a random direction. Either the intensity of the photon or the number of photons emmited at that point is taken to be proportional to the local model luminosity. Next, the optical depth to the next point of scattering is determined. If this point lies outside the model boundaries, the photon escapes, otherwise a scattering angle is determined and the process is repeated. After the photon escapes, its intensity is attenuated using the integrated absoption optical depth along its path. All photons are then binned in solid angles, to produce a series of images of the model as seen from different directions. The signal-to-noise ratio in the resulting images is determined by the number of photons traced, and the ultimate accuracy only depends on the integration step sizes.

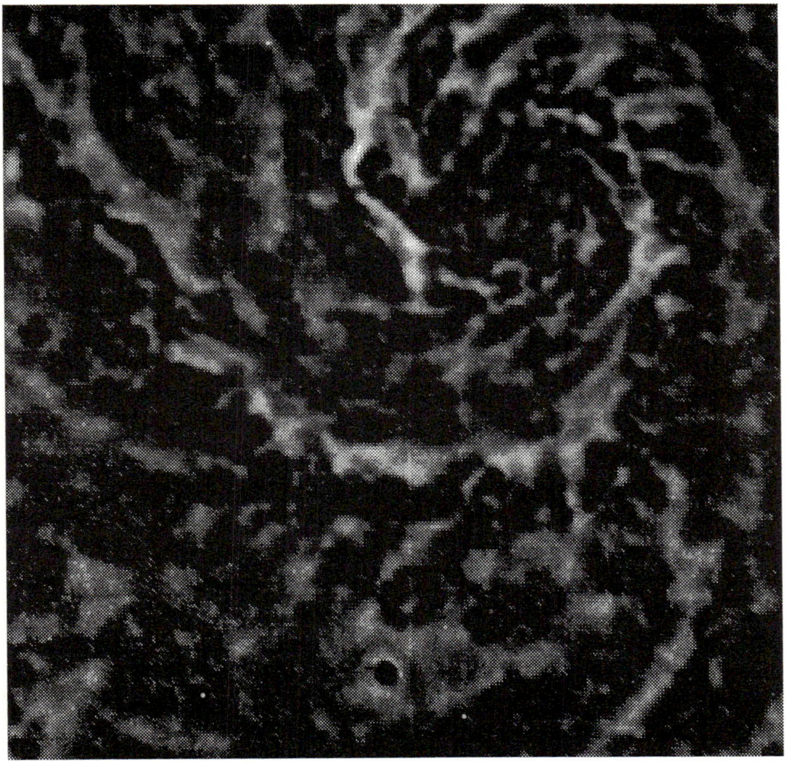

Fig. 2. A map of the filamentary dust distribution (white) in the central region of M51, as derived from optical and NIR colours.

3 The confrontation with reality

Current extinction models assume smooth distributions for both the light and dust components. They actually show a closer resemblance to dusty S0 than to spiral galaxies, which contain structures on all scales, with dust and gas concentrated in 'clumps' ranging from parsec sized globules to Kpc sized Giant Molecular Clouds (GMCs). Not only the distribution of dust, but also that of the light is far from smooth. Young stars, which dominate the UV and blue luminosity, are concentrated in star-forming regions and giant HII complexes along the spiral arms. Only the 'old disk', and population II stars are uniformly distributed, with the exception of an occasional spiral arm, bulge or bar.

As an example of dust structures, Fig. 2 shows the filamentary dust distribution in the inner parts of M51[1]. How does this filamentary dust distribution affect the extinction properties of spiral galaxies? Because of their low covering factor the filaments will contribute little to the total face-on extinction. Never-the-less, as shown by Block et al. (1995) in the case of NGC 4736 and NGC 4826, the cold filamentary dust presents up to 90 per cent of the total dust mass. Also, star-formation occurs in the regions with the highest gas, and therefore highest dust densities. On average, the light from young blue stars will therefore be more absorbed than that of the more uniformly distributed older red population.

Besides the distributions of dust and stars, the dust properties themselves have to be determined. Current extinction models assume an average Galactic extinction law (e.g. Savage and Mathis 1979), and scattering phase function (Henyey and Greenstein 1941). But already in the Galaxy itself, dust properties vary from the diffuse ISM to dense molecular clouds (Kim, Martin and Hendry 1994). Outside the LMC and SMC, the dust extinction law has only been studied in a few other galaxies. In for example NGC 4594, the 'Sombrero' galaxy, Knapen et al. (1991) find an extinction law in agreement with the Galactic one. Kinney et al. (1994), however, find quite a different UV and optical extinction law in a sample of 39 starburst galaxies. Thusfar little attention has been given to the sensitivity of the model results to different dust properties. So, also here, more research remains to be done.

4 Discussion

Why should we still worry about extinction models for spiral galaxies? One may argue that the now widely available near infra-red (NIR) arrays have made extinction models obsolete. On the one hand, the NIR is (almost) free from the effects of *dust extinction* (absorption and scattering) that trouble our observations at optical and UV wavelengths. And on the other hand,

[1] Based on data kindly provided by H.-W Rix (Rix and Rieke 1993)

the NIR is almost free from *dust emission*, which dominates in the mid- and far infra-red. Furthermore, the NIR is free from the blue light of young, massive stars, which dominate the appearances of spiral galaxies at shorter wavelengths (e.g. Rix and Rieke 1993). These factors make the NIR K-band (or K'-band) ideal to study the underlying stellar mass distributions in spiral galaxies, free from the effects of dust and star-formation.

Never-the-less, if we want to study stellar populations through global colours and colour gradients, we still have to know the amounts of reddening suffered by the light in the optical and UV bands. If we want to interpret the data that will hopefully be obtained by the ISO satellite, we have to understand radiative transfer in spiral galaxies; is the dust heated by star-forming regions, active nucleii, or by the general inter-stellar radiation field? How does the evolution of the ISM affect our view of galaxies at higher redshifts? If there has been massive star-formation in the early history of the universe, why don't we observe strong Ly-α emission? Also, the study of internal extinction provides a direct link between the optical appearance of galaxies and their gas contents as determined by HI or CO observations (e.g. Block et al. 1994). Can all the observed extinction be attributed to known components in the ISM or is there a very cold molecular component which is missed in CO observations as suggested by Pfenniger, Combes and Martinet (1994)?

It is clear that the current generation of global extinction models is still too naive to be used as diagnostic tools for spiral galaxies. To address the questions listed above, the models should include atleast some of the details observed in real spiral galaxies. This could be achieved in one of the following ways. First, we can try to introduce structures by trail and error, until the model images reproduce the appearance of spiral galaxies as e.g. observed in the optical and NIR. Secondly, we can use high resolution imaging in Hα, HI, CO, FIR etc. to map the distribution of the ISM and use these data as input to the models. And finally, we can use the results of N-body/SPH models which include a star-formation recipe, as input for more realistic extinction models.

But, in the end, we should always keep in mind that a model is *supposed* to be a simplified abstraction of reality, and that it should only include the essential ingredients. The major challenge will be to identify what these ingredients are.

References

Block, D.L., Witt, A.N., Grosbøl, P., Moneti, A., (1994): A&A **288**, 383
Block, D.L., Witt, A.N., Grosbøl, P., (1995): in *The Opacity of Spiral Disks* (Eds. J.I. Davies and D. Burstein. Kluwer, Dordrecht, Boston, London)
Bruzual, A.G., Magris, C.G., Calvet, N., (1988): ApJ **333**, 673
Burstein, D., Haynes, M.P., Faber, S.M., (1991): Nat **353**, 515
Byun, Y.-Y., Freeman, K.C., Kylafis, N.D., (1994): ApJ **432**, 114

Chołoniewski, J., (1991): MNRAS **250**, 486
Cunow, B., (1992): MNRAS **258**, 251
Davies, J.I., Burstein, D., Eds. (1995): *The Opacity of Spiral Disks*, NATO ASI Series Vol. 469 (Kluwer, Dordrecht, Boston, London)
de Jong, R.S., (1995): Ph.D. Thesis (University of Groningen, The Netherlands)
de Vaucouleurs, G., (1959): Astron. J. **64**, 397
de Vaucouleurs, A., de Vaucouleurs, G., Buta, R.J., Corwin Jr., H.G., Fouqué, P., Paturel, G., (1991): *Third Reference Catalogue of Bright Galaxies* (Springer-Verlag, New York)
Disney, M., Davies, J.I., Phillipps, S., (1989): MNRAS **239**, 939 (DDP)
Elmegreen, D.M., (1980): ApJS **43**, 37
Giovanelli, R, Haynes, M.P., Salzer, J.J., Wegner, G., Da Costa, L.N., Freudling, W., (1994): AJ **107**, 2036
Han, M., (1992): ApJ **391**, 617
Henyey, L.G., Greenstein, J.L., (1941): ApJ **93**, 70
Heidmann, J., Heidmann, N., de Vaucouleurs, G., (1972): Mem. R. Astron. Soc. **76**, 121
Hobson, M.P., Padman, R., (1993): MNNRAS **264**, 161
Holmberg, E., (1958): Medd. Lunds Astron. Obs. Ser. **2**, No. 136
Huizinga, J.E., van Albada, T.S., (1992): MNRAS **254**, 677
Kim, S.-H., Martin, P.G., Hendry, P.D., (1994): ApJ **422**, 164
Kinney, A.L., Calzetti, D., Bica, E., Storchi-Bergmann, T., (1994): ApJ **429**, 582
Knapen, J.H., Hes, R., Beckman, J.E., Peletier, R.F., (1991): A&A **241**, 42
Kylafis, N.D., Bahcall, J.N., (1987): ApJ **317**, 637.
Peletier, R.F., Valentijn, E.A., Moorwood, A.F.M., Freudling, W., Knapen, J.H., Beckman, J.E., (1995): A&A **300**, L1
Pfenniger, D., Combes, F., Martinet, L. (1994): A&A **285**, 79
Regan, M.W., Vogel, S.N., (1995): These proceedings
Rix, H.-W., Rieke, M.J., (1993): ApJ **418**, 123
Rix, H.-W., (1995): in *The Opacity of Spiral Disks* (Eds. J.I. Davies and D. Burstein. Kluwer, Dordrecht, Boston, London)
Savage, B.D., Mathis, J.S., (1979): ARA&A **17**, 73
Tully, R.B., (1972): MNRAS **159**, 35p
Tully, R.B., Fouqué, P., (1985): ApJS **58**, 67
Valentijn, E.A., (1990): Nat **346**, 153
Valentijn, E.A., (1995): in *The Opacity of Spiral Disks* (Eds. J.I. Davies and D. Burstein. Kluwer, Dordrecht, Boston, London)
Witt, A.N. (1977): ApJS **35**, 1
Witt, A.N., Thronson, H.A., Capuano, J.M., (1992): ApJ **393**, 611
Witt, A.N., Lindell, R.S., Block, D.L., Evans, R., (1994): ApJ **427**, 227
Witt, A.N., Lindell, R.S., Block, D.L., Evans, R., (1995): These proceedings

3D Radiative Transfer Models of Galaxies

Barry F. Madore[1]

[1] NASA/IPAC Extragalactic Database (NED), IPAC, JPL,
Caltech, Pasadena, CA 91125, USA

Abstract. A new radiative transfer code is being developed which models three-dimensional galaxies, including a realistic mix of stellar populations, dust and gas, through which the emitted stellar radiation field is allowed to pass. These photons are scattered, absorbed and re-emitted according to the best available prescriptions for the relevant physical processes as a function of wavelength. The galaxy is then viewed along a finite number of preselected lines of sight so as to reduce the computational dimensionality and complexity of the modeling.

1 Introduction

Every crisis in science offers an opportunity to advance the field. And despite Mike Disney's warning, just such an 'opportunity' in extragalactic astronomy occurred when Edwin Valentijn published his 1990 paper in *Nature* discussing the apparent distribution of axial ratios of ESO Sb-Sc galaxies. From these data it was concluded that *spiral galaxies are optically thick over their entire disks.* Not only did Valentijn suggest that galaxies are optically thick everywhere, but that they behaved as being optically thick; this may at first glance appear as a subtle distinction, perhaps; but it is an important one, given that it is legend that spiral galaxies are almost everywhere optically thin and that they are 'known' to act that way (Holmberg, 1958, 1975; de Vaucouleurs, 1976). The reaction from the community was swift and relentless (*e.g.,* see Bosma, 1992; Han, 1992; Byun, 1993; White & Keel, 1992; and also Burstein, Haynes, & Faber 1991, and numerous references therein). Valentijn even went so far as to suggest that the dark matter in the outer reaches of spiral galaxies required to maintain the flat rotation curves was due to this same obscuring material. And while many of the specific claims of Valentijn, Disney and Burstein are not widely accepted as proven, the debate that ensued did sensitize the community to the lack of compelling evidence to the contrary.

Specifically inspired by the controversy, but also in recognition of the general scientific importance of the subject, a NATO Advanced Study Workshop was held in Cardiff in 1994 (Davies & Burstein 1995) devoted to the specific observational and theoretical problems associated with the dust opacities of disk galaxies. At this meeting two things clearly emerged: the first was the age-old plea for 'new data', and the second was a clear need for improved models of multi-component gas-disk galaxies. The interpretation of kinematic, spectral, and broadband data on galaxies as individuals and as statistical ensembles demands an infrastructure of data and theory; but that support structure is still grounded in what Disney amusingly refers to as "the astronomical folklore of the 50's". Our group, in combination and collaboration with other astronomers, is well on the way to furnishing the urgently needed observations; and we are also developing the tools necessary to interpret them.

Huizinger (this volume) has also reviewed the contributions made by many workers in the field to a better understanding of the flow of radiation through dusty galaxies. Most of these seminal papers have been analytic in nature tightly focussed on one or two effects, or addressing a particular data set, but always requiring considerable simplification.

These are the necessary first steps needed to assess the magnitude and sign of including known phenomena, and to test the sensitivity of available observations to known variables. Driven by larger, high-quality, linearized data sets, acquired at new wavelengths, greater and greater demands (and constraints) are being applied to the available models. This increase in information and understanding has raised our awareness to a level of complexity in the real world that truly demands a variety of modifications to existing theoretical models of galaxies that are clearly coupled in their nature and in their level of importance.

2 Dust

Dust undoubtedly dominates as the regulator of the transfer of stellar radiation in those galaxies where it is present. Nevertheless, how and where the dust is actually distributed is probably one of the biggest uncertainties in the study of the apparent structure of galaxies and their subsequently inferred composition. The evolution of those components can only be classified as a distant dream.

The interstellar medium is a complex mix of physics acting on a wide variety of scales. One can no longer treat one wavelength in total isolation. And so, one cannot and should not overly simplify (or ignore) physical processes occurring at other wavelengths. A comprehensive study of attenuation in the optical and ultraviolet cannot overlook scattering any more than it can ignore the existence of wavelength dependence in albedo, absorption and re-emission. Faced with this explosion of wavelength-dependent complexity the situation might seem hopeless. But, it is not.

The relative geometry of stars, dust and gas cannot be smoothed over all scales and simply treated as a fluid continuum: clumping, filling factors and relative geometry are critical factors that must be explicitly considered. Various stellar populations are more or less spatially correlated with the dust. All of them interact with the dust to varying degrees, dependent not only on their intrinsic spatial distributions, but also on our particular viewing angles. This parameter space is also enormous; and it too is to be ignored at our peril. Still an approach to this problem is tractable as well.

For radiative transfer studies, the Hubble sequence is a progression of increasing complexity and therefor it symbolizes the progression of increasing demands on the modeler. Along the Hubble sequence the combination of stellar populations changes from relatively simple in E ans S0 galaxies to a mix of the oldest and youngest stars of high anf low metallicities; the geometry goes progressively from radially symmetric systems, to a blend of spheroids and disks; to highly structured systems with rich morphology including bars, rings, spiral arms and fragments. Finally, the dust content goes from levels of dubious detection to radially dependent mixtures of hot, cool and cold dust and gas in a variety of phases, on a continuum of scales and at densities ranging from diffuse and optically thin to compact and totally opaque.

3. A Cellular Approach

In an attempt to address some of the above-identified problems we have begun constructing a new generation of 3D multi-wavelength models of morphologicallly realistic galaxies designed from the outset to have the richness of form associated with observed galaxies, with the light distribution being generated according to known principles, subject to our current understanding of stellar populations, and obeying the physics of the interstellar medium on scales down to 100 pc.

Our approach differs from all previous approximations in a number of important ways: (1) it is a fully 3-dimensional representation of the stellar and gaseous mix, (2) it is of high spatial resolution (3) it is multi-wavelength and (4) it incorporates all identifiable relevant physics of radiative transfer (including absorption, re-emission, and scattering.) Dust, gas and stellar radiation interactions are dealt with locally (in 100 pc cells) as well as globally. The grid is large enough to assemble and encorporate all of the main morphological features (bars, bulges, disks, spiral arms, etc) of a complete and identifiable galaxy anywhere along the Hubble sequence.

The 'cellular automaton' approach to the star/dust/radiation mix is implemented in a way that is tailored to rapid computation. Models can be constructed relatively rapidly by (1) restricting the number of viewing angles (2) by selectively and progressively following the radiation flow from the highest energy-density cells additively across the entire grid and (3) by using precalculated look-up tables for the 'microphysics' of radiative transfer, light degradation and redistribution (both in wavelength and direction) on the scales assigned to individual cells.

Perhaps the most computationally demanding aspect of incorporating physically realistic radiative transfer in the interstellar medium is in accommodating the continuously variable angular dependence of the scattering by dust. Adopting a pure forward scattering scenario is trivial but unfortunately not realistic. A Henyey-Greenstein formalism is a reasonable parameterization, but its asymmetry term (the so-called 'g-factor') immediately introduces a continuum of angles for the flow of radiation, which when combined with the inevitable mix of multiple scatterings, can lead to computational divergence. Kylafis & Bahcall (1987) suggest a convergent series solution to the problem; we outline here our alternative scenario.

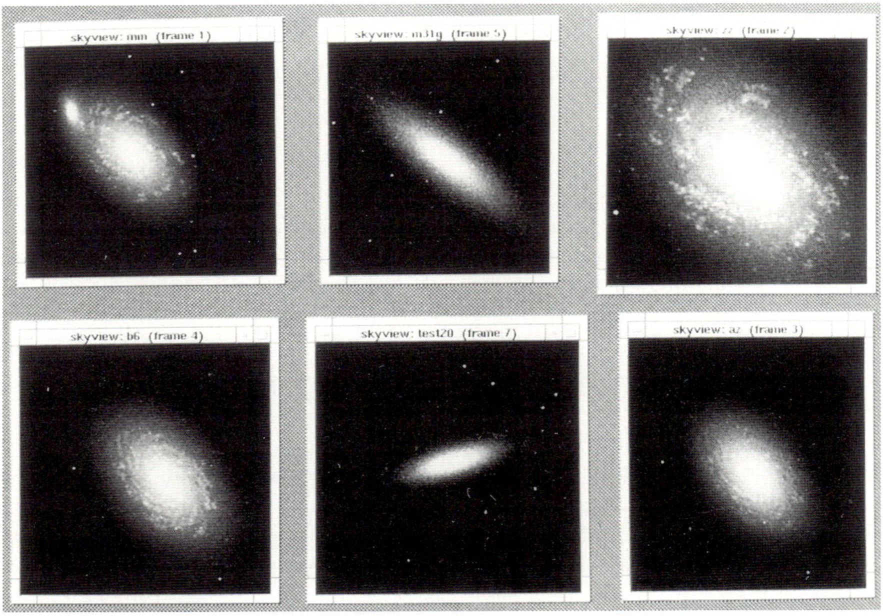

Fig. 1. Examples of early results of the Galaxy Reconstruction Program under development by the author and his colleagues in Pasadena. The images represent a sampling of galaxies of different Hubble types viewed at a variety of inclinations, orientations and wavelengths.

While the scattering of radiation is naturally a continuous function of line-of-sight angle we have decided to limit the computing demands imposed by this term by first considering a finite set of viewing angles and restricting the flow of radiation to these selected directions alone. To do this we have constructed 'lattice crystals' representing cells within which all of the microphysics of energy generation, extinction, scattering and spectral redistribution can occur. The facets of these crystals represent allowed viewing angles. Operationally the facets act as accumulators at one step and then transmission points between contiguous cells at the next. The grid is searched for facets of high energy density. The radiation of these facets is preferentially propagated outwardly to the peripheral cells that then act as final buffers. Each sweep of the grid moves some fraction of the light out of the galaxy, and internally redistributes and redirects the remainder. Newly augmented high-energy facets are again identified and the process iterated until some (user-specified) percentage of the initial energy is fully liberated.

A vast array of complex geometries can be accommodated within the framework of this program. No analytic approximations or overly simplifying assumptions need be made in order to investigate the effects of different distributions of stars, gas and dust. Indeed as new components (a warm interstellar medium?) or geometries (warps?) are identified they can be easily added to the code without major modification.

We are very much aware of two shortcomings intrinsic to **any** attempt to understand the radiative transfer in dusty galaxies at the present time: First of all, there is still a great deal of uncertainty in the simplest parameterization of the attenuating properties of the Galactic interstellar medium (*i.e.*, the variations of g factors, albedo, and extinction coefficients *etc.*, especially as a function of wavelength, but also more speculatively as a function of environmental factors such as metallicity or proximity to high-temperature radiation fields, *etc.* Secondly, there is no obvious minimum physical scale length below which one is totally justified in macroscopically averaging over. After all, the basic units of extinction are grains of micron size, with mean free pathes between photon-grain interaction events ranging from millimeters to kiloparsecs. Clumping of the interstellar medium occurs at all scales from giant molecular clouds on down to comets.

To address the first concern (about the frailty of our current knowledge of the multiwavelength properties of the interstellar medium and its variations) we can only appeal to the best published data, as supported, extended and interpolated by modern theoretical grain models. Some variations in grain properties with metallicity are well documented (*e.g.*, the observed decrease of the $2200\AA$ extinction feature with decreasing metallicity of the gas); and these will be incorporated in the look-up tables associated with the code. As new data and modeling becomes available we will adapt the code to reflect these advances.

The second problem, that of scale size and clumping characteristics of the interstellar medium, can be explored incrementally, indeed with the same program that we are developing for the synoptic study of entire galaxies.

Acknowledgements

This work was undertaken at IPAC with partial support from NASA through grants to the *NASA/IPAC Extragalactic Database* [NED] and the *Astrophysics Data Program* [SDP]. Collaborators on this project include Wendy Freedman (Observatories of the Carnegie Institution of Washington) and Shoko Sakai (Jet Propulsion Laboratory).

References

Bosma *et al.* (1992) *Ap.J.,* **500**, L21
Burstein, D., Haynes, M., & Faber, S.M. (1991) *Nature*, **353**, 515
Byun, Y. (1993) *PASP*, **105**, 993
Davies, J.I., & Burstein, D., Eds. (1995): *The Opacity of Spiral Disks*, NATO ASI Series Vol 469 (Kluwer, Dordrecht, Boston, London)
de Vaucouleurs, G. *et al.,* (1976) *Second Reference Catalogue of Bright Galaxies*, (Univ. Texas Press)
Han,M. (1992) *Ap.J.*, **391**, 617
Holmberg, E.E. (1958) *Medd. Lunds Astr. Obs.,* **2**, 136
Holmberg, E.E. (1975) *Stars & Stellar Systems,* Vol. IX, eds. A. Sandage, M. Sandage, & J. Kristian, (Univ. Chicago Press)
Kylafis, N.D. & Bahcall, J.N. (1987), Ap.J., **317**, 637
Valentijn, E. (1990) *Nature,* **346**, 153
White, R.E. & Keel, W.C., (1992) *Nature*, **359**, 129

Near-IR Colors: The Structure of Dust Lanes in Barred Galaxies

Michael W. Regan & Stuart N. Vogel

Department of Astronomy,University of Maryland,College Park, MD 20742

Abstract. We present a sample of barred galaxies for which we have obtained near infrared and optical photometry and BIMA interferometer observations of CO emission. The combination of infrared and optical data increases the range of sensitivity to dust column density. We obtain the dust extinction and relative scale height of the dust and stars using a radiative transfer model to interpret the photometry. The morphology of the dust extinction regions closely resembles the distribution of CO emission, although there is some evidence that the ratio of CO brightness to dust extinction is higher in the nuclear regions. An analysis of a larger sample of galaxies with IR data indicates that dust column densities appear larger in the nuclear regions than in the bar dust lanes.

1.0 Introduction

There is a discrepancy between what models predict for the dust and gas distribution in barred spirals and what is observed with optical extinction and with millimeter interferometers observing CO emission. The optical observations of strongly barred spirals are dominated by the two straight offset dust lanes along the leading edges of the bar. The majority of the previous CO observations show a "twin peaks" structure (Kenney et al 1992) where the CO emission has two peaks at the inner end of the dust lanes. Recent high resolution models of the gas flow in barred spirals predict that the highest column densities are in a nuclear ring of dust and gas (Piner, Stone, & Teuben 1995).

In this paper we will try to resolve some of the discrepancy by using new CO and near infrared observations of a sample of barred spiral galaxies.

2.0 Observations

The near infrared observations were obtained on three different observing runs. The first set were obtained on the 1.3m at Kitt Peak using the Simultaneous Quad Infrared Imaging Device (SQIID) in January 1994. The second set were obtained on the 1.3m at Kitt Peak using the Cryogenic Optical Bench in March 1994. The third set were obtained on the 1.5m at Cerro Tololo using the CTIO IR Imager (CIRIM).

The millimeter observations were obtained during 1994 and 1995 using the Berkeley- Illinois - Maryland Array observing the CO J=1−0 transition. All the observations were made in three configurations of six antennas each yielding beams of approximately 4 arcseconds.

3.0 Discussion

One reason why the center of the galaxies do not show very much extinction in the optical is because at optical wavelengths the dust optical depth is high enough to absorb most of the light from the stars on the far side of the galaxy. In this case the light near the bulge is dominated by the unreddened stars on the near side of the galaxy leading to unreddened colors. By using colors that have K-band as their long wavelength band this effect can be avoided over a larger range of dust column depth.

By using a collection of colors that have K-band as their long wavelength band as input to a radiative transfer model it is possible to determine some of the geometry of the dust distribution in a galaxy. This was done for the galaxy NGC 1530 (Regan & Vogel 1995), and the resulting dust optical depth is shown in Fig. 1. ¿From our modeling of NGC 1530 we have determined that for both I-K and J-K the reddened regions are good tracers of the dust column density. These colors are useful because neither of these colors get bluer in regions of recent star formation, but both do get redder when there is dust absorption. J-K is better for tracing large column depths of dust since it takes a large amount of dust to saturate J-K. I-K has the advantage of more leverage on the dust extinction since a given amount of dust will have a larger effect on the I-K color than on the J-K color.

We compare the dust column depth as determined by the I-K color to the CO integrated emission in NGC 1530 in Fig. 2. The two correlate quite well showing that there is a relationship between the dust and the CO emission. The two would be expected to correlate since the molecular gas is assumed to form on the surface of grains.

For several other galaxies in our sample we have both CO and near infrared colors. We can use these galaxies to test whether the relationship between CO and dust extinction found in NGC 1530 is repeated in these galaxies.

If we compare the near infrared color as determined from the 1 micron - K-band color to the CO integrated emission (Helfer & Blitz 1995) for NGC

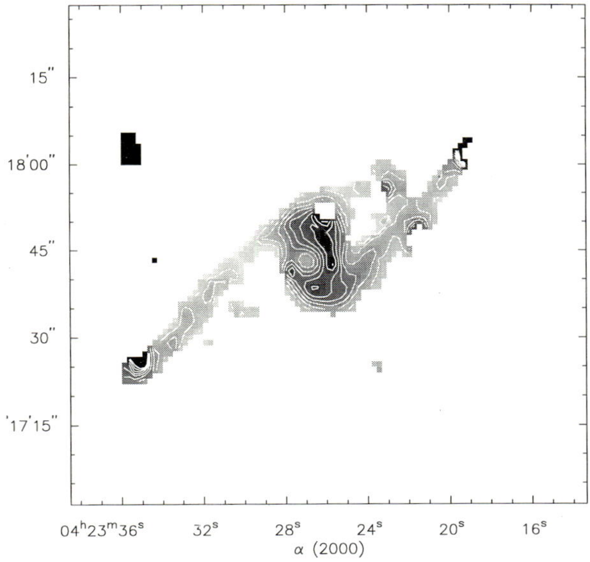

Fig. 1. Total dust optical depth for NGC 1530 as output from the radiative transfer model. A nuclear ring and the concentration of the dust at the inner ends of the dust lanes are the two most prominent features in the image.

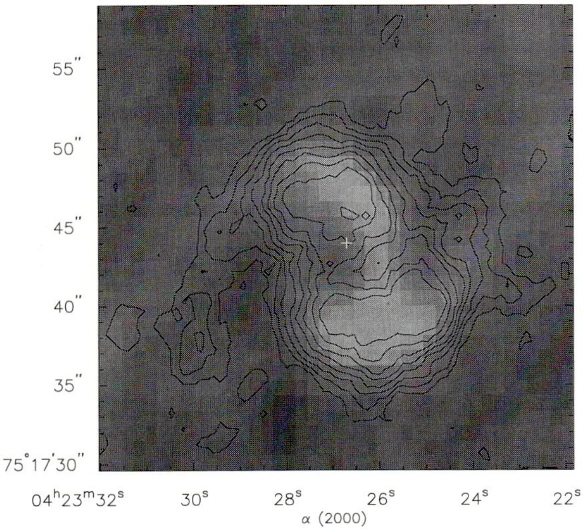

Fig. 2. CO integrated brightness compared with I-K color for NGC 1530. The contours are the BIMA CO (J=1−0) total intensity map and are at two sigma intervals. The beam is 5.2″ by 6.5″ with a position angle of -74°. The gray scale image is the I-K color variation. Here we can see that the I-K color correlates with the CO total intensity although there are regions where there is extinction and no CO.

1068 (Fig 3.), we can see several interesting features. The color map traces two spiral arms that can be traced down to the end of the bars confirming the conclusion of Helfer and Blitz (1995) that the previously identified CO ring is actually the inner extension of two spiral arms. The two arms as seen in the near infrared colors continue to a larger radius than they do in the CO map. This could be due to heating of the CO in the inner region due to the large amount of star formation at that radius. It is also possible that the lack of CO could be an interferometer artifact of the small declination of NGC 1068.

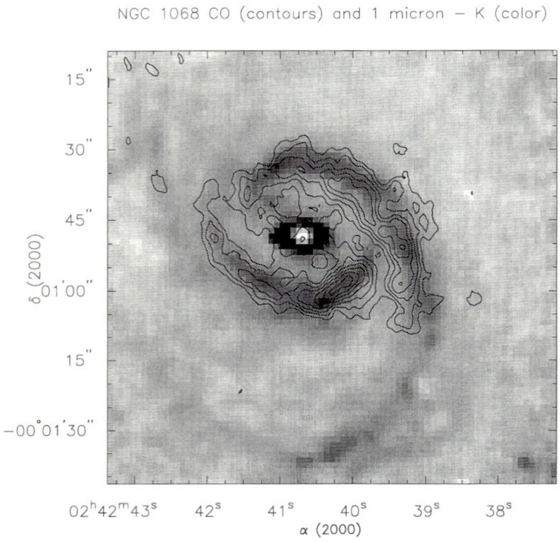

Fig. 3. CO integrated emission (contours) compared with 1 micron-K color (grey scale) for NGC 1068. The two spiral arms are quite striking in the color map and confirm that the inner CO "ring" is really just the inner extension of the spiral arms. Note that the central region was saturated by the Seyfert nucleus so the colors are meaningless in this region.

NGC 5383 is a well studied barred spiral galaxy that shows strong dust lanes in the optical. Our observations show that the CO and J-K colors are well correlated and can best be described as a triple peak structure (Fig. 4). The two strongest peaks are located at the ends of the straight bar dust lanes. The weakest peak is located at the center of the galaxy. This galaxy shows the best overall correlation between CO emission and dust extinction.

NGC 2903 (Fig. 5) is another barred spiral galaxy that shows a strong correlation between dust and CO. In this case the dust reveals a twin peaks structure that is washed out in the CO total intensity map but shows up in the channel maps. The dust extinction traces the dust lanes and seems to be somewhat downstream of the CO near the nucleus. In this galaxy we can also

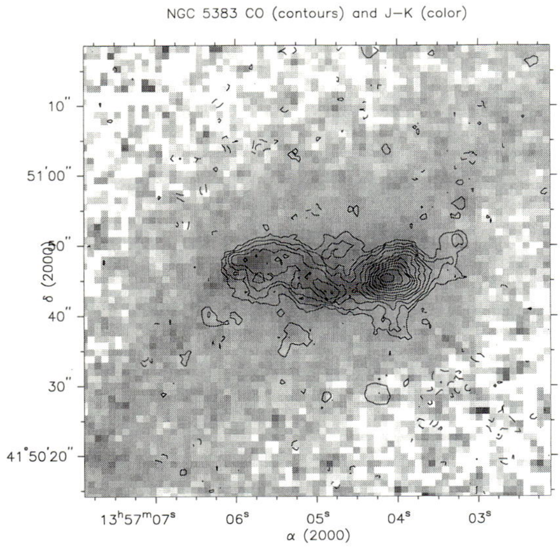

Fig. 4. CO integrated emission compared with J-K color for NGC 5383. The contours are the BIMA CO (J=1–0) total intensity map and are at two sigma intervals. The beam is 4.3". The grey scale image is the J-K color variation from SQIID and COB observations. In this galaxy the correlation between CO and J-K color is very strong.

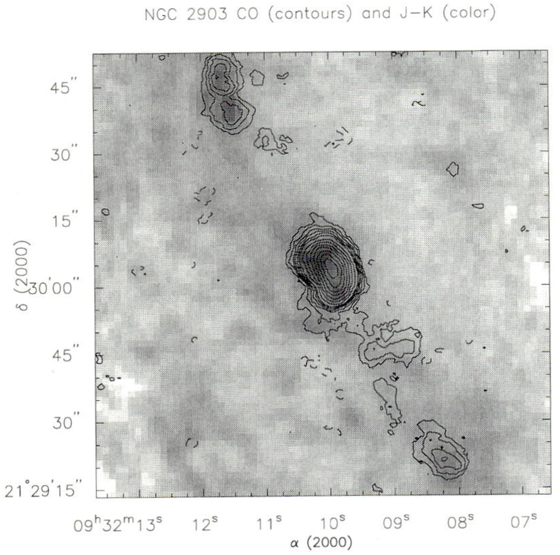

Fig. 5. NGC 2903 CO and J-K The contours are the BIMA CO (J=1–0) total intensity map and are at two sigma intervals. The beam is 5.0". The color image is the J-K color variation. It is possible to see two peaks in the dust extinction in the central region.

detect the CO and dust at the bar ends where many spiral galaxies show an excess of star formation.

For an example of an early type barred spiral we observed NGC 4314. The CO emission and J-K colors are shown in Fig. 6. In this case the CO seems to be more of a "twin peaks" morphology than the extinction since we can clearly see a ring of extinction in the J-K colors but only a faint plateau is visible in the CO map. One end of the dust lanes has a much stronger CO peak than the other one as was seen in the previous galaxies.

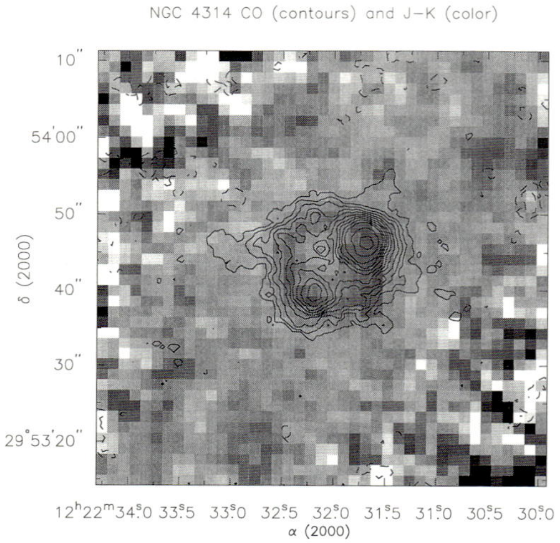

Fig. 6. NGC 4314 CO and J-K. The contours are the BIMA CO (J=1−0) total intensity map and are at two sigma intervals. The beam is 5.5" by 4.8" with a position angle of -82°. The color image is the J-K color variation from SQIID observations. A dust ring can be seen in the J-K color and faintly in the CO. The CO shows two strong peaks of emission with one peak being twice as strong as the other.

¿From these observations we can see that the infrared colors are excellent tracers of the dense dust and gas in galaxies. Thus, we can use them alone to study the dust morphology of galaxies for which high resolution CO observations have not been obtained.

Although J-K does not have a lot of leverage and the SQIID detectors are not very sensitive, it is still possible to detect dust in those galaxies that have large amounts of it. For example, two galaxies that have a lot of dust are NGC 2146 and NGC 3079. These two starburst galaxies are shown in Fig. 7, and the large amount of dust along their major axes is apparent. The large amount of dust in these galaxies shows that the high CO luminosity of the galaxies cannot all be due to heating of the CO by the high rate of current star formation in the galaxies.

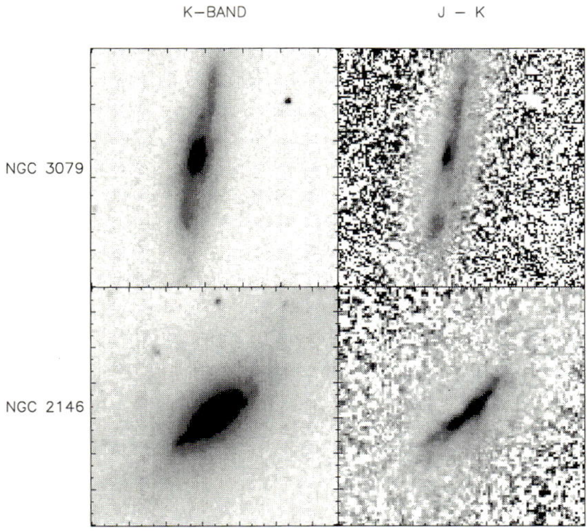

Fig. 7. SQIID 3079 and 2146. The amount of extinction caused by the dust in these two galaxies is large, consistent with the large single dish CO fluxes observed for the galaxies.

With detectors that have higher quantum efficiencies it is possible to get a better signal to noise ratio in the color maps and thus be able to map the morphologies of fainter dust emission. Figure 8 shows two images made using the CIRIM detector at CTIO. NGC 1365 is a prototype of strongly barred spiral galaxies. In this figure we can see how the inner dust structure seems to spiral around the nucleus. It is clearly not a ring and is highly non-symmetrical.

NGC 1672 is classified optically as a weakly barred spiral although from the infrared image it looks very similar to NGC 1365. In fact, their dust morphologies appear to be mirror images of each other.

4.0 Conclusions

We have shown that optical colors do not trace dust column depths due to the fact that at reasonable column depths the dust will obscure the far side stars, leading to unreddened colors. By using K-band as the long wavelength band we show that the near infrared colors do trace the dust column depths over a much wider range of column depths.

We have shown how there is a very high correlation between the dust column depth as measured by near infrared colors and CO emission. Thus, we

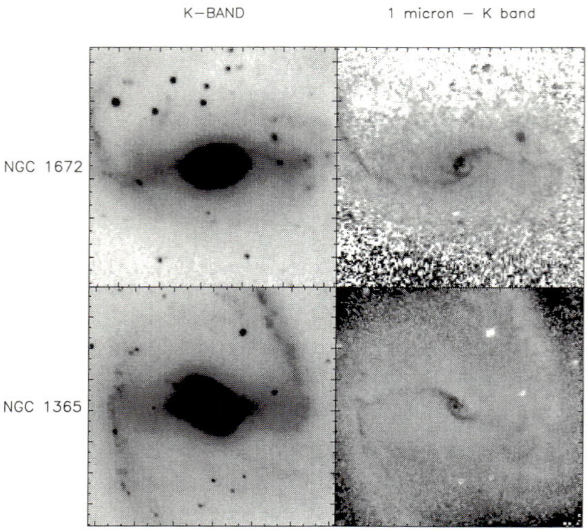

Fig. 8. CIRIM NGC 1672 and NGC 1365.

can determine the dust and gas morphology from near infrared observations alone.

Our observations of barred spirals show that the gas column depths are much higher in the rings and peaks near the nucleus than in the bar dust lanes.

We also see a much stronger m=1 component in the dust and gas distribution than is visible in the stellar potential as traced by the K-band light. This implies that models of barred galaxies can no longer assume that there is plane symmetry between the two sides of the bar.

A "twin peaks" morphology is quite common even though it is not seen in the high resolution models. This means that something must be missing from the models. The most obvious thing that is missing is star formation which could break the symmetry of the gas rings.

References

Helfer, T.T., Blitz, L. (1995), ApJ, 450, 90

Kenney, J.D.P., Wilson, C.D., Scoville, N.Z., Devereux, N.Z., Young, J.S. (1992), ApJ, 395, L79

Piner, B.G., Stone, J.M., Teuben, P.J. (1995), ApJ, 449, 508

Regan, M.W., Vogel, S.N., Teuben, P.J. (1995), ApJ, 449, 576

K′-band observations of the Evil Eye Galaxy and the implied K′-band dust albedo

A.N. Witt[1,2], J.K. Petersohn[1], R.S. Lindell[1], D.L. Block[3], & Rh. Evans[4]

1 Ritter Astrophysical Research Center, The University of Toledo, Toledo, OH 43606, USA
2 European Southern Observatory, Karl-Schwarzschild Str. 2, D-85748 Garching b. München, Germany
3 Dept. of Computational and Applied Mathematics, Witwatersrand University, Wits, 2050, South Africa
4 Colby College, Waterville, ME 04901, USA

Abstract. We are reporting the result of a new analysis of the observations of reddening and attenuation produced by the remarkable dust feature in the galaxy NGC 4826 (Evil Eye Galaxy), published by Witt et al. (1994). This analysis is based on multiple-scattering radiative transfer models for a disk galaxy with an overlying dust lane, and it includes the important feature of disk brightening due to scattering. A revised dust albedo of 0.6±0.1 is derived for the K′-band.

1 Introduction

Among several advantages frequently stated for observing galaxies at near-IR wavelengths is the greatly appreciated fact that the extinction optical depth of dust is much less than that encountered in the optical. In the commonly used K′-band (2.1μm), the interstellar opacity due to absorption and scattering is only about 10% of that in the V-band for the average galactic extinction law (Martin & Whittet 1990). However, in the context of radiative transfer within galaxies (e.g. Witt, Thronson, & Capuano 1992), the attenuation of starlight is determined primarily by the absorption part of extinction, while scattered light is returned to the integrated light of galaxies, albeit with a possible directional change. For this reason, the actual light attenuation at near-IR bands relative to the V-band depends rather critically upon the dust albedo in the near-IR. Very little is known empirically about this dust property in the near-IR in our Galaxy, far less in other galaxies. Here, we report on an effort to determine the effective K′-band dust albedo in the prominent dust lane in the Evil Eye galaxy (M64, NGC 4826, Hubble Type Sab).

2 Observations

The data used for this study were obtained from a K′-band image of NGC 4826, observed by A. Stockton, and a V-image, provided by J. Kormendy. The observational details are described by Witt et al. (1994). Sample surface brightness profiles of NGC 4826 in V and K′ are also shown in Witt et al. (1994), while a reproduction of the V-image and a (V-K′) color map of NGC 4826 are shown by Block et al. (1994). The prominent dust lane, covering the NE half of NGC 4826 and responsible for the nickname of this object, causes severe attenuation in the V-band, but it is essentially invisible in the K′-band where instead the remarkably symmetric disk geometry of NGC 4826 is revealed. (Fig. 2 of Witt et al. 1994)

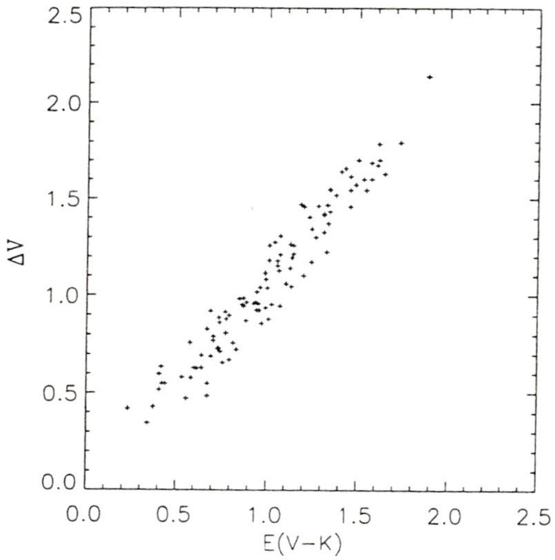

Fig. 1. Observations of NGC 4826. ΔV is the difference in V-surface brightness, in magnitudes, between the dust obscured NE-side of NGC 4826 and points on the unobscured SW-side at equivalent galactocentric distances. E(V-K′) is the corresponding difference in (V-K′) color.

3 Analysis

We took advantage of the exceptional symmetry of the disk of NGC 4826 in order to derive the magnitude of the attenuations in V and in K′. Towards this purpose we extracted V- and K′-band surface brightness profiles by taking a series of $1''.85$ wide cuts through the center of the galaxy, each cut seperated in position angle by 5deg. We folded each surface brightness profile about the center of the galaxy and compared corresponding attenuated and unattenuated positions equidistant from the center in both V and K′. On average, the V-attenuations at a given position were about 15 times larger than the K′-band attenuations at the same location. We display the resulting measurements for a total of 108 matching pairs of positions in Fig. 1 in the form of a ΔV vs. E(V-K′) diagram, where E(V-K′) = ΔV-ΔK′, and where ΔV and ΔK′ are the differences in surface brightness at two diametrically opposite positions on the disk of NGC 4826 respectively, expressed in magnitudes. Witt et al. (1994) analyzed these data in terms of radiative transfer models with an adjustable albedo in the K′-band, using the spherical shell geometry ("nucleus") of Witt, Thronson & Capuano (1992). A best fit was obtained with a K′-band albedo of 0.86, compared to a V-band albedo of 0.6. Several comments are in order:

1. The resultant K′-band albedo is larger by more than a factor of four than the Draine-Lee (1984) grain albedo values for the MRN size distribution of silicate and graphite grains (Mathis et al. 1977), which is frequently used as a reference model.
2. At this stage, we cannot exclude the possibility that the apparent near-absence of dust attenuation in the K′-band is not the result of non-equilibrium near-IR dust emission, commonly observed in galactic reflection nebulae (Sellgren 1984).
3. The Witt et al. (1994) analysis relied upon spherically symmetric models which do not produce the disk-brightening effect expected for disk systems with dust viewed at small inclination angles. This effect is especially prominent at low optical depths (Bruzual et al. 1988), and by being thus more important at K′ than at V, could produce the appearance of a higher effective dust albedo.

For this present contribution we have adopted a new radiative transfer model which simulates the disk of NGC 4826 and the associated dust feature, which covers the NE half of the galaxy. We carried out Monte Carlo simulations of the radiative transfer in a doubly exponential disk of stars and dust, in which an additional dust layer was inserted at a distance of 175 pc above the central plane, to cover half the galaxy's disk. When viewed at high inclination angles, this additional dust layer scatters light received from many directions within the stellar disk below, and it acquires the surface brightness needed to almost compensate for the attenuation at K′ for a lower value of the albedo than was the case with the WTC models. We are displaying these new model predictions together with the observations of NGC 4826 in Fig. 2. The best fit to the observations occurs now for an albedo at K′ of $a_{K'} = 0.6$, with only very little dependence on the phase function asymmetry, which we estimate to lie in the range $0.1 \leq g_{K'} \leq 0.6$.

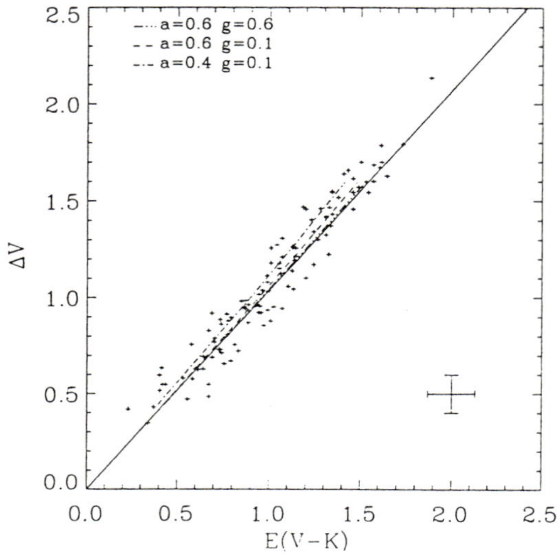

Fig. 2. Comparison with model predictions, base on a doubly exponential disk with an overlying dust lane. The continuum solid line is a leastsquares fit to the data, the endpoints of the model lines correspond to an extinction optical depth of 3.0 at V. The best fit is obtained for $a_{K'}=0.6\pm0.1$, with only weak dependence on the phase function asymmetry.

4 Discussion

Even with the effect of disk brightening included, we find that the effective dust albedo at K′ is as large as the dust albedo at V. For dust features which appear as heavy obscurations in the visible, such as the dark lane in the Evil Eye galaxy, the lower optical depth at K′ combined with a high albedo and aided by the disk brightening effect, makes the corresponding obscuration in the K′-band almost disappear through the return of scattered light. A test of this interpretation might be obtained through polarization observations which should show the integrated light of NGC 4826 in the location of the dust lane to be partially polarized in the K′-band.

The possibility still remains as well that what we perceive as a high effective albedo at K′ is really the result of non-equilibrium dust emission in the near-IR, as observed in Galactic reflection nebulae (Sellgren 1984). This possibility could also be tested observationally by careful surface brightness photometry of NGC 4826 in the 3.0-5.0μm wavelength region. If the near-IR emission of reflection nebulae is a guide, the dust lane in NGC 4826 may actually appear in emission above the integrated light of the stellar disk in this spectral range. However, there are independent indications from observations of scattered light in star-forming regions (Pendleton, Tielens, & Werner 1990; Sellgren 1992) that a higher dust albedo at near-IR wavelengths needs to be considered. This would be an indication that the upper size limit in the size distribution of interstellar grains is substantially larger than the value of $0.25\mu m$ arbitrarily adopted for the MRN model. A recent derivation of a more realistic size distribution, using the maximum entropy method applied to the observed wavelength dependence of extinction by Kim, Martin, & Hendry (1994), has indeed resulted in a size distribution which has still an appreciable number of grains of $1.0\mu m$ radius and larger. Consequently, Kim et al. (1994) predict a dust albedo of about 0.4 for the K′-band, based on a grain mixture of silicate and graphite grains similar to the one assumed for the MRN model.

Furthermore, the analysis of observations of the Thumbprint nebula by Lehtinen and Mattila (1995) in the J, H and K′ bands has also led to albedo values in the range of 0.6 to 0.7.

We plan to extend the present approach of measuring the relative obscuration by strong dust features in galaxies in V and in K′ to other objects and test the general validity of the present results.

5 Conclusions

Scattering by dust of star light originating elsewhere in the disk of near face-on spiral galaxies, so called disk brightening, is an important effect in the K′-band. Reasons contributing to the importance are the relatively low opacity of interstellar dust in the near-IR, the higher than expected dust albedo, and a phase function less forwardly directed than that applicable to optical wavelengths (e.g. Bruzual et al. 1988). As a consequence of this disk brightening effect, the presence of dust lanes becomes almost undetectable in the K′-band, as scattered light replaces almost completely that lost by extinction. The dust albedo required to produce this phenomenon in NGC 4826 has a value of 0.6 ± 0.1. The value lies substantially above that expected from standard models of galactic interstellar dust, and it signifies the presence of dust with sizes in excess of 1 micron.

References

Block, D.L., Witt, A.N., Grosbøl, P., Stockton, A., and Moneti, A. 1994, A&A **288**, 383
Bruzual, G.A., Magris, G.C., and Calvet, N. 1988, ApJ **333**, 673
Draine, B.T., and Lee, H.M. 1984, ApJ **285**, 89
Kim, S.-H., Martin, P.G., and Hendry, P.D. 1994, ApJ **422**, 164
Lehtinen, K., and Mattila, K. 1995, A&A, in press
Martin, P.G., and Whittet, D.G.B. 1994, ApJ **357**, 113
Mathis, J.S., Rumpl, W., and Nordsieck, K.H. 1977, ApJ **217**, 425
Pendleton, Y.J., Tielens, A.G.G.M., and Werner, M.W. 1990, ApJ **349**, 107
Sellgren, K. 1984, ApJ **277**, 623
Sellgren, K. 1992, ApJ **400**, 238
Witt, A.N., Lindell, R.S., Block, D.L., and Evans, Rh. 1994, ApJ **427**, 227
Witt, A.N., Thronson, H.A., and Capuano, J.M. 1992, ApJ **393**, 611

Probing the optical depth of spirals using multi–waveband observations

Rhodri Evans [1]

[1] Colby College, Waterville, Maine, 04901 USA

Abstract. We have investigated whether the reddening of stars can be used to determine the amount and extent of obscuring dust in galaxies. We find evidence for significant optical depths in the central regions of these galaxies, but optical depths of less than unity at ~ 2 scale lengths from the centre.

1 Introduction

The question of whether a significant fraction of the Holmberg diameter of normal spiral galaxies are optically thick or thin is still not resolved. Opinions range from the conventional view that they contain very little dust (eg. Holmberg 1958, de Vaucouleurs, de Vaucouleurs and Corwin 1976) to work which claims to show that they are optically thick out to large radii, with only an upper layer of stars being visible (Valentijn 1990a,b; Burstein 1991 etal.). In this paper we present results comparing the reddening predicted by realistic geometries (dust mixed with the stars and dust “sandwiched” by stars) to the observed colours of individual regions of a sample of spiral galaxies. Using a simple stellar population model we investigate whether the observed colours and colour gradients are due to stellar population changes or changes in observed colour due to the reddening effects of dust.

2 The models and results

We have investigated the reddening produced by different relative geometries of the stars and dust, as well as with different amounts of dust. Throughout the work we have assumed a Rieke and Lebofsky (1985) extinction law, although using a different one would not alter our results significantly. We find that for a screen of dust lying between the observer and the stars the reddening increases linearly as the optical depth of the dust increases. However, such a geometry is clearly incorrect when we consider stars in external galaxies. In these systems the stars are going to be mixed with the dust, and so we have also considered a "slab" geometry, where the stars and dust are uniformly mixed, and a "sandwich" geometry, where the stars sandwich the dust. The reddening produced with such geometries is very different, in the case of the slab geometry the reddening reaches a maximum value, and for sandwich geometries the reddening actually decreases as the optical depth increases.

Because of the non-linearities mentioned above the reddening vectors on colour–colour plots will no longer be linear. They neither increase linearly as the optical depth is increased, nor do they follow a straight line on such a plot. We have investigated whether there are combinations of colours which lead to these reddening vectors deviating in direction from the direction taken by a changing stellar population. The stellar population model we have chosen is a very simple one, a combination of OB type stars, AF type stars and GKM giants. We find that whereas some colour–colour plots are not good discriminators between stellar population effects and dust reddening effects, combinations of wavebands which include near the infrared can be used to suggest reddening due to dust. We have compared the observed colours of 8 spiral galaxies to those predicted by our models, including the direction reddening would move stars on the colour-colour plane. (see Evans 1992,1995).

References

Burstein, D., Haynes, M.P. and Faber, S.M. (1991)*Nature,***353**, 515.
Evans, Rh. (1992) *PhD Thesis* – University of Wales.
Evans, Rh. (1995) *Mon. Not. Royal Ast. Soc.* – submitted
Holmberg, E. (1958) *Medn.Lunds astr.Obs.* Ser. 2, No. 136
Rieke, G.H. and Lebofsky, M.J. (1985) *Astroph. J.***288**,618.
Valentijn, E.A. (1990a) *Nature,* **346**, 153
Valentijn, E.A. (1990b) *IAU Symposium No. 144.*
de Vaucouleurs, G., de Vaucouleurs, A. and Corwin, H. (1976) *Second Reference Catalogue of Bright Galaxies,*–(RC2)

Dust extinction from Paschen-Balmer lines

L. Petersen, T. Christensen and P. Gammelgaard

Institute of Physics and Astronomy, University of Aarhus, Denmark

Abstract. We have studied the dust extinction of spiral galaxies by determing the attenuation of the emission from giant extragalactic H II regions. The aim is to push CCD observations as far into the near-IR as possible to compare the near-IR Paschen lines with blue Balmer lines separated by a wide wavelength interval including the corresponding multiplet lines P_δ/H_ϵ and P_γ/H_δ, which originate at the same upper atomic level.

Introduction. The visual extinction, A_V can be calculated from the observed and predicted ratio of two emission lines at λ_1 and λ_2 as

$$A_V = \frac{2.5 \log(R_o/R_p)}{A_{\lambda_2}/A_V - A_{\lambda_1}/A_V} \tag{1}$$

assuming a foreground absorbing dust screen with the normalized extinction law A_λ/A_V given by Cardelli et al. (1989). A_V has been derived both by applying Eq. (1) to the corresponding multiplet lines and as an average value based on all observed lines (Fig. 1). Spectra were obtained at the NOT with the two-channel Low Dispersion Spectrograph covering λλ3950–4900 Å and 8500–11000 Å simultaneously to ensure precise relative spectrophotometry.

NGC 5461 is an extensively studied H II region in NGC 5457 (M101) bright enough for observations of the near-IR P_γ emission line. Figure 1 illustrates the advantage of a wide wavelength interval for extinction determination and from a weighted least-squares linear fit of all observed line fluxes represented by the dashed line we derive $A_V = 1.12 \pm 0.07$. For P_δ/H_ϵ and P_γ/H_δ we get $A_V = 1.46$ and 1.38, respectively. A comparison with literature data based on H_α/H_β or synthesis models (Rosa & Benvenuti 1994) shows a considerable scatter amounting to more than 1 mag but with the newer results in favour of the high dust extinction also found in the present study.

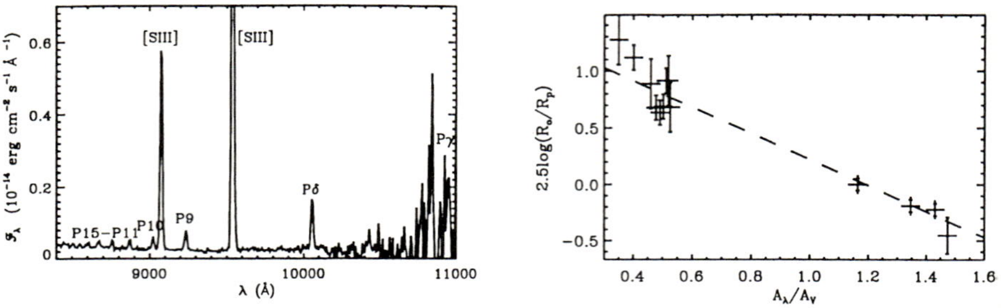

Fig. 1. Left, near-IR spectrum of NGC 5461 with the Paschen lines indicated. Right, observed and predicted fluxes relative to H_β . When displayed in this way they will show a linear relation with a slope of $-A_V$ according to Eq. (1).

NGC 628 and NGC 2403. 7 H II regions were observed in these two spiral galaxies with the results shown in Table 1. A detailed analysis of these data can be found in Petersen & Gammelgaard (1995). Except for the extreme value of region NGC 628, 27 by McCall et al. (1985) the values of A_V derived in this study agree well with the previous published results based on the H_α/H_β ratio but are somewhat higher. For NGC 628 the derived extragalactic extinctions are in the interval 1.4–1.6, while for NGC 2403 we find A_V (Exgal) = 0.7–1.3. By applying extinction corrections from Table 1 to the near-IR S[III]9069,9531 lines and our empirical calibration based on Garnett (1989) we establish a sulphur gradient in NGC 2403 of -0.073 ± 0.021 dex/kpc.

Table 1. Visual extinction derived from Paschen-Balmer line ratios. For comparison, values of A_V based on H_α/H_β from MRS and Fierro et al. (1986) are listed in columns 4 and 5. A_g denotes the Galactic foreground extinction

H II reg ID#	A_V, P_δ/H_ϵ	A_V, lin. fit	A_V (MRS)	A_V (FTP)	A_g
NGC 628, 27	1.57	1.61 ± 0.17	2.12 ± 0.28		0.09
NGC 628, 79	1.78	1.58 ± 0.11	0.99 ± 0.28		0.09
NGC 628, 83	1.65	1.47 ± 0.10	1.03 ± 0.14		0.09
NGC 2403, 1	0.84	0.83 ± 0.02	0.71 ± 0.14	0.77 ± 0.22	0.12
NGC 2403, 2	1.05	0.81 ± 0.09		0.55 ± 0.22	0.12
NGC 2403, 3	1.06	1.00 ± 0.03		0.87 ± 0.22	0.12
NGC 2403, 4	1.59	1.45 ± 0.10		0.77 ± 0.22	0.12

References

Cardelli J.A., Clayton G.C., Mathis J.S., 1989, ApJ 345, 245
Fierro J., Torres-Peimbert S., Peimbert M., 1986, PASP 98, 1032 (FTP)
Garnett D.R., 1989, ApJ 345, 282
McCall M.L., Rybski P.M., Shields G.A., 1985, ApJS 57, 1 (MRS)
Petersen L., Gammelgaard P., 1995, A&A (accepted)
Rosa M.R., Benvenuti P., 1994, A&A 291, 1

VI

NUCLEAR ACTIVITY:

CLUES FROM THE NEAR–IR

Evidence for Biased IMFs

G. H. Rieke[1]

[1] Steward Observatory, University of Arizona, Tucson, AZ 85721, USA

Abstract. The accuracy of evolutionary models applied to starburst galaxies is estimated by comparing results from different groups; key parameters appear to be determined only to the factor of two level. Potential errors in the observational constraints for these models are also discussed. The evidence for an IMF biased toward high mass stars seems to be outside plausible errors in either models or observational constraints, at least for a small number of galaxies including M82, NGC 253, NGC 1614, and NGC 7552. Improvements in infrared spectrometers will advance our understanding of this behavior, for example by improving measures of the dynamical masses and by allowing use of HeI lines to constrain the temperatures of the hot stars.

1 Introduction

The initial mass function (IMF) is a fundamental boundary condition for theories of star formation. It is difficult to measure the IMF directly over a large mass range except in the solar neighborhood. A "local IMF" has been determined extending from 0.1 $M_\odot$ to $50 - 100$ $M_\odot$, albeit with substantial uncertainties at both the low and high mass ends. On a small scale – tens of parsecs – it is obvious that there are substantial deviations from the local IMF. It has been argued that on a larger scale – hundreds of parsecs – these deviations should average out to yield a single, universal form for the IMF. Given the difficulties in detecting low mass stars even 100 pc from the sun, this hypothesis might seem immune to test.

However, a test is possible in exceptional circumstances, where very luminous starbursts take place in galaxies with low mass nuclei. If the properties of these starbursts are fitted with models based on the local IMF, the required mass appears to exceed the upper limit determined by the dynamics. Since the low mass stars contribute nothing to the observable properties of the starburst, this result suggests that star formation in these roughly kpc-sized regions is biased toward massive stars. By far the strongest case for this phenomenon has been made for M82 (Rieke et al. 1980; Puxley et al. 1989; Bernlöhr 1992; Rieke et al. 1993; Doane and Mathews 1993).

The hypothesis of a biased IMF has been challenged by Lester et al. (1991) and Satyapal et al. (1995). Both groups criticize the derivation of the absolute K magnitude of M82. Unfortunately, they both also mis-represent the previous measurements in ways that greatly exaggerate the difference between their derived M_K's and those used to constrain the evolutionary models. Nonetheless, their claims emphasize the importance of possible errors in the chain of arguments requiring a biased IMF.

We will review this situation in the remainder of this paper, both to assess the robustness of the evidence for biased IMFs and to suggest general improvements that are now possible in starburst modeling.

2 Accuracy of Starburst Models

Tinsley & Spinrad (1971) and Tinsley (1972) demonstrated how to use stellar evolution as a useful constraint on galaxy evolutionary models. The conventional technique of using an evolutionary stellar population model to synthesize the composite optical spectrum of the stars is not easily applicable to starbursts because of the large amounts of interstellar reddening. However, an evolutionary model can be used to predict other parameters that are less reddening sensitive. This approach has been applied to starbursts by many authors, such as Rieke et al. (1980), Bernlöhr (1992), Rieke et al. (1993), Krabbe et al. (1994), Leitherer & Heckman (1995), and Genzel et al. (1995). The most important boundary conditions are: 1.) dynamical mass; 2.) ionizing continuum strength; 3.) maximum temperature of stars responsible for the ionizing continuum; 4.) near infrared stellar luminosity; and 5.) bolometric luminosity. The case for a biased IMF therefore is equivalent to a statement that the measured values of these five parameters cannot be fitted within the errors by any pattern of star formation using a local IMF and assuming "normal" stellar evolution.

All these models are dependent on the theoretical evolutionary tracks for massive stars. Rieke et al. (1993) show the effects on the models of using alternate evolutionary tracks; the model predictions seem relatively robust to large changes in the tracks. However, Chiosi (private communication) has emphasized that improvements in calculating hydrogen and helium burning lifetimes make it desirable to convert the models to the recent evolutionary calculations of Meynet et al. (1994).

The models use a variety of observational and theoretical approaches to convert the theoretical stellar parameters (e.g., luminosity and temperature) into the observed parameters (e.g., colors and magnitudes). In addition, various numerical techniques are used to combine the outputs of the stars into the composite parameters of the starburst population. The discreteness of the calculated evolutionary tracks can lead to non-physical oscillations in the outputs. These oscillations can dominate the differences from one set of models to another; to maximize the accuracy, it is useful to interpolate smooth behavior between the tracks and possibly to use numerical smoothing on the outputs of the models.

The agreement among starburst models for a given set of input stellar tracks, and to some extent for different tracks, can be determined from four sets of calculations published with adequate detail for meaningful comparisons: 1.) Rieke et al. (1980); 2.) Bernlöhr (1992); 3.) Rieke et al. (1993); and 4.) Leitherer and Heckman (1995). The first three have been compared in detail by Rieke et al. (1993). We have run a set of models specifically to compare their models with those of Leitherer and Heckman (1995), and find good agreement for solar metallicity. The overall errors judging from these four sets of models are, however, larger than generally realized. The predicted M_K's could be in error by $\pm$ 0.6, the bolometric luminosities by a factor of 2, and the UV fluxes by a factor of $\sim$ 1.6. Further improvements in the models are highly desirable, since many of these discrepancies seem to arise from the conversion of theory to observables and from the oscillations due to the discreteness of the tracks.

Most of the starburst properties of interest for infrared modeling have only a weak dependence on metallicity (Leitherer & Heckman 1995). However, the calculated properties of red supergiants are strongly dependent on metallicity (Meynet et al. 1994), with a correspondingly strong effect on the K luminosities in Leitherer and Heckman's models. The observed blue-to-red supergiant ratio is in reasonable accord with theory near solar metallicities, but it behaves in the *opposite* sense to the theoretical calculations as the metallicity is reduced (Langer & Maeder 1995). Consequently, the models should be reasonably accurate only around solar metallicity. Until improved stellar tracks are available, it must be assumed that the near infrared properties of low metallicity galaxies are only very roughly constrained.

3 Observational Constraints

We will discuss the uncertainties in the observational constraints using M82 as an example. We can then compare directly with arguments for a biased IMF in this galaxy.

3.1 Dynamical Mass

Near infrared spectrometers have advanced sufficiently to use the 2-0 CO bandhead at 2.3μm to determine the dynamical masses in starbursts (e.g., Gaffney et al. 1993; Shier et al. 1994). This technique has many advantages. If the mass is determined by infrared measures of the stellar dynamics, then it should be little influenced by extinction or by non-gravitational forces on the gas. Another advantage is that the CO bearing stars are old enough to give a true measure of the rotation. The youngest populations of CO bearing stars are red supergiants that are typically $\sim 10^7$ years old; at a typical velocity of 100 km/s they will have travelled 1 kpc, about one circuit of the galactic nucleus for a starburst of typical dimensions. In contrast, the ionized gas can trace very young complexes of ionizing stars that are not well distributed around the nucleus (Krabbe et al. 1994).

A variety of techniques have been used to measure the dynamics of the stars and gas around the nucleus of M82, virtually all giving consistent results. These measurements have been summarized by Götz et al. (1990), whose mass model yields a total of 7×10^8 $M_\odot$ within the 30″ diameter starburst region.

A more difficult problem is to estimate what portion of this total mass can take part in the starburst. One approach is to assume that the gas flows continuously into the galaxy nucleus until it becomes unstable and triggers a starburst. When the gas becomes a major factor in determining the overall gravitational field, the system should become unstable because the gas can dissipate energy and hence collapse into clumps under its self-gravity. Thus, the mass in the starburst (stars plus gas) should be no more than half the total from the dynamics, or no more than 3.5×10^8 $M_\odot$ in the case of M82. Allowing 1×10^8 $M_\odot$ for the remaining gas, no more than 2.5×10^8 $M_\odot$ can have formed into stars.

Another estimate of the available mass can be obtained by extrapolating the 2μm surface brightness of the galaxy toward the nucleus from outside the starburst region. The large scale image of Ichikawa et al. (1995) is useful for this purpose. The estimated M_K for the underlying stellar population obtained in this fashion can be converted to a mass according to the ratio of K luminosity to mass for the solar neighborhood (Thronson & Greenhouse 1988) (or the many similar relations suggested during this conference). The result is that all of the dynamical mass could easily be accounted for in terms of the old stellar population. That is, the mass in newly formed stars is likely to be well below the upper limit of 2.5×10^8 $M_\odot$ derived from stability considerations.

3.2 Ionizing Continuum

In the case of M82, a definitive measurement of the ionizing continuum has been made by Puxley et al. (1989) using the H53α line. This line is not affected by extinction, and it is at a sufficiently high frequency that no corrections are needed for stimulated emission (see, e.g., Seaquist et al. 1985). By comparing with the 3.3mm continuum strength, Puxley et al. are also able to constrain T_e. They find that $N_{LyC} = 1.1 \times 10^{54}\ s^{-1}$ and $T_e = 5000K$.

Measurements of H53α are feasible only for exceptionally bright starbursts. However, infrared hydrogen recombination lines are measurable in many starbursts and should have only modest sensitivity to extinction. The values of N_{LyC} and T_e from Puxley et al. provide predictions of the infrared line strengths to compare with the infrared data for M82. As a result, we can test the infrared technique, which is more sensitive to extinction than H53α and does not give an independent estimate of T_e.

The most thorough study of the near infrared recombination lines in M82 is by Satyapal et al. (1995). They demonstrate the shortcomings of previous measurements using aperture spectrometers, presumably because of uncertainties in placing the instrument aperture on a source with such complex structure. Their imaging line measurements yield from Paβ and Brγ a value of $N_{LyC} = 0.8 \times 10^{54}$ s^{-1} taking a canonical value of T_e = 10,000K and N_{LyC} = 0.6×10^{54} s^{-1} taking T_e = 5000K.

It appears that estimates based purely on the infrared lines underestimate N_{Lyc} because of regions of large optical depth that are not accounted properly in the extinction estimates. This hypothesis is supported by McLeod et al. (1993), who show that the infrared lines are compatible with the H53α results if, for example, the dust is mixed homogeneously with the ionized gas with a maximum A_V = 55 through the galaxy, or with a two-screen model with maximum A_V = 20. Other indications of regions of heavy extinction in M82 are the depth of the 10μm silicate feature, implying $A_V = 20 \pm 4$ (Aitken & Roche 1984) and the CO (J = 1 - 0) line, with indicates a peak extinction of $A_V \sim 400$ with a standard CO/H_2 conversion or $A_V \gtrsim 40$ with any plausible conversion (Satyapal et al. and references therein).

It is encouraging, however, that estimating the UV continuum from the extinction-corrected Brγ emission gives a result that is within a factor of two of the correct value for M82. This galaxy should represent a worst case situation since it is so nearly edge-on, so a similar procedure should work well for many other starbursts.

In all starburst modeling to date, the UV flux derived as above has been used as a constraint under the assumption that every UV photon is useful in creating hydrogen recombination lines. In such dusty regions, a significant portion of the ionizing flux is likely to be absorbed by dust. The extent of this effect needs to be taken into account in future starburst models. Doing so will significantly increase the demands made on the stellar population and hence will tend to drive the models to more highly biased IMFs.

3.3 Hot Stellar Temperature

In a heavily obscured region such as M82, the stellar temperature can be constrained by the infrared fine structure lines of Ne II, Ar III, and S IV (Roche et al. 1991). The measurements of these lines by Achtermann & Lacy (1995) yield $T_{eff} \lesssim 35{,}000$K for M82. The optical lines, such as O III, give a higher value for T_{eff} (McLeod et al. 1993). This difference could result if the stars on the front side of the starburst are younger and hotter than the integrated population, a possibility consistent with suggestions that the starburst is propagating outward from the center.

Virtually all starburst galaxies appear to have relatively low ionizing stellar temperatures (Roche et al. 1991). The uniformity of behavior raises the question of whether dust in the emitting regions, high metallicity in the hot stars, or some other cause affects the measured temperatures. In the case of M82, the metallicity of the interstellar gas seems to be near or slightly higher than solar (e.g., O'Connell & Mangano 1978; Puxley et al. 1989; McLeod et al. 1993; Achtermann & Lacy 1995). One would expect the stellar metallicity to be less than that of the gas, since the stars should have formed when there had been less time to enrich the gas. Consequently, there is no reason to ascribe the low effective temperature to high metallicity in M82.

Interstellar dust makes the UV spectrum appear *harder* than it would be in a dust-free environment (Aanestad 1989; Shields & Kennicutt 1995). This behavior arises because: 1.) the dust removes metals from the gas, raising T_e; 2.) the grain opacity as a function of wavelength hardens the spectrum; and 3.) photoelectrons emitted by the grains can heat the gas. The tendency for the optical lines to give higher temperatures than the infrared fine structure lines, as noted above, may result from the effects of dust rather than from geometric effects. In any case, there is no evidence that the effective temperature of the hot stars could exceed 35,000K.

Measuring the ionizing stellar temperature with mid-infrared fine structure lines and the total UV flux from near infrared lines may result in systematic biases because of the differing levels of extinction in the two spectral regions. One of the most straightforward means of constraining the stellar temperature within HII regions is by measurement of the ionized fraction of helium. Helium-ionizing photons will be exhausted at a point interior to the hydrogen Strömgren radius for stellar $T_{eff} \lesssim 40{,}000$K, resulting in a zone of neutral He within the nebula; for hotter stars, however, helium is ionized throughout the HII region (Osterbrock 1989). Measurements of the relative strength of helium and hydrogen recombination lines can thus be used to estimate T_{eff} in the case of partial He ionization, and to place a lower limit on stellar temperature for full ionization.

At infrared wavelengths, the strongest He recombination lines are the HeI $2P^o \rightarrow 2S$ triplet and singlet transitions at 1.083 and 2.058μm, respectively. Unfortunately the fluxes of these lines are difficult to interpret, due to radiative transfer effects (e.g., Clegg 1987, Shields 1993). Vanzi et al. (1995) show that the ratio of the $4^3D \rightarrow 3^3P^o$ transition of HeI at 1.70 μm to Br10 is a useful diagnostic of temperature in the 35,000 to 40,000K region of interest for starbursts. The ratio is extinction independent and depends only weakly on T_e. M82 shows no evidence for the 1.70μm HeI line, in agreement with the temperature upper limit from fine structure lines.

3.4 Near Infrared Stellar Luminosity

The near infrared stellar luminosity has been a very controversial parameter for M82 because of the large extinction corrections that must be applied. Satyapal et al. (1995) make these corrections from maps of the ratio of Paβ to Brγ line strengths, but their approach only works if the ionized gas and stars have identical distributions in the galaxy. In fact, they predict regions where the intrinsic stellar colors are too blue for stars with strong CO absorption bands, a result that appears to contradict 2μm spectroscopy of the galaxy and probably indicates that this assumption is incorrect. Previous estimates were based on the colors of the reddened stellar continuum. Although free of the problems with the approach of Satyapal et al., these estimates have their own shortcomings because of ad hoc assumptions about the distributions of stars and dust and the possibility of distortions in the colors due to thermal reradiation by dust.

To avoid these problems, an improved stellar-based dereddening approach is needed. Fortunately, to the necessary degree of accuracy, nearly all stars have roughly the same H $-$ K colors, ($0 \lesssim H - K \lesssim 0.5$), regulated by H$-$ opacity in cool stellar atmospheres. This "standard stellar color" could be distorted if there is sufficient emission by hot dust to increase the K flux above the stellar contribution, causing an overestimate of the extinction resulting in an overestimate of the stellar luminosity. Fortunately, for M82 it has been shown that this problem does not arise (McLeod et al. 1993). Two groups have determined M_K from point-by-point dereddening assuming standard stellar colors, with close agreement: $M_K = -22.7$ (Telesco et al. 1991) and $M_K = -22.5$ (McLeod et al. 1993).

Estimates of the near infrared luminosity can be improved further. The role of emission by hot dust can be determined by comparing the strengths of the second overtone CO bands (in the H window) with the first overtone bands (in K). The relative strengths of these bands are well behaved in individual stars (Origlia et al. 1993), and dust emission at a level sufficient to change the photometric colors will be easily detectable.

Assuming that the H $-$ K color is not contaminated, then it can be dereddened to estimate the bolometric magnitude of the stars that dominate the near infrared emission. There is a partial accidental cancellation between the K bolometric correction (Elias et al. 1985) and the error that would result in the dereddened K magnitude from a population with a color that deviates from the average (Rieke 1993; Haller, Rieke, & Rieke 1995). Hence, point-by-point dereddening from H and K images to derive the bolometric luminosity of the red stars should provide accurate results. If the evolutionary starburst models are adjusted to predict the bolometric luminosities of the stars in the near infrared, then they can be constrained in a similar manner to previous studies but with much greater confidence.

3.5 Bolometric Luminosity

Fortunately, estimation of the bolometric luminosity of a starburst galaxy is relatively uncontroversial. Most starburst models that succeed in matching the four other criteria have little trouble also matching the predicted luminosity; hence, luminosity estimation is both relatively easy and non-critical.

4 Biased IMFs

In the case of M82, Rieke et al. (1993) showed that no successful starburst models could be constructed using either the Scalo (1986) or Basu and Rana (1992) versions of the local IMF. The discrepancies were a factor of three to four in the predicted outputs in both K-band and ionizing fluxes.

This work made very conservative estimates of the boundary conditions. They assumed all the UV photons were available to ionize gas; accounting for absorption by dust could increase the shortfall in ionizing flux by a further factor of 2 or 3. They assumed half the dynamical mass in the nuclear region is available for the starburst; we have seen that the mass of old, underlying stars may be larger than half, leaving correspondingly less for recent star formation. They also used the mass in the most efficient possible manner – a short burst of star formation to produce the present-day red supergiants, a quiescent period, and another short burst to produce the ionizing flux. More realistic rates of star formation with time could double the mass consumption. As a result, the evidence for a biased IMF falls well outside the modeling and potential observational errors.

The difficulty in fitting the parameters can be removed if the IMF in M82 includes far fewer stars of 3 $M_\odot$ or less, compared with the local IMF. However, the exact form of the IMF in M82 is clearly very poorly constrained, so long as it is modified to provide a significantly larger proportion of massive stars than we deduce locally (Scalo 1986; Basu & Rana 1992).

Recent measurements indicate a number of other galaxies with very low values of M/L and which also are likely to require IMFs biased toward massive stars. Examples are NGC 253 (Engelbracht, private communication), NGC 1614 (Shier et al. 1995), and NGC 7552 (Oliva et al. 1995).

Slightly biased IMFs are also required to fit optical measurements of disk galaxies (Kennicutt, Tamblyn, & Congdon 1994). It is not clear in these cases whether the difference is due to errors in determination of the "local IMF", or whether we are seeing a true variation in the disk of the Milky Way compared with those of other galaxies. However, in M82, NGC 253, and similar galaxies, it appears that the deviation lies outside reasonable errors in the determination of the local IMF. Formation in the gravitational well of a galactic nucleus, or some other environmental factor, evidently can have a strong effect on the process of star formation.

5 Summary

We have reviewed the arguments for an IMF biased toward high mass stars in M82. We find that both for this galaxy and a number of others (NGC 253, NGC 1614, NGC 7552), the evidence for such a bias seems well outside likely observational or modeling errors. Although it is not clear how universal the effect is, we conclude that conditions in a nuclear starburst can substantially modify the process of star formation as we know it in the solar neighborhood.

We have also noted shortcomings in the current approach to starburst modeling and have suggested a number of improvements:

1. Output parameters of the models are likely to have errors as large as factors of two in the case of solar metallicity, and much larger in the case of the infrared behavior of starbursts in low metallicity environments. Further improvements in stellar tracks and incorporation of them in improved evolutionary models are both important.

2. Use of the first overtone CO bandhead to measure stellar dynamics promises to yield accurate dynamical masses even in dusty starbursts. However, we need new approaches to estimating the portion of this mass that is locked up in the underlying old stellar population vs. the portion that can take part in a starburst.

3. Use of Brγ line strengths, after correcting for reddening by comparison with Paβ, appears to provide a reasonably accurate measure of the strength of the ionizing UV continuum

4. Work is needed to estimate the portion of potentially ionizing photons absorbed by dust in starbursts and reradiated in the infrared.

5. The 1.700μm $4^3D \rightarrow 3^3P^\circ$ transition of HeI can be useful as a near infrared indicator of the maximum stellar temperature.

6. The accuracy of the starburst constraints from near infrared photometry can be improved in many cases if the $H - K$ color of the stars is used to calculate the bolometric luminosity of the cool stellar population. This technique can be applied safely for galaxies where the relative strengths of the first and second CO overtone bands indicate the K photometry is not contaminated by emission by hot dust.

Hitherto, infrared starburst modeling has been largely based on photometry with a modest infusion of constraints from spectra. The improved capabilities of infrared spectrometers can provide tremendous improvements, such as use of the CO lines to determine dynamical masses, of H recombination lines to measure the ionizing fluxes, and of faint HeI lines to set limits on the maximum stellar temperatures. We can therefore expect rapid progress in understanding these events.

This work was supported by American Airlines, the European Southern Observatory, and the U.S. National Science Foundation under grant AST91-16442.

References

Aanestad, P. (1989): ApJ **338**, 162
Achtermann, J.M., & Lacy, J.H. (1995): ApJ **439**, 163
Aitken, D. K., & Roche, P.F. (1984): MNRAS **208**, 751
Bernlöhr, K. (1992): A&A **263**, 54
Basu, S. & Rana, N.C. (1992): ApJ **393**, 373
Clegg, R.E.S. (1987): MNRAS **229**, 31P
Doane, J.S., & Matthews, W.G. (1993): ApJ **419**, 573
Elias, J.H., Frogel, J.A., & Humphreys, R.M. (1985): ApJS **57**, 91
Gaffney, N.I., Lester, D.F., & Telesco, C.M. (1993): ApJL **467**, L57
Genzel, R., Weitzel, L., Tacconi-Garman, L.E., Blietz, M., Cameron, M., Krabbe, A., Lutz, D., & Sternberg, A. (1995): ApJ **444**, 129
Götz, M., McKeith, C.D., Downes, D., & Greve, A. (1990): A&A **240**, 52
Haller, J., Rieke, M.J., & Rieke, G.H. (1995): to be submitted to ApJ
Ichikawa, T., Yanagisawa, K., Itoh, N., Tarusawa, K., Driel, W. van, & Ueno, M. (1995): AJ **109**, 2038
Kennicutt, R.C., Tamblyn, P., & Congden, C. W. (1994): ApJ **435**, 22
Krabbe, A., Sternberg, A., & Genzel, R. (1994): ApJ **425**, 72
Langer, N., and Maeder, A. (1995): A&A **295**, 685
Leitherer, C., and Heckman, T.M. (1995): ApJS **96**, 9
Lester, D.F., Carr, J.S., Joy, M., & Gaffnet, N. (1990): ApJ **352**, 544
McLeod, K.K., Rieke, G.H., Rieke, M.J., & Kelly, D.M. (1993): ApJ **412**, 111
Meynet, G., Maeder, A., Schaller, G., Schaerer, D., & Charbonnel, C. (1994): A&AS **103**, 97
O'Connell, R.W. & Mangano, J.J. (1978): ApJ **221**, 62
Oliva, E., Origlia, L., Kotilainen, J.K., & Moorwood, A.F.M. (1995): A&A, in press
Origlia, L., Moorwood, A.F.M., & Oliva, E. (1993): A&A **280**, 536
Osterbrock, D. E. 1989: *Astrophys. of Gaseous Neb. and Active Galactic Nuclei* (Univ. Sci. Books)
Puxley, P.J., Brand, P.W.J.L., Moore, T.J.T., Mountain, C.M., Nakai, N., and Namashita, T. (1989): ApJ **345**, 163
Rieke, G.H., Lebofsky, M.J., Thompson, R.I., Low, F.J., & Tokunaga, A.T. (1980): ApJ **238**, 24
Rieke, G.H., Loken, K., Rieke, M.J., and Tamblyn, P. (1993): ApJ **412**, 99
Rieke, M.J. (1993): AIP Conf. Proc. **278**, 37
Roche, P.F., Aitken, D.K., Smith, C.H., & Ward, M.J. (1991): MNRAS **248**, 606
Satyapal, S., et al. (1995): ApJ **448**, 611
Scalo, J.M. (1986): Fund. Cos. Phys. **11**, 1
Seaquist, E.R., Bell, M.B., & Bignell, R.C. (1985): ApJ **294**, 546
Shields, J.C. (1993): ApJ **419**, 181
Shields, J.C., & Kennicutt, R.C. (1995): ApJ, in press
Shier, L.M., Rieke, M.J., and Rieke, G.H. (1994): ApJL **433**, L9
Shier, L.M., Rieke, M.J., and Rieke, G.H. (1995): to be submitted to ApJ
Telesco, C.M., Campins, H., Joy, M., Dietz, K., & Decher, R. (1991): ApJ **369**, 135
Thronson, H., & Greenhouse, M. (1988): ApJ **327**, 671
Tinsley, B.M. (1972): A&A **20**, 383
Tinsley, B.M., & Spinrad, H. (1971): Ast. & Sp. Sci **12**, 118
Vanzi, L., Rieke, G.H., Martin, C.L., & Shields, J.C. (1995): submitted to ApJ

IR Spectra and the L/M Ratios in Seyfert and Starburst Galaxies

A.F.M. Moorwood[1], L. Origlia[2], J. Kotilainen[3] and E. Oliva[4]

1 European Southern Observatory, Karl-Schwarzschild-Str. 2, D-85748 Garching
2 Osservatorio Astronomico di Torino, Strada Osservatorio 20, Torino, I-1002
3 Tuorla Observatory,University of Turku, Väisäläntie 20,Piikkiö, FIN-21500
4 Osservatorio Astrofisico di Arcetri, Largo E. Fermi 5, Firenze, I-50125

Abstract.

In a study of the stellar populations in galaxy nuclei, infrared spectra at R$\simeq$2000 in the H(1.65μm) and K(2.2μm) windows have been obtained of 27 normal, starburst and Seyfert galaxies with IRSPEC at the ESO NTT. The equivalent widths measured for the CO (1.6 and 2.3μm) and SiI(1.59μm) absorption features are consistent with stellar continua dominated by late type stars and have been used to estimate the non-stellar continuum contributions in the Seyfert nuclei. Equivalent widths alone have been found insufficient to definitively establish the presence of red superginats and hence recent star formation. However, L_H/M ratios have also been determined from the same spectra by combining the 1.6μm stellar continuum fluxes with the velocity dispersions measured from the absorption features. Compared with elliptical galaxies the L_H/M ratios found in starburst galaxies are a factor $\simeq$ 6 times larger which is consistent with the presence, as expected, of a population of red supergiants. The average L_H/M ratio of the Seyfert 1's is comparable to the normal spirals and only slightly larger than the ellipticals. There is thus no evidence for red supergiants in the most extreme AGN type in our sample. Seyfert 2's form an intermediate class with an average L_H/M ratio $\simeq$ 2.5 times larger than the ellipticals - indicative of starburst activity. Both the L_H/M and Brγ EW's are consistent with an increase in starburst age along the sequence starburst, Seyfert 2, Seyfert 1. These results are discussed in the context of recent starburst models and starburst - Seyfert evolutionary scenarios.

1 Introduction

Although many Seyfert galaxies exhibit evidence for starburst activity at or around their nucleii it remains unclear as to whether this is more prevalent in active galaxies or if the two phenomena are related (c.f Heckman, 1991). Nevertheless, two essentially different scenarios have been advanced in recent years for a possible evolutionary connection. One is based on the 'standard' AGN model in which the Seyfert characteristics are attributed to the presence of a black hole which is formed and/or fuelled by the remnants of a nuclear or circumnuclear starburst. This has been discussed in most detail in relation to the so-called IRAS Ultraluminous Galaxies which Sanders et. al (1989) have proposed are in the transition phase between starburst galaxies and quasars. The alternative suggestion is that extremely hot stars ('warmers') and SN produced by starbursts in early type spirals can account for the observed Seyfert characteristics without invoking a black hole (Terlevich & Melnick 1985; Terlevich et al. 1992). Amongst other evidence for this, Terlevich, Diaz and Terlevich (1990) have drawn attention to the fact that optical spectra of several Seyfert nucleii exhibit deep CaII triplet absorption features which could be produced by a population of red supergiant stars.

The primary aim of the present work was to detect starburst activity in Seyfert galaxy nucleii. Because, in evolutionary scenarios, it is to be expected that the starburst is more evolved and probably , ultimately absent, in Seyfert galaxies than those classified as pure starburst galaxies our approach was primarily based on detecting red supergiants which start to appear after about 10^7 years in a starburst i.e when the most massive O stars have already disappeared. In Seyfert galaxies, direct detection of young, cool stars by means of their stellar absorption features is also less ambiguous than deducing the presence of OB stars by their associated nebular emission lines (e.g Brγ) which may be contaminated by the narrow line region excited by the AGN. The target spectral features were CO 1.6μm and SiI 1.59μm selected because (i) dilution by the non-stellar AGN continuum is much lower around 1.6μm than at the wavelength of the more commonly observed CO band at 2.3μm and (ii) our discovery during compilation of a stellar library of H band spectra that the CO/SiI ratio, which is unaffected by any non-stellar continuum, depends on the spectral type and, to a lesse extent luminosity class of cool stars (Origlia, Moorwood and Oliva, 1993) and (iii) the lower extinction at these wavelengths compared to the visible. The results presented here are based on infrared spectra obtained with IRSPEC at the ESO NTT of 27 galaxy nuclei comprising Seyfert 1 and 2's, starburst, normal spirals and elliptical galaxies. Most of the observations and a preliminary discussion has already been published elsewhere (Oliva, Origlia, Kotilainen and Moorwood, 1995 referred to as OOKM(1995) hereafter). Here we draw attention to our conclusion that the L_H/M ratios derived by combining the observed continuum emission with velocity dispersions derived from the features provides a more sensitive and reliable means to distinguish red giants and supergiants than the depths of the features themselves. We also discuss the relationship between L_H/M ratio and Brγ EW's and starburst ages in the light of recent models and comment on the significance of these results for the evolutionary scenarios mentioned above.

2 Stellar Populations from IR Spectra

2.1 CO and SiI in Red Giants/Supergiants

The near infrared stellar continuum in galaxies is expected to be dominated by red giants and supergiants and it has been long known that these stars exhibit CO absorption bands around 2.3μm (1st overtone) and 1.6μm (2nd overtone) which are sensitive to both spectral type (temperature) and luminosity.

In the present work we have observed both CO bands plus SiI (1.59μm) but have based most of the analysis on the H band spectra both because of the lower dilution by non-stellar continuum and to exploit the CO/SiI ratio which is sensitive to spectral type and also independent of any non-stellar continuum because of the closeness of these features in wavelength (Origlia, Moorwood and Oliva, 1993).

As an example, Fig. 1 shows H band spectra of the Seyfert galaxy NGC1068 both on and off the nucleus together with that of an M supergiant. All three spectra show the CO and SiI absorption features which are almost as deep in the off nucleus galaxy spectrum as the star but shallower, presumably due to non-stellar dilution, on the nucleus. Fig. 2 is a plot of CO (1.62μm) EW versus CO/SiI(1.59μm) showing individual stars, star clusters, ellipticals, normal spirals and the programme Seyfert and starburst galaxies. Apart from the more extreme Seyfert galaxies, these objects define a relatively tight sequence which depends primarily on temperature. Within this the galaxies show only a relatively small scatter which, expressed in spectral type, corresponds roughly to the range K5-M3. As the luminosity dependence is small and as the observed features could, in general, be attributed to red supergiants or slightly cooler red giants the EW's alone are not considered a reliable basis for the detection of red supergiants as evidence for recent star formation. The 'depression' of the Seyfert 1's is attributed to the presence of non-stellar dust or synchrotron emission associated with the AGN and its magnitude has been used to estimate the fractional dilution which reaches a maximum of $\simeq$0.5 at 1.6μm. In contrast, the non-stellar continuum around 2.3μm estimated from the position of the galaxies in the EW CO(1.62μm) versus CO(1.62μm)/CO(2.29μm) diagram can exceed 0.9.

2.2 L_H/M as a Giant/Supergiant Discriminator

Although, as argued above, CO and SiI equivalent widths alone are an unreliable means for discriminating giants and supergiants the fact that the L_H/M ratios of the latter are two orders of magnitude larger provides an alternative method based on the same spectra. Details are given in OOKM(1995).

In summary, the H band stellar luminosities have been derived from the continuum fluxes after correction for any non-stellar contribution as described above and the masses have been determined using the velocity dispersions measured

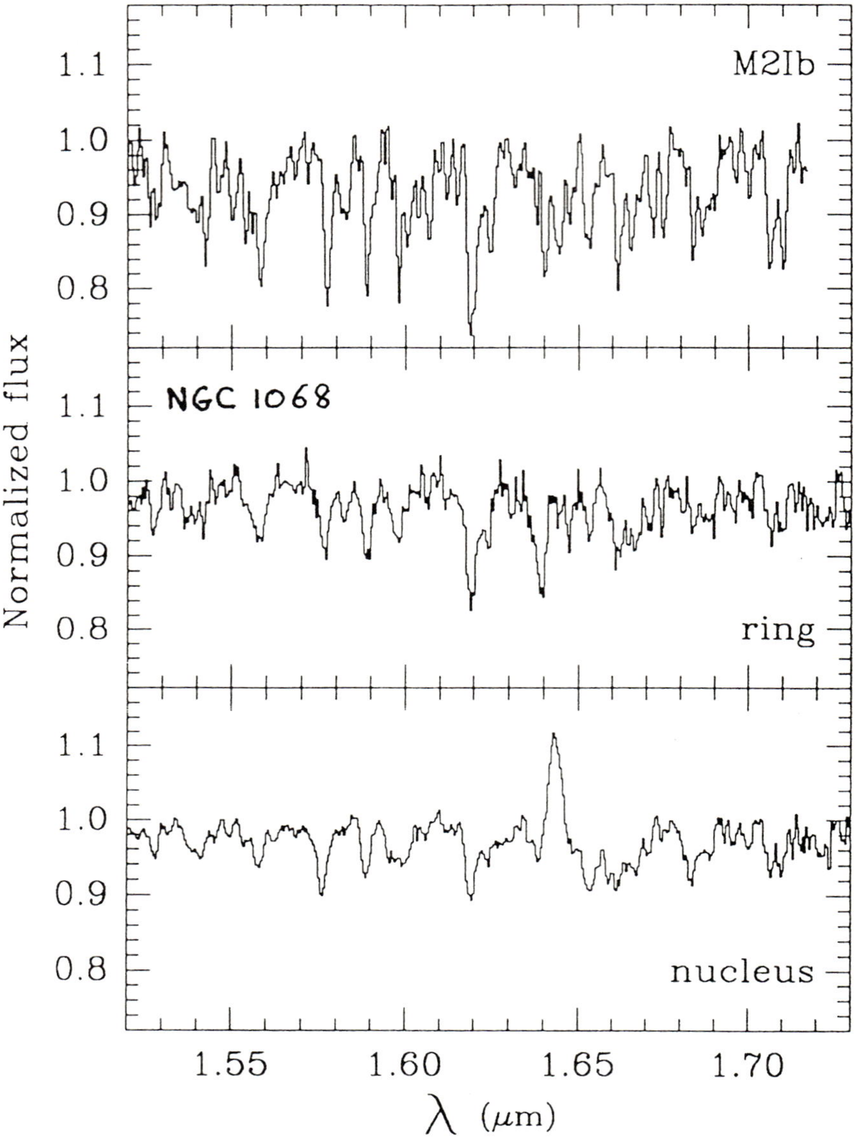

Fig. 1. H band spectra on and off the nucleus of the Seyfert galaxy NGC1068 together with that of an M supergiant star. The CO and SiI absorption features characteristic of cool stars are deepest in the star, only slightly shallower in the off nucleus galaxy spectrum and shallowest on the nucleus due to dilution by non-stellar continuum emission. The emission line present only on the nucleus is [FeII].

from the absorption features. We thus have a very clean method for determining both L_H and M of the stars alone within the same aperture. As discussed below, the resulting L_H/M ratios do exhibit a strong dependence on galaxy type which we attribute to the relative contribution of red supergiants.

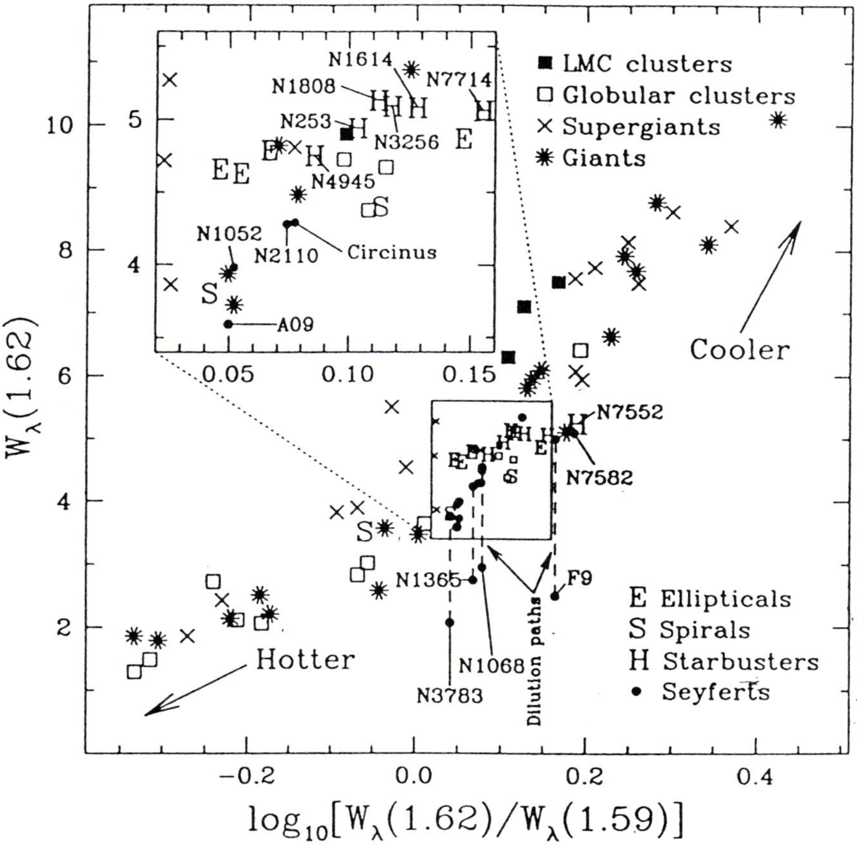

Fig. 2. Plot of CO(1.62μm)EW versus CO/SiI(1.59μm) ratio showing stars, star clusters, ellipticals, normal spirals, starburst galaxies and Seyferts. All objects form a relatively tight sequence except the Seyfert 1's whose EW's are reduced by non-stellar continuum emission.

2.3 Brγ from OB Stars

The measurement of hydrogen recombination line emission from gas ionized by hot stars is a standard technique for detecting and quantifying starbursts in galaxy nuclei. We have observed most of the programme galaxies around Brγ(2.17μm) which is much less affected by extinction than the Balmer lines in the visible. With regard to tracing starbursts in AGN, the limitations of this technique are that the recombination line emission only traces stars formed within the last $\simeq 10^7$ years and it may be difficult to separate the starburst and narrow line region contribution. In practice the EW's measured for this line (corrected for non-stellar dilution using the CO (2.3μm) EW's) have proved to be a useful complement to the L_H/M ratios and provide probably the most reliable means for dating the nuclear starbursts.

3 Results and Discussion

Fig. 3 is a plot of L_H/M versus Brγ EW for the observed galaxies. More detailed information on each of the galxies can be found in OOKM (95) except for the Seyfert 1's IC4329A, A0556-38, NGC4593 and Mkn 1095 added more recently.

With regard to the Brγ EW's it should be noted (i) that none of the elliptical and normal spirals and only one of the Seyfert 1's (NGC1365 which is well known to exhibit circumnuclear star formation) were detected and (ii) the observed EW's have been corrected for the non-stellar continuum contribution as described above i.e the plotted values apply to the stellar populations and are not e.g diluted by non-thermal or hot dust emission in the Seyferts. The Brγ EW's therefore measure the ratio of UV ionizing flux to the stellar continuum, presumed dominated by red giants/supergiants which is strongly related to starburst age as noted previously by various authors (Doyon et al. 1994; Krabbe et al. 1994).

Our contention here is that the L_H/M ratios depend primarily on the supergiant to giant ratio which will also depend on the nature and age of the starburst activity but over a somewhat longer period of time than the Brγ EW's. The sequence defined by the galaxies in Fig. 3 therefore can be understood primarily as one of starburst age and the striking features are its tightness and relatively clear separation of the different galaxy types (for the Seyferts it should be noted that the EW's must be regarded as upper limits because of the possible narrow line region contribution). The segregation is emphasized by the mean L_H/M and Brγ EWs for each of the main galaxy types summarized in Table 1. As reference points we presume that the ellipticals are dominated by red giants and the high L_H/M ratios and Brγ EW's of the starburst galaxies are attributed to the presence of red supergiants and O stars. On this basis, the fact that the Seyfert 1's are essentially indistinguishable from ellipticals implies that their nuclei have not been sites of recent star formation, with important consequences for evolutionary models, whereas some of the Seyfert 2's do lie in the region populated by starburst galaxies.

Table 1. Summary of L_H/M ratios and Brγ EW's

Galaxy Type	N	L_H/M	Brγ EW (Å)
Starburst	7	9.6 +/-3.4	15.3+/-6.8
Seyfert 2	6	4.1 +/-2.2	5.6 +/-5
Seyfert 1	7	2.1+/-0.5	≤2.5
N. Spirals	3	2.5+/-0.3	
Ellipticals	4	1.6+/-0.4	

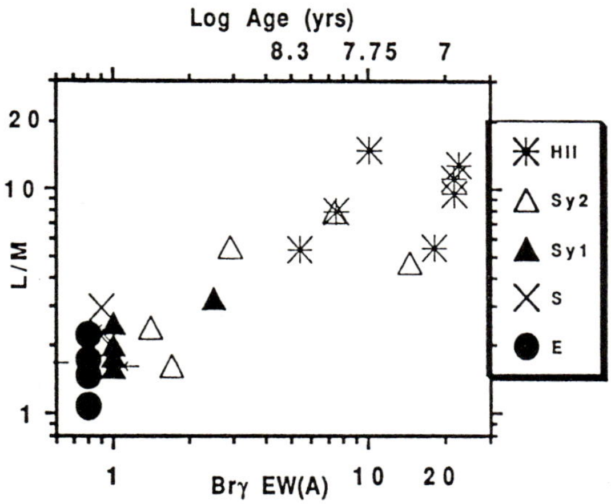

Fig. 3. L_H/M versus Brγ EW. The upper scale indicates the age of a continuous starburst corresponding to the Brγ EW's as derived from the models of Leitherer and Heckman 1995.

The above interpretation of the L_H/M ratios and Brγ EW's is qualitatively in agreement with starburst models. The most recent and most comprehensive are those by Leitherer and Heckman (1995) which cover a wide range in both input and output parameter space and make use of the latest stellar data. These clearly show the effect of red supergiant formation at around 10^7yrs in a starburst as a rapid brightening of the near infrared continuum by several magnitudes and a corresponding decrease in the Brγ EW.

In general, quantitative comparison with the observations is difficult because of the effect on both L_H/M and the Brγ EW's of the underlying old stellar population which is not accounted for in the predicted starburst properties. However, in those galaxies exhibiting substantially enhanced L_H/M ratios it is a reasonable assumption that the luminosity is dominated by the starburst and that the observed Brγ EW's do provide a measure of the starburst age. Detailed predictions then depend on the adopted starburst characteristics. For various reasons it appears more appropriate in the case of galactic nuclei to consider continuous as opposed to single burst models. In this case the most important parameters remaining are those describing the IMF. Amongst those computed by Leitherer and Heckman the models assuming a Salpeter function and lower and upper mass cutoffs at 1 and 30 M$\odot$ leads to ages $\simeq 10^7$yrs for the starburst galaxies in the upper right region of Fig. 3 to $\sim eq 10^8$yrs for the Seyfert 2's at the lower left and $\geq 10^{8.5}$yrs for the Seyfert 1's, normal spirals and ellipticals. Steeper IMFs and/or increasing the upper mass cut-off to 100M$\odot$ increases the ages while assuming single starbursts reduces them. In all cases,however, the relative age sequence which is of most interest here is preserved.

Star formation rates (SFR) estimated from the Brγ luminosities alone cover a wide range from 0.05 - 20 M$\odot$/yr and with a complete overlap of the starburst and Seyfert galaxies i.e the older starburst ages deduced for the Seyferts are not coupled necessarily with lower SFR's.

The most significant result in relation to evolutionary models is probably our failure to detect either O or red supergiant stars in all but one of the Seyfert 1 galaxies. Taken at face value this could be most simply interpreted as evidence that the Seyfert properties are unrelated to starburst activity, at least of the type detected in the starburst and Seyfert 2 galaxies. It is also, however, consistent with the 'classical' evolutionary model if formation/fuelling of the black hole occurs after or actually terminates the starburst phase by gas accretion. The fact that several of the Seyfert 2's contain starbursts which appear to be older than those in the pure starburst galaxies is also suggestive of a starburst, Seyfert 2, Seyfert 1 evolution. Our sample is too small to definitely establish this preferential association of starburst and Seyfert 2 activity in any statistically meaningful sense. Such an evolution is thus a highly speculative suggestion and one which would clearly have consequences for the standard unification model in which the difference in Seyfert 1 and 2 characteristics (mainly the absence of broad lines in the latter) is attributed solely to the combined effect of viewing angle and presence of an obscuring torus. Terlevich has also argued at this meeting that such an evolution is unlikely on the grounds that the Hubble types of starburst galaxies are generally later than those of Seyferts. In our particular sample, however, it should be noted that more than half of the starburst galaxies and several of the Seyferts are classified as Sb.

Irrespective of any evolutionary aspects, the stronger association of starburst activity with Seyfert 2's than Seyfert 1's does also support the suggestion of Cid Fernandez Jr and Terlevich (1995) that starburst activity could account for the blue continuum observed in Seyfert 2's which is otherwise difficult to explain in the standard torus model. Based largely on infrared imaging observations the existence of a torus at all in the prototype case of NGC1068 has also been questioned by Cameron et. al. (1993) who have argued that its broad line region is more probably obscured by a molecular cloud(s) in the line of sight.The relative concentration of molecular clouds and associated starburst activity rather than the presence of a torus therefore remains an interesting possibility for explaining the differences between Seyfert 1's and 2's.

6 Conclusions

L_H/M ratios have been derived from spectra covering the CO and SiI bands around 1.6μm of a sample of 27 normal, starburst and Seyfert galaxy nuclei. A clear segregation has been found with starburst galaxies exhibiting elevated values ($\simeq$ 6) compared with normal ellipticals and spirals which we attribute to the presence of red supergiants. This is consistent with the observed Brγ equivalent widths and recent starburst models which indicate ages $\simeq 10^7$ years which correspond to the expected onset of the red supergiant phase. In contrast, neither the L_H/M or Brγ EW's of all but one of the Seyfert 1's indicates the presence of starburst activity. Several of the Seyfert 2's in the sample do exhibit both Brγ EW's and elevated L_H/M ratios consistent with starburst activity and also ages which are significantly older than in the pure starburst galaxies.

Taken together, these results suggest a possible evolutionary scenario in which starburst galaxies evolve via Seyfert 2's to Seyfert 1's due to the formation of a black hole or at least accretion of gas from the starburst region. Irrespective of whether or not the evolutionary aspect is correct, the apparently stronger association of starbursts with Seyfert 2 than Seyfert 1 activity found here supports independent suggestions (Cameron et al. 1993; Cid Fernandez Jr and Terlevich 1995) that this rather than the obscuring torus postulated in the classical unification model could equally well if not better account for the observed difference between Seyfert 1 and 2 galaxies.

References

Cameron, M., Storey, J., Rotaciuc, V., Genzel, R., Verstraete, L., Drapatz, S., Siebenmorgen, R., and Lee, T.J. (1993): Ap.J.,419,136

Cid Fernandez Jr, R., Terlevich, R.(1995): MNRAS, 272, 423

Doyon, R.,Joseph, R.D., Wright, G.S. (1989): Proc. 22nd ESLAB Symposium on IR Spectroscopy in Astronomy, ESA SP-290, 269

Heckman, T.M.(1991): Massive Stars in Starburst Galaxies eds. C. Leitherer, N. Walborn, T. Heckman, C. Norman (Cambridge University Press), 289

Krabbe, A., Sternberg, A., Genzel, R. (1994): Ap. J., 425, 72

Oliva, E., Origlia, L., Kotilainen, J.K, Moorwood, A.F.M. (1995): A&A,301,55

Leitherer, C., Heckman, T.M. (1995): ApJS, 96, 9

Origlia, L., Moorwood, A.F.M., Oliva, E.(1993):A&A, 280, 536

Sanders, D.B., Soifer, B.T., Elias, J.H., Madore, B.F., Matthews, K., Neugebauer, G., Scoville, N.Z. (1988): Ap.J., 325, 74

Terlevich, R., Melnick, J. (1985): MNRAS, 213, 841

Terlevich, E., Diaz, I., Terlevich, R. (1990): MNRAS, 242, 271

Terlevich. R., Tenorio-Tagle, G., Franco, J., Melnick, J. (1992): MNRAS, 255, 713

Near Infrared Emission Lines in the Active Nuclei of Spiral Galaxies

Rodger I. Thompson[1]

1 Steward Observatory, University of Arizona, Tucson, AZ 85721, USA

Abstract. This contribution surveys the dominant near infrared spectral features in AGNs and the physical regions associated with them. In particular those features which separate the central engine contributions from star formation activity are emphasized. The galaxies studied in this work display a range of main energy contributions from central engine dominated to star formation dominated. All galaxies, however, display contributions from both energy sources. Two of the galaxies, NGC 1068 and NGC 7469, display emission from the high ionization species Si X. This is the first observation of Si X in AGNs.

1 Introduction

Although galaxies with active nuclei are often singularly classified, Seyfert 1, Seyfert 2, Starburst, etc. they most often display a mixture of spectral features from each class of galaxy. If we distinguish not by spectral type but by energy generation mechanism we can rank galaxies by the proportion of the mechanism operating in each. For the purposes of this contribution we consider only two mechanisms, a black hole central engine and massive star formation. We do not consider here whether black holes actually exist as the driving mechanism, we simply use it as the standard model for a centrally concentrated luminosity source. The other main mechanism, stellar luminosity is of course present but not part of the AGN nature of the galaxy. Some spectral components should be present in both types of energy generation and are listed as intermediate. Implicit in this discussion is the widely held paradigm of dust obscuration determining whether broad line emission from the! central engine is observable.

In general we attribute the classical broad and narrow line regions to central engine sources. Starburst or star formation activity is associated with regions of molecular emission and high stellar continuum from red giant and supergiant stars. Coronal line emission may come from either luminosity source based on our lack of certainty of the production mechanism. Also, even though the forbidden Fe II emission is generally discussed in conjunction with the narrow line region, it is probably not coincident with it as is discussed in Section 2.2.1.

One of the key aspects in performing these studies is the availability of contemporaneous wide spectral range spectra. The components described here range from 0.88 to 2.5 mm, a little less than a factor of 3 in wavelength. All of the spectra shown here were obtained with the GRIS (Thompson 1994) which is a cold cross dispersed echelle spectrometer.

Another example of the usefulness of broad range spectra is the discovery of the emission lines of [Si X] at 1.4299 mm. This spectral region is not generally observed with limited range spectrometers due to the heavy telluric absorption at nearby wavelengths.

2 Central Engine Features

Classical Seyfert 1 galaxies display both broad and narrow line spectral components, often in the same line. Here we consider the presence of broad emission lines of hydrogen and helium as well as narrow emission lines of S III and other species not generally observed in H II regions as evidence for the existence of a black hole central engine. Figure 1. below shows both the broad and narrow line components of the He I 1.0830 mm emission from the Seyfert 1 galaxy NGC 4151. For comparison the spectra of NGC 1068 and NGC 7469, which display only narrow line emission are also displayed.

2.1 Broad Line Component

The most direct spectral signal of a central engine is broad emission from either atomic hydrogen or helium. This is a defining characteristic of Seyfert 1 nuclei such as NGC 4151. The Seyfert 1.5 and 2 nuclei of NGC 1068 and NGC 7469 in fig. 1 only have narrow line components. Near infrared broad lines are generally He 1.0830, Paschen b, Brackett g, and at high enough redshifts to clear the telluric absorption, Paschen a. However, the weakness of Brackett g often makes it difficult to see the broad line component of the line.

2.2 Narrow Line Emission

Narrow line emission in hydrogen and helium is well observed in most AGNs but is hard to separate from similar emission in H II regions associated with star formation. One of the most ubiquitous sets of narrow lines is the emission from [S III]. These lines at 0.9068, 0.9531, and 1.032 mm are excellent indicators of the narrow line region. Figure 2. shows this emission in several AGNs.

2.2.1 Forbidden Fe II Emission

Although not generally seen at high intensity in H II regions the forbidden Fe II emission at 1.644 mm is a ubiquitous feature in the spectra of AGNs. It is also a very strong line in the spectra of supernovae (Graham et al. 1990). It is also thought that [Fe II] emission should be strong in the partially ionized regions created by high energy photons (Dennefeld and Pequignot 1983). Mouri et al. 1993, argues strongly that [Fe II] 1.644 mm emission is associated with nuclear activity in the central engine. Unfortunately, until there is a better understanding of the production mechanism for Fe II emission it can not be classified as being

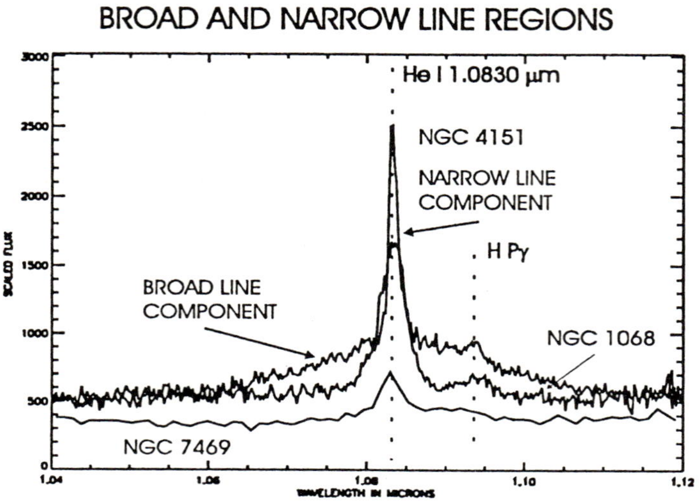

Fig. 1. He I broad and narrow line emission in NGC 4151, NGC 1068, and NGC7469

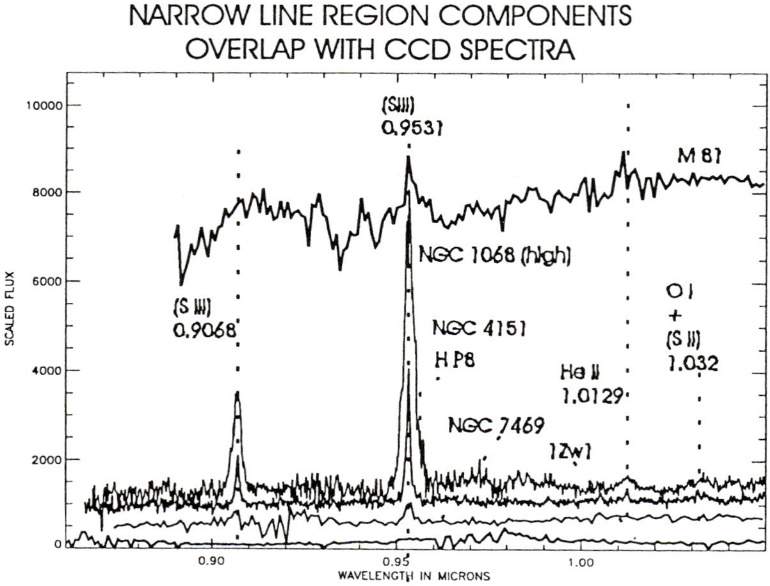

Fig. 2. Narrow line [S III] emission

indicative of a central engine or of the later supernova production from a massive starburst. It is probable, however, that it is produced in a region independent of the standard narrow line production region.

3 Intermediate Features

Intermediate features are spectral signatures that are expected from both central engine and starburst energy production. The first of these are the Coronal lines of high ionization species.

3.1 Coronal Lines

Coronal lines involve high ionization states which may be achieved by high energy photon emission from either a central engine or supernovae. In addition it may be possible to produce the high ionization with high energy shocks from either galaxy mergers or winds from a central engine. Coronal lines from [Si VI] and [Si VII] have been known to exist in AGNs for some time (Oliva & Moorwood 1990). The spectra of NGC 1068 and NGC 7469 shown in Fig. 3 show for the first time (Thompson 1995b) emission from [Si X] which has an ionization potential of 401.4 eV (Greenhouse et al. 1993). Previous to this the line had only been observed in the solar corona (Penn and Kuhn 1994). The line is at 1.4299 mm, a region subject to extensive telluric absorption and therefore not heavily observed. These spectra were taken during an unusually cold and dry night at Kitt Peak, which accounts for their observability.

Other possibilities for this line are the H2 (4,2) Q(5) line at 1.42958 mm and the H2 (9,6) O(4) line at 1.431123 mm, although they would have to greatly exceed their expected flux relative to the H2 (1,0) S(1) line to account for the observed flux. InNGC 4151 the coronal lines appear to have a wider line width than the molecular or even narrow line region widths (Thompson 1995a) which indicates that at least in this galaxy the lines are more likely associated with the central engine than any star formation activity.

4 Star Formation Features

Two principal spectral features are associated with the presence of star formation in AGNs, molecular hydrogen emission and enhanced stellar absorption features from giant and supergiant stars. The molecular hydrogen emission indicates that there are extensive molecular clouds capable of star formation. The giant and supergiant stellar features indicate that star formation has recently occurred in the galaxy.

4.1 Molecular Hydrogen Emission

Since H2 can be excited by either shocks or UV fluorescent mechanisms, the primary indicator is not the excitation mechanism but the presence of molecular clouds capable of star formation. An important example is the very strong H2 emission in merger galaxies where the presence of starbursts is well documented. The ubiquitous presence of H2 emission is strong evidence of high star formation rates even in AGNs dominated by the central engine. The primary observed line is the (1,0) S(1) line at 2.12183 mm along with the (1,0) S(3) line at 1.9575 mm. These lines are among the strongest lines emitted in shock excitation of molecular clouds. UV excitation stimulates the (1,0) S(0) line at 2.2227 mm along with lines at shorter wavelengths such as (3,1) Q(1) 1.3138 mm and (4,2) S(1) at 1.3112 mm.

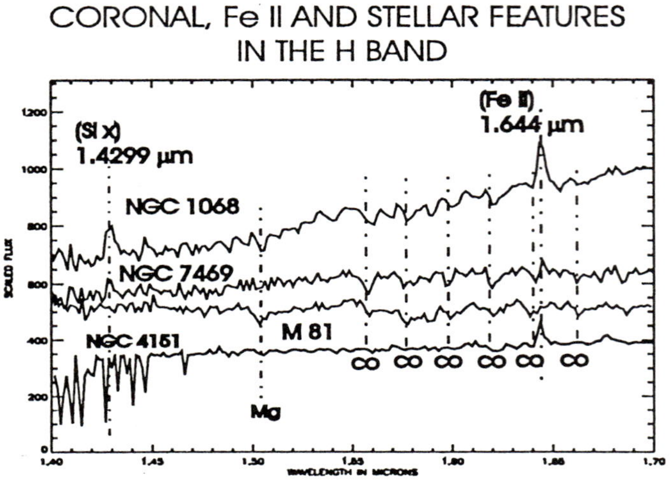

Fig. 3. Coronal, Fe II, and Stellar Features

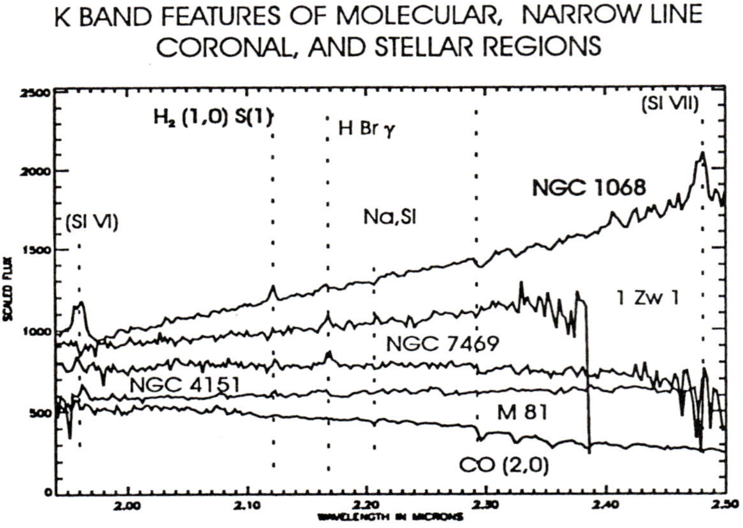

Fig. 4. Stellar Absorption Features in the K band

4.2 Stellar Absorption Features

Stellar absorption features are a sensitive measure of the relative contributions of the central engine versus star formation contributions. Well known are the first overtone CO band features at 2.3 mm which are luminosity sensitive. Perhaps even more sensitive are the CO and other giant star features in the H band around 1.5 to 1.7 mm. If we take M 81 to be representative a galaxy with "normal" stellar populations then Fig. 3 shows that the starburst dominated galaxy NGC 4769 has significantly stronger features than M 81, indicating an enhance giant and supergiant population due to the increased star formation rate. The K band spectra shown in Fig. 4 are not nearly as discriminating. The ratio of the second overtone CO band absorption strength to the 1.644 mm [Fe II] emission strength may be an important indicator of the relative star formation to central engine ratios.

5 Ranking the Observed Galaxies

With criteria we have developed in this paper we can construct a crude grid evaluating the ratio of star formation activity to central engine activity. Table 1 below presents that ranking for the small number of galaxies in this sample. The galaxy 1 Zw I is an anomaly with very few spectral features. It is placed in this ranking but may have more SO like features than AGN features. The table ranks the spectral features with strong central engine features on the left and galaxies toward the top. Star formation features increase toward the right and starburst galaxies toward the bottom

Table 1. Galaxy Ranking by Luminosity Source

Object	Broad Line Strength	Narrow Line Strength	FeII Line Strength	Coronal Line Strength	Molecular Line Strength	Stellar Abs Line Strength
NGC4151	strong	strong	strong	strong	moderate	weak
NGC1068	strong	strong	strong	strong	moderate	very weak
1 Zw I	not visible	strong	not visible[a]	not visible	not visible	not visible
M81	none	very weak	none	none	normal	
NGC7469	none	moderate	weak	strong	strong	very strong

[a] strong in optical

6 Conclusions

The main conclusions of this contribution are threefold. First, most AGNs are a combination of black hole central engine and starburst luminosity. Second, wide spectral range near infrared spectra are excellent tools for determining the relative contributions of central engine and starburst activity. Finally, a new coronal line of [Si X] at 1.4299 mm was detected in NGC 1068 and NGC 7469.

References

Dennefeld. M., & Pequignot, D. 1983, A&A, 127, 42.
Graham, J.R., Wright, G.S., & Longmore, A.J. 1990, Ap.J., 352, 172.
Greenhouse, M.A., Feldman, U., Smith, H.A., Klapisch, M., Bhatia, A.K., & Bar-Shalom, A. 1993, Ap. J. Supp., 88, 23.
Mouri, H., Kawara, K., & Taniguchi, Y. 1993, Ap. J., 406, 52.
Oliva, E. & Moorwood, A.F.M. 1990, Ap.J. (Letters), 348, L5.
Penn, M.J., and Kuhn, J.R. 1994, Ap.J., 434, 807.
Thompson, R.I. 1994, P.A.S.P., 106, 94.
Thompson, R.I. 1995a, Ap.J., 445, 700.
Thompson, R.I. 1995b, (in preparation)

Near-IR Properties of Quasar and Seyfert Host Galaxies

K.K. McLeod[1] and G.H. Rieke[2]

[1]Smithsonian Astrophysical Observatory, 60 Garden St. MS 12,
Cambridge, MA, 02138, USA
[2]Steward Observatory, University of Arizona, Tucson, AZ 85721, USA

Abstract. We have obtained deep, near-IR images of nearly 100 host galaxies of nearby quasars and Seyferts. We find the near-IR light to be a good tracer of luminous mass in these galaxies. The Seyferts are found in galaxies of type S0 to Sc. The low-luminosity quasars live in similar kinds of galaxies spanning the same range of mass centered around L*. However, for the most luminous quasars there is a correlation between host-galaxy mass and nuclear B-band luminosity. Some of the luminous quasars in our sample have now been imaged using WFPC2 aboard HST. We compare our IR images to the HST images and find that, contrary to early reports, there is not a large discrepancy between ground- and space-based results. The host galaxies of these quasars are likely smooth, and luminous early type galaxies. We discuss the advantages of near-IR imaging for detecting the hosts of luminous quasars.

1 Introduction

Historically, most of the emphasis in the study of Active Galactic Nuclei (AGN) has been placed on the "N." However, concentrating on the "G" gives us the potential to answer important questions relevant to the understanding of galaxy formation and evolution as well as the AGN phenomenon. By studying the host galaxies of Seyferts and nearby quasars, we can hope to find out: Can any galaxy host an AGN? Do all galaxies go through an active phase? How does a galaxy feed a "monster" in its middle? How does the AGN affect the host galaxy?

To this end, we undertook a project to obtain deep, near-IR images of $\sim$ 50 nearby quasars and $\sim$ 50 Seyferts. The results have been presented in McLeod and Rieke (1994a, b, and 1995a). We describe in these papers the relationships between host galaxy luminosities (and masses) and nuclear properties; the existence of substantial obscuration coplanar with the disks of host spirals of Seyferts; and the search for signs of disturbances that could aid the flow of fuel towards the centers of the galaxies. The reader is encouraged to consult these papers for details; here we will give only a brief description of the data and present a few results. We will also compare our images to visible images recently obtained by several groups using WFPC2 aboard HST.

2 The IR Images

We obtained deep near-IR images for nearly 100 AGN using a 256x256 NICMOS array camera on the Steward Observatory 2.3m telescope. We chose two samples of AGN, selected to avoid some of the biases in previous studies, that allowed us to investigate host galaxy properties over 10 B magnitudes in nuclear luminosity. For low-luminosity AGN we used the CfA Seyfert sample, which is selected on the basis of the nuclear spectrum and which has roughly equal numbers of Sy1's and Sy2's. For high-luminosity AGN, we chose the lowest redshift ($z < 0.3$) quasars from the PG sample, to ensure a sample selected on the basis of nuclear properties and close enough so that the host galaxies would be resolved.

There are several advantages in using near-IR imaging for AGN host galaxy studies. Of course, there are the usual arguments that near-IR light shows the galaxy's mass-tracing, red stars while suffering less extinction by dust. Also, near-IR images can be combined with visible images to investigate host galaxy colors. However, the huge advantage in obtaining near-IR images is illustrated in Fig. 1. Especially in the case of quasars, the big problem in detecting the host galaxies is that there is an overwhelmingly bright point source in the middle. Luckily, nature has conspired to make the host galaxy starlight brightest at the same wavelength where the nuclear light is at a local minimum. Thus, near-IR wavelengths allow us a view of the host with less contamination from the nucleus than is possible at visible wavelengths.

3 Host Galaxy Properties

One of the most basic questions to be addressed in the study of AGN host galaxies is: what kinds of galaxies are the hosts? In the case of the Seyferts, all of the host galaxies are disk galaxies. They range in type from S0 to Sc, with a median type of Sab. Many have bars, some discovered in the IR images. However, as a class the Seyferts do not have significantly more bars than non-active galaxies of the same Hubble type. There is no apparent difference in galaxy morphology between Sy1's and Sy2's.

For the closest and very lowest-luminosity quasars, the hosts also appear to be spiral galaxies. However, for the more distant and high-luminosity quasars in our sample, we cannot tell from our images whether the hosts are

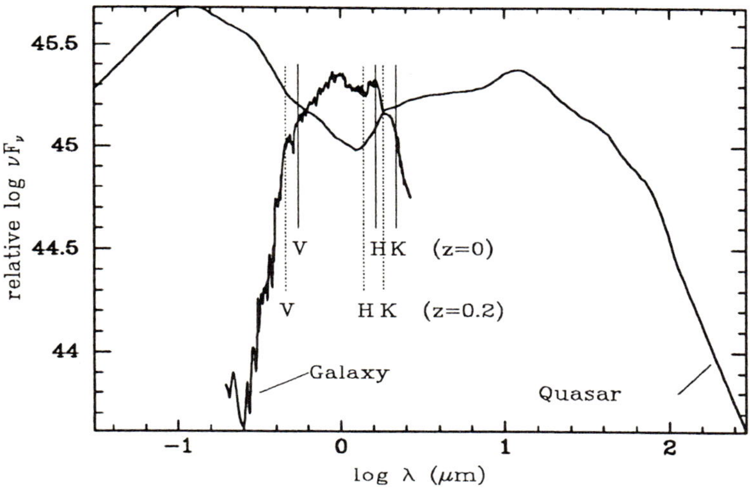

Fig. 1. Energy distributions of quasars (Elvis et al. 1994) and galaxies. Observing in IR light provides much better contrast of galaxy starlight to nuclear light. Note: normalization is arbitrary (for luminous quasars, the nucleus would be much brighter relative to the galaxy).

spirals or ellipticals. Ground-based images do not have enough spatial resolution to investigate the 2-D morphologies, and we are not able to distinguish from 1-D radial profiles whether the hosts are fit by exponential disk laws, de Vaucouleurs laws, or neither. In fact, we have learned from the nearby Seyferts that profile fits are ambiguous even for disk galaxies.

Another probe of a host galaxy is its near-IR luminosity, a good measure of the luminous mass in these systems. The host IR magnitudes are plotted against the nuclear blue magnitudes in Fig. 2. We see from this figure that the host spirals of low-luminosity AGN (nuclear magnitudes $M_B > -23$ mag, $H_0 = 80$ km/s/Mpc) span a range of masses centered around that of an L^* galaxy (i.e. a galaxy at the knee of the Schechter luminosity function). Some of the nearby Seyferts are even found in dwarf galaxies. However, for more luminous AGN, there is a correlation between host mass and nuclear luminosity. Apparently, there is a minimum host mass required to create or sustain very high activity.

3 Hubble Space Telescope Results

As Fig. 2 shows, the highest luminosity quasars are generally not found in galaxies with masses less than that of an L^* galaxy. It therefore came as a big surprise when early reports from the repaired HST indicated that host galaxies were not visible for many of the nearby quasars imaged with WFPC2. Simulations suggested that, depending on the type of galaxy, the hosts could

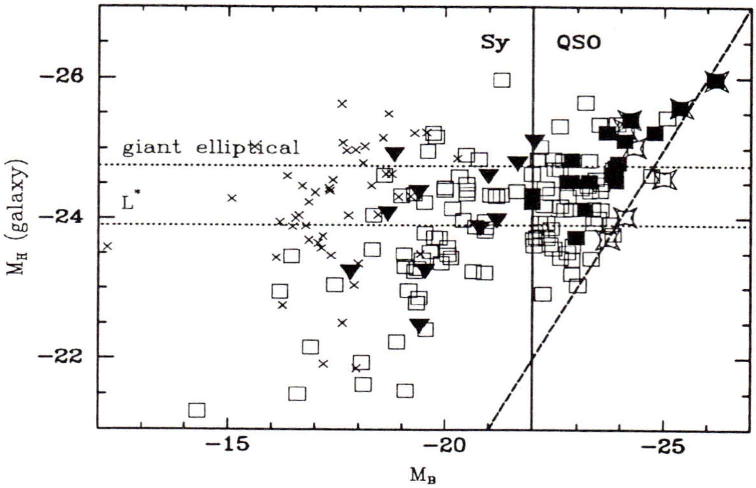

Fig. 2. Nuclear v. host galaxy luminosity for Seyferts and quasars (see McLeod and Rieke 1995a for details). The highest luminosity quasars are generally not found in low-luminosity galaxies.

have been detected down to about 1 mag fainter than the V magnitude of an L^* galaxy (Bahcall, Kirhakos, and Schneider 1994, 1995a).

To try and reconcile the apparent conflict between the HST and our near-IR data, we have taken a closer look at the eight objects in Bahcall et al. (1995a–these objects are denoted by stars in our Fig. 2). We already had H-band images for all eight of these objects. We have also obtained new, deep K-band images for two of the quasars. Together with our IR data, we have analyzed some of the HST data using images from the HST Archive and radial luminosity profiles provided by John Bahcall and his collaborators. The full results of our analysis may be found in McLeod and Rieke (1995b). Here we can briefly summarize the results as follows:

1) HST's complicated point spread function (saturation within in the central 0.5" radius, bleeding, diffraction spikes) makes removal of the bright nucleus difficult in the WFPC2 images. Bahcall et al. have done an excellent job of testing the PSF removal; from their tests, we conclude that the residual light does indeed show the host galaxies.

2) The HST images do not go as deep as the near-IR images; it is more difficult to see the faint, outer parts of the galaxies in the HST images. However, if we stretch the contrast to the limit, we find that the fuzz shape in the HST images shows good correspondence to that in the near-IR images.

3) If the HST images are smoothed via a 1-D radial profile analysis to boost the signal-to-noise in the faint, outer parts of the galaxies, the resulting

host magnitudes are consistent with those determined in the near-IR for normal to slightly red galaxies (caveat: there are large uncertainties in both!).

4) The H-K colors of the fuzz are also consistent with starlight and inconsistent with scattered nuclear light.

5) The fact that HST did not see sharp, high-surface-brightness features (e.g. spiral arms) for these galaxies, combined with our results suggest that most of these quasars live in **luminous, smooth, early-type** galaxies.

6) Recently, several groups have obtained images of several more quasars with WFPC2: Bahcall et al. (1995b,c); Hutchings and co- workers (Hutchings et al. 1994, Hutchings and Morris 1995); and Disney et al. (1995). While a variety of host galaxies are detected (one spiral and several peculiar systems), the data indicate a strong preference for the highest-luminosity AGN to live in early-type galaxies. **Near-IR imaging provides an excellent way to detect these smooth hosts.**

References

Bahcall, J. N., Kirhakos, S., and Schneider, D. P. (1994): HST Images of Nearby Luminous Quasars. ApJL. **435**, L11

Bahcall, J. N., Kirhakos, S., and Schneider, D. P. (1995a): HST Images of Nearby Luminous Quasars II: Results for Eight quasars and Tests of the Detection Sensitivity. ApJ. in press

Bahcall, J. N., Kirhakos, S., and Schneider, D. P. (1995b): PKS 2349−014: A Luminous Quasar with Thin Wisps, a Large Off-Center Nebulosity, and a Close Companion Galaxy. ApJL. **447**, L1

Bahcall, J. N., Kirhakos, S., and Schneider, D. P. (1995c): The Apparently Normal Galaxy Hosts for Two Luminous Quasars. ApJ. submitted

Disney, M. J. et al. (1995): Interacting Elliptical Galaxies as Hosts of Intermediate-Redshift Quasars. Nature. **376**, 150

Elvis, M. et al. (1994): Atlas of Quasar Energy Distributions. I. The Data. ApJS. **95**, 1

Hutchings, J. B., Holtzman, J., Sparks, W. B., Morris, S. C., Hanish, R. J., and Mo, J. (1994): HST Imaging of Quasi-Stellar Objects with WFPC2. ApJL. **429**, L1

Hutchings, J. B., and Morris, S. C. (1995): Imaging of Low-Redshift QSOs with WFPC2. AJ. accepted

McLeod, K.K., and Rieke, G.H. (1994a): Near-Infrared Imaging of Low-Redshift Quasar Host Galaxies. ApJ. **420**, 58

McLeod, K.K., and Rieke, G.H. (1994b): Near-Infrared Imaging of Low-Redshift Quasar Host Galaxies. II. High-Luminosity Quasars. ApJ. **431**, 137

McLeod, K.K., and Rieke, G.H. (1995a): Near-Infrared Imaging of CfA Seyfert Galaxies. ApJ. **441**, 96

McLeod, K.K., and Rieke, G.H. (1995b): Luminous Quasars in Luminous Early-Type Host Galaxies. ApJL. submitted

NIR view of the nuclear region of NGC 1808

J.K. Kotilainen[1], D.A. Forbes[2], A.F.M. Moorwood[3], P.P. van der Werf[4] and M.J. Ward[5]

[1]Tuorla Observatory, University of Turku, FIN–21500 Piikkiö, Finland
[2]Lick Observatory, University of California, Santa Cruz, CA 95064, U.S.A.
[3]European Southern Observatory, D–85748, Garching bei München, Germany
[4]Leiden Observatory, P.O. Box 9513, NL–2300 RA Leiden, The Netherlands
[5]Astrophysics, Nuclear Physics Building, Keble Road, Oxford OX1 3RH, UK

Abstract.

We present high spatial resolution near-infrared imaging of the nuclear region of the starburst galaxy NGC 1808. The continuum emission is smooth throughout the circumnuclear region and shows no morphological change with wavelength. Most of the continuum is produced by the evolved bulge population, while red supergiants and hot dust only make a small contribution. The circumnuclear line emission arises from hot spots in a ring-like structure. From comparison of Brγ and Hα fluxes, we derive extinctions between A_V of 3 and 5 towards the hot spots. From analysis of the line and continuum luminosities, using an evolutionary starburst model, we derive for each hot spot star formation rate 0.1–0.6 $M_\odot$ yr^{-1} and supernova rate 0.4–11 x 10^{-3} yr^{-1}. The age of the burst is between 8 and 17 Myr in the circumnuclear region, but much older, ~40 Myr, in the nucleus. The line ratios of the nucleus are similar to Seyfert nuclei, and the multiwavelength evidence for and against hidden Seyfert activity in NGC 1808 is discussed.

1 Introduction and observations

The nuclear region of the starbursting spiral galaxy NGC 1808 (16.4 Mpc, $1'' \sim 80$ pc) contains hot spots and a superwind into the halo (Phillips 1993). The nucleus is surrounded by compact radio sources (Collison et al. 1994) powered by supernova remnants (SNR). NGC 1808 has a central molecular ring (Koribalski et al. 1993) at the Inner Lindblad Resonance (ILR). Moorwood & Oliva (1988) first detected the near-infrared (NIR) Brγ , H_2 1-0

S(1) 2.121 μm and [Fe II] 1.644 μm lines from the nucleus of NGC 1808. Krabbe, Sternberg & Genzel (1994, hereafter KSG) first imaged the Brγ and H_2 lines and 2.15 μm continuum. Here we present higher spatial resolution ($\sim$1.0″ FWHM) images of NGC 1808 in the broad-band JHK continuum and Brγ and H_2 lines (observed with the IRAC2B camera on the 2.2m telescope at ESO, La Silla, in January 1994) and the [Fe II] line (observed with the IRIS on the 3.9m AAT in January 1994), and analyse the morphology, extinction, star formation and the possible presence of Seyfert activity in the nuclear region of NGC 1808. For full discussion, see Kotilainen et al. (1995).

2 Results

The NIR continuum emission is smooth, with no counterparts to the optical hot spots. There is no change in morphology from J to K band, indicating that red supergiants (RSG) and hot dust do not dominate the NIR emission. The Brγ , H_2 , [Fe II] and 6cm radio maps (Fig. 1) all show hot spots throughout the circumnuclear region, but there are differences in the detailed morphology, indicating that some peaks are dominated by SNRs while others are HII regions. Our Brγ map generally agrees with the map of KSG, but there are some differences, probably due to our higher spatial resolution.

There is little correlation between the Brγ and Hα peaks due to high extinction. It was estimated by assuming that it is caused by dust absorption in front of the emission regions. Using standard interstellar dust, case B recombination, and the observed Hα and Brγ fluxes, the resulting A_V is between 2.6 and 4.8 for the hot spots. For the continuum, we compared the observed J-H colour of the hot spots of NGC 1808 to the mean colour of normal galaxies and attributed any deviation to extinction. We derive A_V between 0.5 and 4.1 for the hot spots, lower than for the emission lines. Most likely, the intrinsic colours of the hot spots are bluer than in normal galaxies. In what follows, we correct both the line emission and the continuum for the extinction derived from the Brγ /Hα ratio.

3 Discussion

3.1 Star formation activity in NGC 1808

We have applied the Brγ and radio luminosities to the evolutionary starburst models of Leitherer & Heckman (1995, hereafter LH95) to study the star forming properties and ages of the hot spots. First, the Brγ luminosities give the number of ionizing photons in each hot spot. The star formation rate (SFR) was then derived from comparison of these values to Fig. 38 in LH95. Assuming constant star formation with IMF slope 2.35, and upper mass limit 30 $M_\odot$, we derive SFR = 0.1 – 0.6 $M_\odot yr^{-1}$ per hot spot. Using the relation between the nonthermal radio emission and SN rate (Condon &

Yin 1990), and the 5 GHz fluxes and spectral indices (Collison et al. 1994), we derive SN rates between 0.4 and 11 x 10^{-3} yr^{-1} per hot spot. These values are upper limits, since part ($<$ 25%, based on Brγ) of the radio emission may be due to thermal emission. From the SN rates, with the above assumptions, we estimated the age of the burst for each hot spot (Fig. 6 of LH95). All the circumnuclear hot spots are young, between 8 and 17 Myr, whereas the nucleus is much older, $\sim$40 Myr. Note that the L/M ratio and the Brγ equivalent width of the nucleus of NGC 1808 (Oliva et al. 1995) are in the lower end of starburst galaxies, providing independent support to a decaying nuclear burst.

Our SFRs are in good agreement with those derived by KSG for the nucleus and most hot spots, while our SN rates are always lower. Our age estimate agrees well with KSG for the youngest hot spots and the nucleus, but is lower for the other hot spots. Note that even if we have overestimated the SN rates (see above), the burst ages are only overestimated by 1–2 Myr. On the other hand, if the hot spots were much older than 10 Myr, RSGs should dominate the continuum, whereas they contribute not more than 10 % in the K band (Tacconi-Garman et al. 1995).

3.2 The nuclear component

While the circumnuclear hot spots are located in the area occupied by starbursts in the [Fe II] /Brγ vs. H_2 /Brγ diagram, the nucleus lies in the area of Seyfert nuclei. This may be simply due to relative lack of hot ionizing stars and implies the presence of an old, evolved, SN-rich starburst (see above). The line ratios may also result from higher density or dust content in the nuclear ISM compared to the circumnuclear region. The ionizing photons, instead of being radiating in hydrogen recombination lines, would be absorbed by dust grains or escape altogether. Alternatively, there may be additional H_2 and [Fe II] emission from shocked gas and streaming motions (Hollenbach & McKee 1989).

On the other hand, an obscured Seyfert may contribute to the excitation of H_2 and [Fe II] in the nucleus. The possibility of a weak Seyfert nucleus in NGC 1808 was suggested by Véron-Cetty & Véron (1985) based on broad Hα profiles. Phillips (1993) attributed them to the strong underlying Balmer absorption and claimed that the nuclear activity is better explained by young stars and SNRs. Against the Seyfert hypothesis, the nucleus falls onto the correlation (Forbes & Ward 1993) between [Fe II] and radio emission for starbursts, indicating that the nuclear emission is due to SNRs, as in the hot spots. However, while the strong soft X-ray emission from NGC 1808 is extended, the hard X-ray source is compact and highly absorbed (Awaki & Koyama 1993), indicating a Seyfert. Another argument for the presence of a Seyfert is the presence of hot dust in the K band light of the nucleus. This is common in Seyfert nuclei (Kotilainen et al. 1992) but in the starbursts model the dust is heated by hot stars, while we have argued above that there is a deficiency of hot stars in the nucleus.

The real nature of the nucleus of NGC 1808 is still open. Whether an old decaying starburst or an obscured Seyfert, there are evolutionary implications. The nuclear starburst may have triggered the younger burst in the circumnuclear region through shocks from the superwind. Alternatively, the formation of a dust obscured Seyfert may be the result of feeding the nucleus by gas driven inward by the starburst at the ILR.

4 Conclusions

We have presented high spatial resolution NIR line and continuum imaging of the circumnuclear region of the starburst galaxy NGC 1808. The continuum emission shows smooth morphology throughout the circumnuclear region, with no morphological change with wavelength. Most of the continuum is produced by the old bulge stellar population, and red supergiants and hot dust can only make a small contribution.

All emission line images show extended emission produced in several regions, most clearly resolved in Brγ . Both Brγ and [Fe II] are well correlated with the radio emission, but there are differences in their detailed morphology that can differentiate between hot spots dominated by SNRs and HII regions. From analysis of the Brγ and Hα fluxes, we derive foreground extinctions of A_V between 3 and 5 for the hot spots.

We have used the line and continuum luminosities with evolutionary starburst models to derive for each hot spot star forming rate 0.1–0.6 $M_\odot$ yr^{-1}, and SN rate 0.4–11 x 10^{-3} yr^{-1}. The age of the current burst is constrained to be ~40 Myr in the nucleus, and much younger, between 8 and 17 Myr, in the circumnuclear region. The circumnuclear hot spots lie in the area of starbursts in the [Fe II] /Brγ vs. H_2 /Brγ diagram, whereas the nucleus has similar line ratios to Seyfert nuclei. The multiwavelength evidence does not decisively prove for or against the presence of a hidden Seyfert nucleus. Finally we discuss the evolutionary implications of an old nuclear burst or a hidden Seyfert.

References

Awaki,H, Koyama,K., 1993, Adv. Space Res. 13, 221
Collison,P.M., Saikia,D.J., Pedlar,A. et al. , 1994, MNRAS 268, 203
Condon,J.J., Yin,Q.F., 1990, ApJ 357, 97
Forbes,D.A., Ward,M.J., 1993, ApJ 416, 150
Forbes,D.A., Boisson,C., Ward,M.J., 1992, MNRAS 259, 293
Hollenbach,D., McKee,C.F., 1989, ApJ 342, 306
Koribalski,B., Dickey,J.M., Mebold,U., 1993, ApJ 402, L41

Kotilainen,J.K., Ward,M.J., Boisson,C., DePoy,D.L., Smith,M.G., 1992, MNRAS 256, 149
Kotilainen,J.K., Forbes,D.A., Moorwood,A.F.M., van der Werf,P.P., Ward,M.J., 1995, A&A submitted
Krabbe,A., Sternberg,A., Genzel,R., 1994, ApJ 425, 72 (KSG)
Leitherer,C., Heckman,T.M., 1995 ApJS 96 38 (LH95)
Moorwood,A.F.M., Oliva,E., 1988, A&A 203, 278
Oliva,E., Origlia,L., Kotilainen,J.K., Moorwood,A.F.M., 1995, A&A 301, 55
Phillips,A.C., 1993, AJ 105, 486
Tacconi-Garman,L. et al, 1995, these proceedings
Véron-Cetty,M.P., Véron,P., 1985, A&A 145, 425

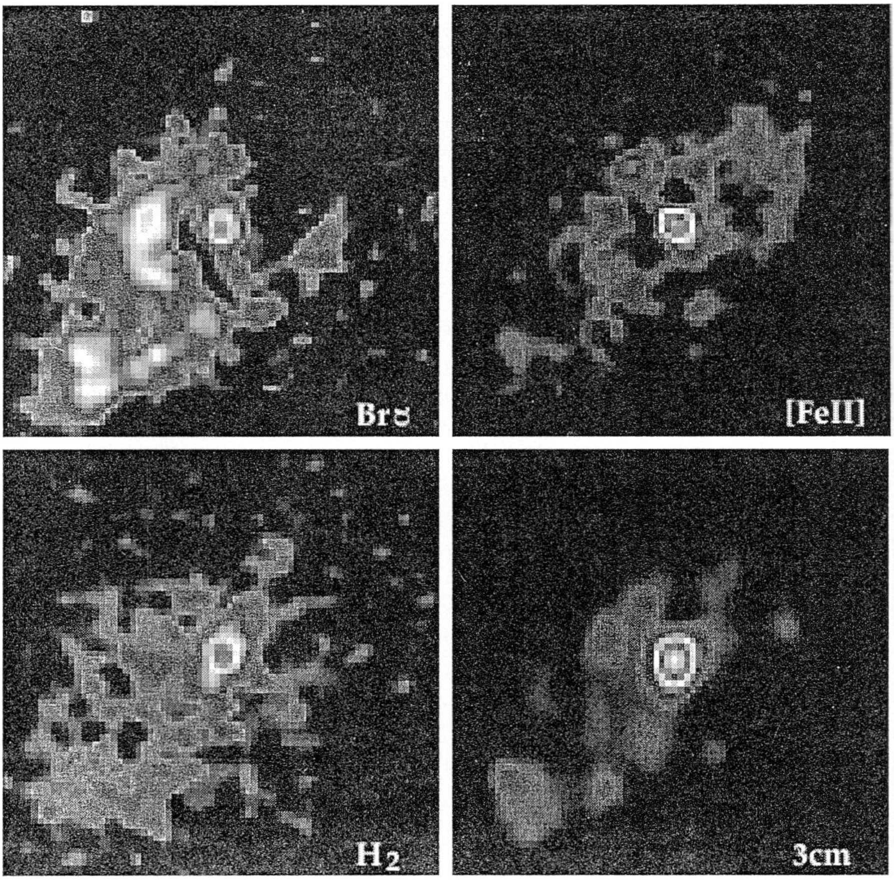

Fig. 1. Images of the nuclear 20 arcsec (~1.6 kpc) region of NGC 1808. **a):** The Brγ 2.166 μm line. **b):** The H_2 1-0 S(1) 2.121 μm line. **c):** The [Fe II] 1.644 μm line. **d):** The 5 GHz (6cm) radio emission (Collison, priv. comm).

Revealing the origin of nuclear starbursts in late-type spirals: IC 342

T. Böker, N. Förster, R. Genzel, A. Krabbe, L.E. Tacconi-Garman and M. Tecza

Max-Planck-Institut für Extraterrestrische Physik (MPE),
Giessenbachstrasse, 85740 Garching, Germany

Abstract. We have used the new MPE imaging spectrometer 3D to observe the central 12 arcsec of the nearby Scd spiral IC342 in K-band. The data, with a resolution of 1.2 arcsec spatially and 300km/s spectrally show rich structure and compare well with radio and sub-mm maps. Linemaps in Br γ, H_2 $(1-0)$ and 2.06 μm HeI reveal the circumnuclear ring / spiral arm system which consists of various hot spots, presumably regions with enhanced star formation. These measurements seperate between regions of different starburst evolution.

1 Introduction and results

At the MPE a program has been started to investigate many of the problems related to the detailed history and evolution of the centers of spiral galaxies, some of which are:

1. What triggers the starburst activity?
2. Is the whole central starburst region always composed of many smaller star-forming sites?
3. How does the continuum emission split between stars and thermal emission from dust and/or gas?
4. What is the exact composition of the stellar population in the starburst(s)?

As a start to this program we have chosen the nearby Scd spiral IC342. With a linear size of 30 kpc and an absolute blue magnitude of $M_B = -21.4$, IC342 is very similar to the Milky Way. Because of its small distance of 1.8 Mpc (McCall 1989) and its almost face-on orientation, IC342 is well suited for a high-resolution view in its inner 100 pc.

The data presented here were taken in January 1995 at the Calar Alto 3.5 m telescope. We took 45 frames with 100 s on-source integration time each. Standard methods of correction for non-linearities, bad pixels and atmospheric absorption were applied, the absolute flux calibration was done by using the value given by Becklin (1980).

The quantitative results are presented in Table 1, while Fig. 1 shows the spatial distribution of gas emission as well as some spectra of specific boxes in the nuclear vicinity. A more detailed analysis of these results will be given in a future paper (Böker 1996), together with a complete dataset in H-band, also taken with 3D. However, one can already say that the quality of our spectra will enable us to put strong constraints on the stellar population in the central cluster. This is demonstrated in Fig. 3 where a comparison is made between the spectra of the central cluster in IC342 and that of the K5Ib star HR8726 taken from Kleinmann & Hall (1986). The striking agreement shows that the contribution to the continuum light from gas and dust is negligible in this region.

Table 1. Quantitative results of three of the four regions indicated in Fig. 1d. The spectroscopic CO-index CO_{SP} was defined by Doyon et al. (1994)

Region	Center	East	West
Offset (arcsec)	(0,0)	(-0.5,2.5)	(-0.75,-3.75)
Aperture size	(1.25 x 1.25)"	(2.0 x 2.75)"	(2.75 x 2.0)"
total integration time	4500s	2700s	1800s
K-flux (2.0-2.4 μm)	$6.5x10^{-15}$	$2.1x10^{-15}$	$2.3x10^{-15}$
Spectral index ($F \propto \lambda^{\beta}$)	-2.40	-1.91	-2.03
$Br\gamma$ flux (Wm^{-2})	$< 5.2x10^{-19}$	$6.6x10^{-18}$	$1.8x10^{-17}$
H_2(1-0) flux (Wm^{-2})	–	$2.5x10^{-18}$	$4.1x10^{-18}$
HeI flux (Wm^{-2})	–	$2.0x10^{-18}$	$1.1x10^{-17}$
CO_{SP}	0.307	0.182	0.183

References

Böker, T. et al., 1996, in preparation
Becklin, E.E. et al., 1980, ApJ, **236**, 441
Doyon, R., Joseph, R.D. and Wright, G.S., 1994, ApJ, **421**,101
Ishizuki, S. et al., 1990, Nature, **344**, 224
Kleinmann, S.G. and Hall, D.N.B., 1986, ApJ,**62**, 501
McCall, M.L., 1989, AJ, **97**, 1341
Turner, J.L. and Ho, P.T.P., 1983, ApJ, **286**, L79

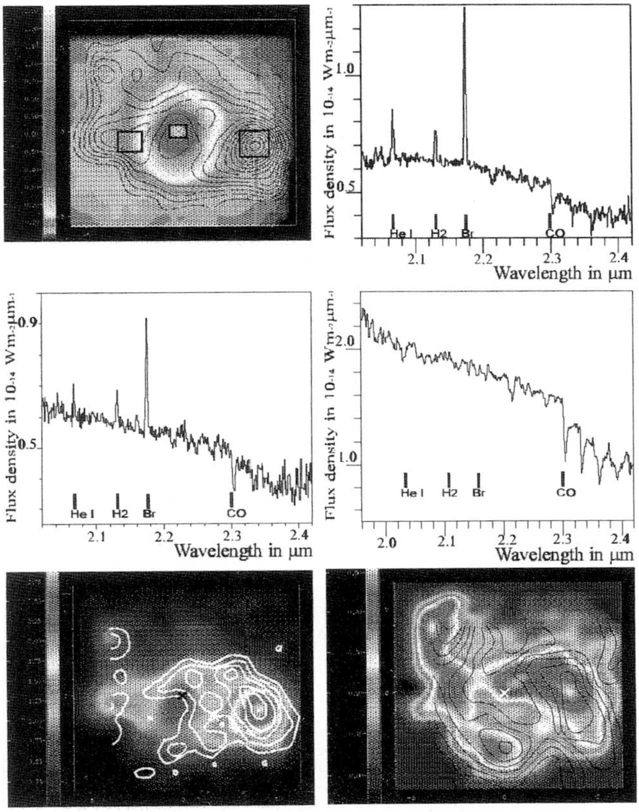

Fig. 1. top left: K-band image of the central 12" of IC342 (colour) vs. Br γ emission (contours). Crosses indicate the dynamical center (Becklin, 1980) top right and middle: Spectra of regions indicated by the boxes in the K-band image bottom left: Br γ emission (colour) vs. 2cm radio continuum map (Turner & Ho, 1983) (contours) bottom right: H_2 $(1-0)$ emission (colour) vs. ^{13}CO (Ishizuki et al. 1990) (contours)

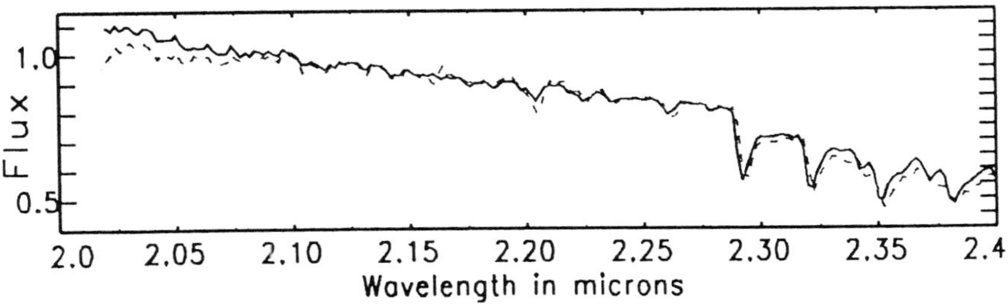

Fig. 2. Comparison between the spectrum of the central 1.25" of IC342 (dashed line) and that of a K5Ib template from the Kleinmann & Hall catalogue, convolved to 3D's resolution and multiplied with a black-body curve of T = 4200 K (solid line).

The nature of nuclear starbursts: M82

N. M. Förster, T. Böker, A. Krabbe, N. Thatte, H. Kroker and R. Genzel

Max-Planck-Institut für Extraterrestrische Physik
Giessenbachstraße, 85740 Garching, Germany

Abstract. We have obtained new K-band observations of the nuclear regions of the starburst galaxy M82, with a spatial resolution of 1.5 $''$ and a spectral resolution of 300 km/s. Maps of the Brγ and H_2 $1-0$ $S(1)$ line emission and of the CO index are presented, as well as spectra of the nucleus and two bright Brγ sources.

1 Introduction

As part of a program to determine the physical conditions of the gas and to characterize the stellar population in M82, we have observed this starburst galaxy in the K-band using the new MPE imaging spectrometer 3D. With a spectral resolution of $\lambda/\Delta\lambda \sim 1000$ and a spatial resolution of 1.5 $''$, the data allow a detailed spectroscopic study on scales as small as 25 pc (at the distance of M82, 3.3 Mpc (Freedman and Madore 1988), $1'' \approx 15$pc). More specific questions concerning the evolution and history of the star formation activity in the nuclear regions of M82 can now be addressed, for example:

· Is the whole central starburst region composed in fact of many smaller individual starburst regions?
· What is the contribution of hot dust emission in the starbursting regions?
· What is the composition of the starburst(s) stellar population?

2 Results

Fig. 1a shows the K-band image of the central regions of M82 that were observed, with superposed contours of the Brγ line emission at 2.166 μm. The morphology of the Brγ emission, tracing the most intense star-forming regions, differs clearly from that of the broad-band emission in K to which late-type stars contribute the most. It shows several clumps offset from the center of M82 and relatively faint emission is detected at the nucleus. The HeI $\lambda 2.058$ μm emission, also associated with regions of active star formation (not shown here), follows closely that of Brγ as expected. In Figs. 1b, 1c and 1d, spectra taken at the nucleus and at the two brightest Brγ clumps are presented. The most striking differences are seen in the CO bands longwards of 2.3 μm. The quality of the data allows comparison with catalogues of stellar spectra such as the Kleinmann and Hall (KH 1986) atlas. To illustrate this, the spectrum of a K5Ib star (KH 1986) is plotted as a dotted line in Fig. 1b along with that of M82's nucleus; both compare remarkably well.

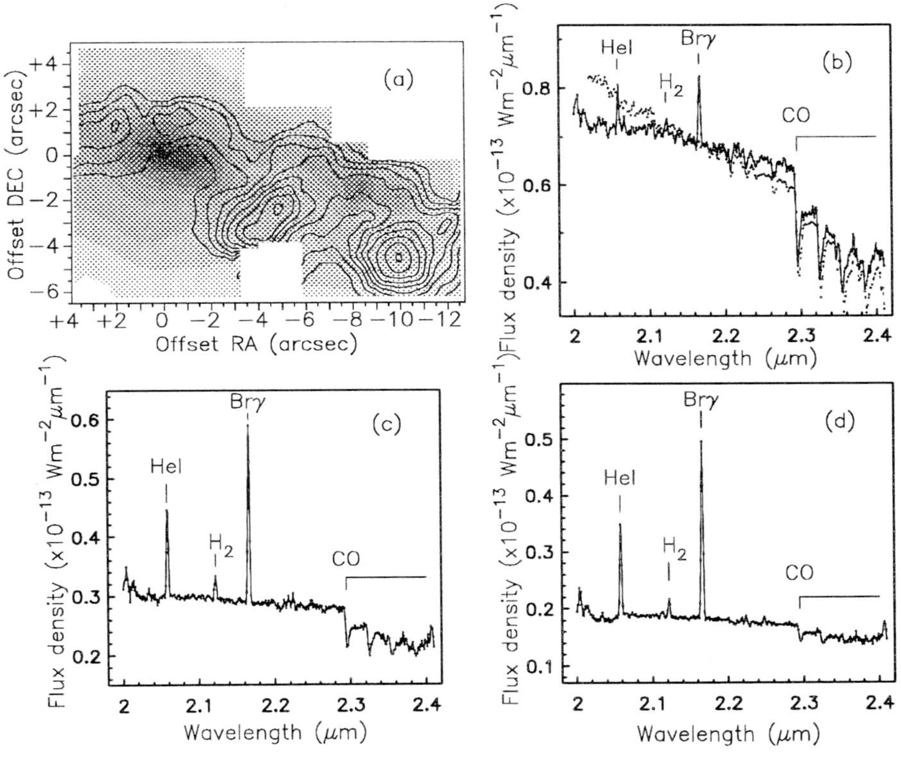

Fig. 1. (*a*) *K*-band image plotted from 0.5 (white) to 5.0 (black) ($\times 10^{-15}$Wm^{-2} μm^{-1}arcsec^{-2}) with Brγ contours in steps of 1.5 starting at 7.5 ($\times 10^{-18}$Wm^{-2} arcsec^{-2}). The nucleus is at (0,0). (*b*) Spectrum of the nucleus taken in a 2$''$ aperture (full line), plotted along with the spectrum of a K5Ib star (KH 1986) normalized to match the M82's nuclear spectrum at 2.2 μm (dotted line). (*c*), (*d*) Spectra of the Brγ sources located at (-4.5,-2.5) and (-10,-4.5), in a 2$''$ aperture.

Fig. 2a shows the H_2 $1-0$ $S(1)$ line emission at 2.122 μm overlaid with Brγ contours. The H_2 emission appears mostly diffuse and differs from the broad-band, the Brγ and the HeI emission. In particular, it can be noted that little H_2 emission is detected at the location of the brightest Brγ clump. In Fig. 2b, the map of the CO spectroscopic index (CO_{sp}, Doyon *et al.* 1994) is presented along with Brγ contours. The CO_{sp} map, which share similarities with the K-band image but significant differences with the Brγ linemap, clearly shows spatial variations, with an important area of abnormally high CO index ($CO_{sp} > 0.30$ mag) reaching 0.33 mag at the nucleus.

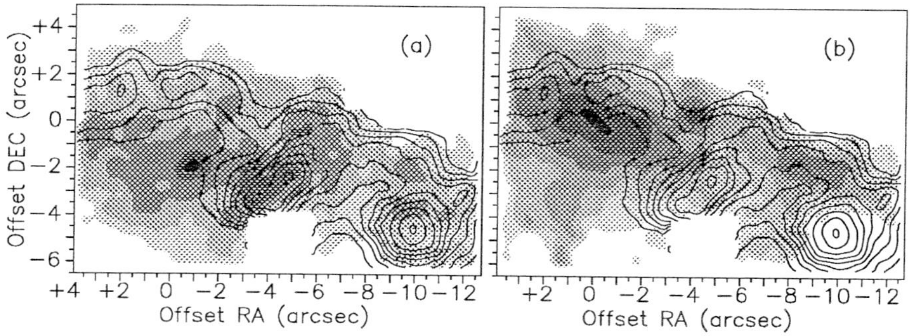

Fig. 2. (a) H_2 $1-0$ $S(1)$ linemap plotted from 1.0 (white) to 2.6 (black) ($\times10^{-18}$W m^{-2}arcsec^{-2}), with Brγ contours (levels as in Fig. 1a). (b) CO_{sp} map plotted from 0.16 mag (white) to 0.31 mag (black), with Brγ contours (levels as in Fig. 1a).

The irregular morphology of the Brγ and HeI emission, associated with the most intense star-forming regions, of the CO_{sp}, which is higher for red giants and supergiants, and the broad-band emission in K, as well as the differences between these distributions are probably due in part to non-uniform extinction effects and contamination by hot dust emission. They could also be explained in terms of age effects and thus point out at a complex starburst activity. However, extinction and hot dust emission have to be carefully accounted for before a definitive interpretation is made. More details will be presented in a subsequent paper (Förster *et al.* 1995).

References

Doyon, R., Joseph, R.D. and Wright, G.S., 1994, ApJ, **421**,101
Förster, N.M. *et al.* 1995, in preparation
Freedman, W.L. and Madore, B.F. 1988, ApJ **332**, L63
Kleinmann, S.G. and Hall, D.N.B. 1986, ApJS **62**, 501 (KH 1986)

Near-IR Imaging of the Central Regions in NGC 1068 and NGC 7552

A. Quirrenbach, E. Schinnerer, F. Prada, A. Eckart, M. Risse

Max-Planck-Institut für Extraterrestrische Physik

The Nuclear Bar in NGC 1068

NGC 1068 is the closest (d = 18 Mpc) and brightest Seyfert 2 galaxy. The infrared emission from NGC 1068 is dominated by a bright compact nucleus (500 mJy at 2.2 μm, $\geq 1.5 \times 10^{11}$ $L_\odot$), and a prominent stellar bar at a position angle of 45°, which extends out to radii of about 16″ (Scoville et al. 1988, Thronson et al. 1989).

We have imaged the central 19″ of NGC 1068 in the near-IR K-band with the MPE SHARP1 speckle camera in August 1992 at the ESO NTT in La Silla, Chile. A total of 4000 frames with 0.5 s exposure time were obtained; individual frames were coadded after recentering on the brightest pixel. An analysis of these “shift-and-add” images shows that the nuclear source is smaller than 0.″1. Our K-band image of NGC 1068 suggests that at radii larger than about 8″ the bar structure turns into two arm-like features. This radius is coincident with the inner radius of the circumnuclear gas / dust / star forming ring (e.g. Telesco and Decher 1988), suggesting an interpretation in terms of spiral arms that have formed at the inner Lindblad resonance near the end point of the bar. To further analyze the structure of the stellar bar, we have fitted ellipses to the isophotes. A decrease of the ellipticity towards smaller radii, and an isophote twist by about 20° are apparent. Similar isophote twists have been observed within the central regions of the barred Seyfert galaxies NGC 1097 and NGC 5728 (Shaw et al. 1993). Such structures, called bars within bars, have been proposed as a possible gas feeding mechanism in active galactic nuclei (Shlosman et al. 1989, Friedli and Martinet 1993). A knot about 0.″6 east of the nucleus is found in an image deconvolved with a modified Lucy algorithm (Hook et al. 1994). The knot is coincident with a peak in the 10 μm emission (Cameron et al. 1993), and with extended emission of excited H_2 (Blietz et al. 1994); it may thus be due to warm dust associated with the narrow-line clouds.

The Circumnuclear Starburst in NGC 7552

NGC 7552 is a southern infrared-bright galaxy at a distance of about 32.7 Mpc ($H_0 = 50\,km\,s^{-1}\,Mpc^{-1}$) (Forbes et al. 1994a). It is described as a nearly face-on ($i = 28°$) barred galaxy (Feinstein 1990, Forbes et al. 1994b) and classified in the RSA catalog as an object of type SBbc(s), and as a HII-region galaxy (Ward et al. 1980) or a LINER (Durret & Bergeron 1987). Forbes et al. (1994a) found in radio maps a 1 kpc starburst ring, which is not visible in single color optical images mainly because of dust. They estimated that the starburst occurred about 5×10^7 yrs ago. A large concentration of molecular gas in the central 45″ was observed in the $^{12}CO(1\text{-}0)$-line by Claussen et al. (1992), also suggesting a nuclear starburst.

We have obtained near-infrared JHK images with the MPE SHARP1 speckle camera in August 1992 at the ESO NTT. The images show a bright nuclear component, with spiral arms indicated. The color maps very clearly show the red, extincted circumnuclear starburst region. The color maps from previous NIR-data (Forbes et al. 1994a) did not show a NIR-bright reddened core on the scale of about 1″, which is usually expected for galaxies with nuclear activity. In our J-K color map we can for the first time detect a weakly reddened central component. The colors of individual knots in the starburst ring indicate that they can be interpreted as stellar clusters with normal intrinsic NIR colors. A knot 2.″05 south and 1.″90 east of the nucleus is of special interest. At 2.2 μm it is the brightest knot and its colors can be explained by a combination of reddened light from late type stars with a strong contribution of hot dust. A closer analysis of this starburst nucleus using data from our NIR imaging spectrometer 3D is in progress.

References

Blietz, M. et al., 1994, ApJ 421, 92
Cameron, M. et al., 1993, ApJ 419, 136
Claussen, M.J., Sahai, R., 1992, AJ 103, 1134
Durret, F., Bergeron, J., 1987, A&A 173, 219
Feinstein, C., Vega, I., Mendez, M., Forte, J.C., 1990, A&A 239, 90
Forbes, D.A., Norris, R.R., Williger, G.M., Smith, R.C., 1994a, AJ 107, 984
Forbes, D.A., Kotilainen, J.K., Moorwood, A.F.M.,1994b, ApJ 433, L13
Friedli, D., Martinet, L., A&A 277, 27
Frogel, J.A., Persson, S.E., Aaronson, M., Matthews, K., 1978, ApJ 220, 75
Hook, R., Lucy, L., Stockton, A., Ridgway, S., 1994, ST-ECF Newsletter, 21, 17
Scoville, N.Z., Matthews, K., Carico, D.P., Sanders, D.B., 1988, ApJ 327, L61
Scoville, N.Z. et al., 1985, ApJ 289, 129
Shaw, M.A., Combes, F., Axon, D.J., Wright, G.S., A&A 273, 31
Shlosman, I., Frank, J., Begelman, M.C., Nature, 338, 45
Telesco, C.M., Decher, R., 1988, ApJ 334, 573
Thronson, H.A. et al., 1989, ApJ 343, 158
Ward, M., Penston, M.V., Blades, J.C., Turtle, A.J., 1980, MNRAS 193, 563

Near IR and mm Imaging Spectroscopy of the Nuclear Region of NGC 1068

N. Thatte, L. Tacconi, H. Kroker, A. Krabbe, L.E. Tacconi-Garman, R. Genzel

Max-Planck-Institut für extraterrestrische Physik,
Postbox 1603, D-85740 Garching, Germany

Probing the central engine: New results from imaging spectroscopy in the millimeter and near infrared

The central few arc seconds of the prototypical Seyfert 2 galaxy, NGC 1068, have been the subject of intense study ever since Antonucci and Miller (1985) first reported the presence of broad lines in the polarized optical spectrum, corroborating the existence of a hidden Seyfert 1 component. However, emission at optical wavelengths is heavily affected by differential extinction effects, which make it difficult to determine the intrinsic distribution of emitting material.

We present sub arc-second imaging spectroscopy in the K band (in the lines of [Si VI] (1.962 μm) and the H_2 (S(1) transition, 2.122 μm)) obtained with the MPE instruments 3D and ROGUE, (Weitzel et al. 1993,), as well as high spatial and velocity resolution interferometric data in the lines of HCN (J=1$\rightarrow$0) and ^{13}CO from the IRAM interferometer. (Sternberg et al. 1996, Tacconi et al. 1994)

The high sensitivity ^{13}CO data clearly show that the "starburst ring" in fact consists of two tightly wound spiral arms (fig. 1a). The HCN J=1$\rightarrow$0 line, which traces dense molecular gas, shows a high concentration of molecular material in the central few arc seconds (fig. 1a). We have aligned the millimeter and near infrared reference frames, based on the relative justification of Gallimore et al. (1995) between the HST optical continuum and the centimeter radio, and astrometry (Thatte et al. 1996) done with the MPE SHARP II camera and the ESO COME-ON+ system. Figure 1b shows an

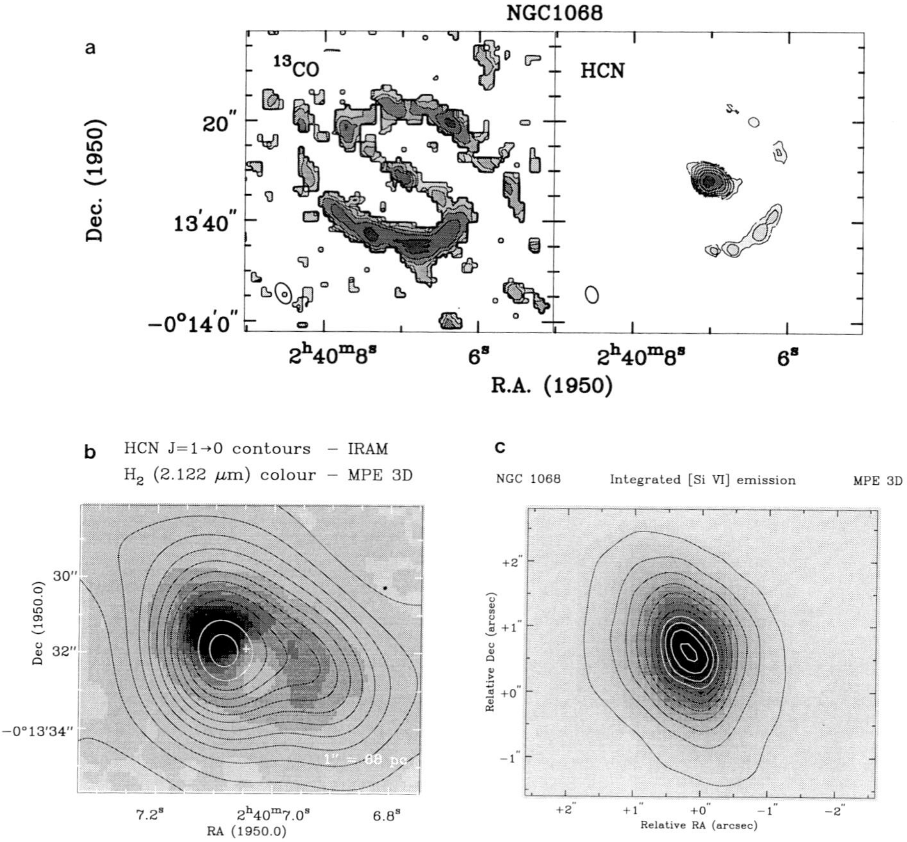

Fig. 1. Following page (a) Velocity integrated emission maps in the lines of ^{13}CO and HCN (J=1→0), obtained with the IRAM interferometer. The interferometer beam is shown in the lower left hand corner of each sub-panel. **(b)** Gray scale rendering of the emission in the H_2 S(1) line at 2.122 μm with superposed contours of the millimeter HCN emission. The K band continuum peak is marked by a +. **(c)** Integrated emission in the [Si VI] line at 1.962 μm (contours and gray scale). The K continuum peak is marked with a +. The emission is extended in the NE direction.

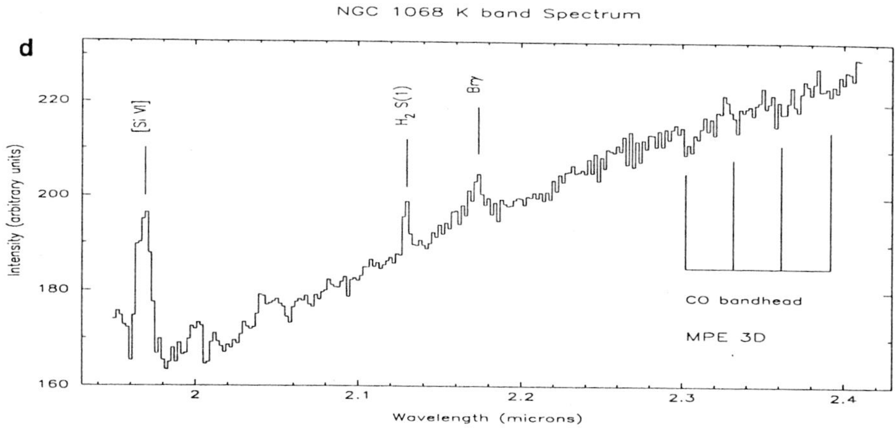

(d) Nuclear spectrum of NGC 1068 obtained with the MPE 3D imaging spectrometer. The various features are marked.

overlay of the dense molecular material traced in the H_2 S(1) line and in the millimeter HCN transition. The position of the K band peak, which should indicate the hidden nucleus (Thatte et al. 1996), is marked with a +. Both tracers indicate that the molecular material is asymmetrically distributed w.r.t. the nucleus.

We have detected extended coronal line emission in [Si VI] in the ENLR of NGC 1068. The spatial extent of the emission is shown in figure 1c, where a + marks the location of the K continuum peak. The morphology of the [Si VI] emission is similar to that of [O III] emission (Macchetto et al. 1994), lending credence to models in which the coronal line gas is excited by photoionizing continuum from the central engine. The [Si VI] line is also spectrally resolved, with kinematic structure similar to that of the [O III] emission.

References

Antonucci, R.R.J. & Miller, J.S. (1985): ApJ, **297**, 621.

Gallimore, J.F., Baum, S.A. & O'Dea, C.P. (1995): ApJ, submitted

Macchetto, F., Capetti, A., Sparks, W.B., Axon, D.J. & Boksebberg, A. (1994): ApJ, **435**, L18

Sternberg, A. et al. (1996): in preparation.

Tacconi, L.J., Genzel, R., Blietz, M., Cameron, M., Harris, A.I. & Madden, S. (1994): ApJ,**426**,L77

Thatte, N. et al. (1996): in preparation.

Weitzel, L., Cameron, M., Drapatz, S.,Genzel, R. & Krabbe, A. (1993): *Proceedings of the Los Angeles Conference* on " Infrared Astronomy with Arrays: The Next Generation", UCLA, July 1993. *Experimental Astronomy*, editors Ian. S. McLean and George Brims, (Dordrecht:Kluwer) 1994.

NIR Imaging Spectroscopy of F10214: Evidence for a Starburst Around an AGN at Z = 2.284

H. Kroker, R. Genzel, A. Krabbe, L.E. Tacconi-Garman, M. Tecza, N. Thatte

MPI für extraterrestrische Physik (MPE), Garching, Germany

Abstract. We present 1" imaging spectroscopy of the galaxy F10214. ¿From the distribution of the Hα and [NII] lines we find strong evidence for a starburst around the point like Seyfert 1.9 nucleus.

1 Introduction

F10214 seemed to be one of the most luminous objects currently known. The total luminosity of $> 3 \cdot 10^{14} L_{\odot}$ (e.g. Rowan–Robinson et al. 1991) can be explained by the magnification effect of a gravitational lens (Broadhurst & Lehár 1995). We observed F10214 in the K-band with 3D, the MPE imaging spectrometer. 3D obtains 256–channel spectra simultaneously for everyone of a 16x16 pixel field on the sky (0."5/pixel, R=1100; more details in Genzel et al. 1995).

2 Results and Discussion

Beside the bright Hα+[NII] lines, some weaker lines are visible the spectrum (Fig 1). We find different morphology of Hα, [NII] and the line–free K–continuum (Fig. 2): K–continuum appears in two components. The southern source is about 3 times brighter. We could not detect any line in the northern source. Hα is more extended in the EW direction than [NII]. This becomes more prominent in [NII]/Hα line ratios of increasing spatial regions. We deconvolved the spectra of the blended Hα+[NII] lines with four single gaussians for 6 spatial regions: the central 0."75, 1."25, 1."75, 2."25 and

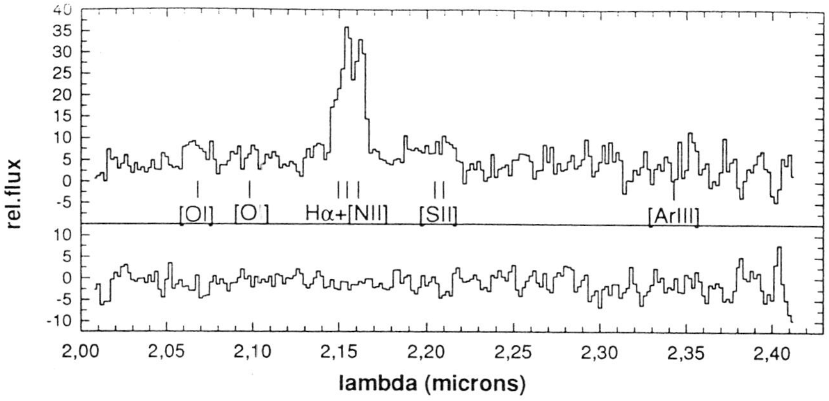

Fig. 1. K-band spectra of the central 2". Some lines for z=2.284 are marked. The lower spectrum is a region off source and represents the noise in the data.

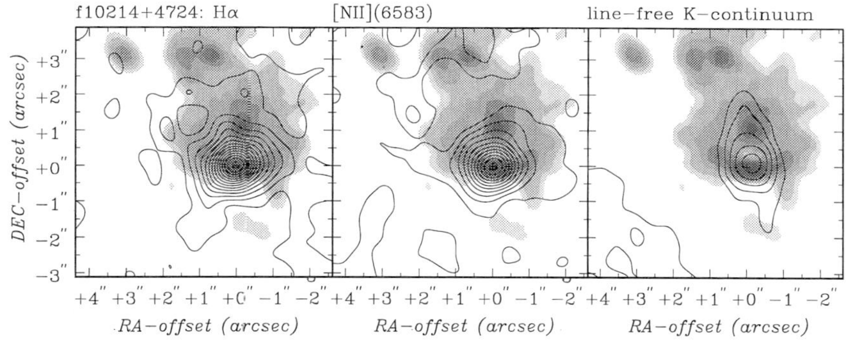

Fig. 2. Channel maps of three spectral regions: a) Hα b) [NII](6583) c) line-free K-continuum. The contours represent our data in linear scale, the grayscale represent the K-band map of Graham et al. (1995) in log scale.

two ring-like regions containing the flux from the inner 2."25, but without the central 0."75 and 1."25, respectively. It was not possible to fit 3 lines to the data with a [NII](6583)/[NII](6548) line ratio of 3:1 that is given from atomic transition probability. Therefore it was necessary to assume a flux contribution to both [NII] lines from a broad Hα line. To decrease the amount of free parameters the fit was constrained by fixing the relative position of all lines and forcing the narrow lines to have the same width (Fig. 3). ¿From this fit we derived a substantial contribution of a broad Hα line with a FWHM of 2400km/s (FWZP=3500km/s) for all regions. The width of the narrow lines is $600 \pm 100 km/s$. The line ratio [NII](6583)/$H\alpha_{\rm narrow}$ drops from 1.7 ± 0.5 (0."75 region) to 1.2 ± 0.4 (2."25 region) and ends with a value of 0.7 ± 0.45 (2."25 − 1."25 ring). This implies a higher flux of $H\alpha_{\rm narrow}$

in the outer regions, thus more extension than [NII]. For all spatial regions the [NII](6583)/$H\alpha_{broad}$ line ratio stays constant, so $H\alpha_{broad}$ has the same extension as [NII].

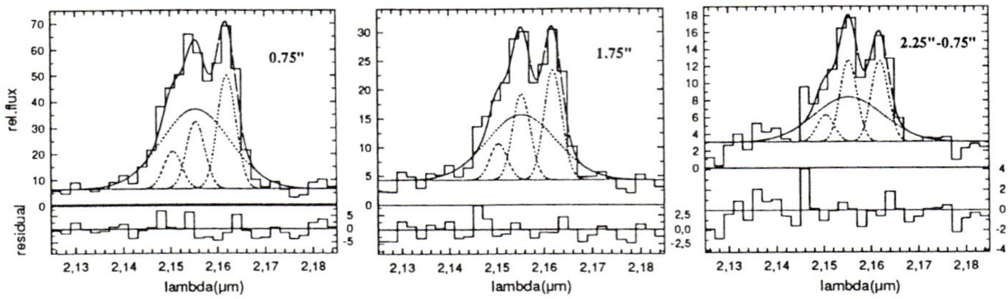

Fig. 3. Spectral line fits to the $H\alpha$ line system for three regions of different spatial size: a) 0."75 b) 1."75 c) 2."25 – 0."75. The columns represent the data, the dashed lines are single gaussian fits, the sum is drawn solid. The residuals are plotted below.

Graham et al.(1995) resolved an arc like structure in the K-band of 0."3 radial thickness and a EW elongation of 2" with a brighter inner arc. Following the gravitational lens hypothesis of Broadhurst & Lehár (1995), this size is due to a real source radius r of 0."2 and a magnification factor m of 5. For r = 0."01 the magnification encreases to m~50 and results in a arc length of 0."5. We attribute the [NII] and broad $H\alpha$ line emission to a pointlike and therefore highly magnified Seyfert 1.9 nucleus that forms the central 0."5 arc. The more extended emission of $H\alpha_{narrow}$ contributes to the 2" arc. Although still diluted by AGN emission in our data, it shows more characteristics of a star forming region. This implies a circum–nuclear starburst region of a size of several pc. The extension of the starburst region is comparable to nearby objects that show enhenced star forming activity in a ring like structure around a Seyfert nucleus like NGC 1068 (2kpc ring), or NGC 7469 (1kpc ring) (Genzel et al, 1995). A more detailed explanation will be given in Kroker et al. (1995).

References

Broadhurst, T., Lehár, J. (1995) Ap.J. (Let.) in press

Genzel, R., Weitzel, L., Tacconi-Garman, L.E., Blietz, M., Cameron, M., Krabbe, A., Lutz, D., Sternberg, A. (1995) Ap.J. **444**, 129

Graham, J.R. and Liu, M.C. (1995) Ap.J. **449** L29

Kroker, H. et al. (1995), in preparation

Rowan–Robinson, M. et al. (1991) Nat, **351**, 719

K-Band Imaging of Arp's Interacting Spirals

C.F. McCain[1], K.C. Freeman[1] and P.J. Quinn[2]

[1]Mt. Stromlo & Siding Spring Observatories, Private Bag,
Weston Creek PO, Weston, ACT 2611, AUSTRALIA
[2] European Southern Observatory,Karl-Schwarzschild-Str. 2
D-85748 Garching b. München, Germany

Abstract. We present K-band integrated photometry of a class of interacting galaxies with a spheroidal and irregular components. These systems are Arp 118 (NGC 1143/44), Arp 140 (NGC 0274/75), Arp 142 (NGC 2936/37), Arp 144 (NGC 7828/29) , Arp 146 and Arp 147 (IC 298/IC 298A). Two of these systems are found to have very high internal velocities within the irregular component: Arp 118 has an internal velocity of ~1100 km/s (Hippelein, 1989; McCain et al., 1995), while Arp 142 has ~900 km/s.

1 Introduction

We are interested in the kinematics and dynamics of a sample of interacting galaxies, comprising a spheroidal and an irregular component. In the blue images, the irregular compponents are severely affected by internal extinction. In order to know their old population masses and distributions, we have observed them in the Kn-band (2.0-2.3μ). The data were taken during two observing runs at the 3.9m Anglo-Australian Telescope, using the IRIS infrared camera at the f/15 Cassegrain focus. The pixel scale is 0.61 $arcsec$/pix.

2 Results

Our K-band images reveal the underlying structure hidden in the B images: eg. in the K images, secondary nuclei appear on the rings of Arp 146 and Arp 147, and in the central region of Arp 144's inner ring. NGC 2936, the irregular component of Arp 142, shows a slightly distorted face-on spiral structure in the K image (Fig. 1); NGC 2936 has an internal velocity of ~900 km/s. Integrated magnitudes are given in Table 1.

Table 1. Integrated K-band magnitudes for the six interacting systems.

Object		$f_K(mJy)$ (integrated)
Arp 147	ring	10.9±0.16
	spheroidal	4.9±0.28
Arp 146	ring	2.9±0.22
	loop	2.9±0.15
Arp 144	NGC 7828	27.5±0.48
Arp 140	NGC 0275	21.2±1.53
Arp 118	NGC 1144	63.4±1.27
Arp 142	NGC 2936	26.6±1.05
	NGC 2937	26.7±0.23

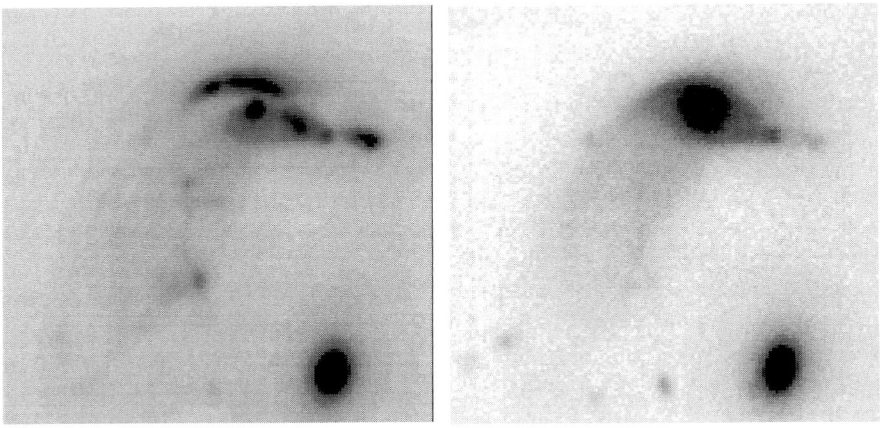

Fig. 1. Arp 142 in (a) B-band, and (b) K-band.

References

Hippelein, H.H (1989): A&A **216**, 11
McCain, C.F., Freeman, K.C., Quinn, P.J., 1995, in prep.

Starburst Episodes Near the Galactic Center

Dieter H. Hartmann[1]

[1] Department of Physics and Astronomy, Clemson University, Clemson, SC29634

Abstract. It was suggested that the Galactic center experienced a giant explosion about 1.5 10^7 yrs ago, releasing $\sim 10^{56}$ ergs, perhaps due to a starburst involving more than $\sim 10^5$ supernovae. This scenario can be tested through a search for the resulting γ-ray line emission or a search for the radio pulsars produced during the starburst

1 Introduction

That the center of the Galaxy may experience a series of explosive events was suggested by Oort (1977) and further studied by Loose, Krügel, & Tutukov (1982) and Sofue (1989). It is conceivable that large scale X-ray structures, such as the north polar spur (NPS), might be explained with shocks induced by these explosions (Sofue 1994), although several arguments suggest that the NPS is a local feature. The shock model requires an energy release of $\sim 3\ 10^{56}$ ergs, and an age of the starburst of $\sim 10^7$ years. It is also conceivable that a massive black hole near the Galactic center could release such a large amount of energy. A recent overview of the activities near the Galactic nucleus can be found in Genzel & Harris (1994).

A typical Type II supernova injects $\sim (1-2)\ 10^{51}$ ergs into the ISM (e.g., Woosley & Weaver 1994), and winds may double this amount (Leitherer *et al.* 1992). About 10^5 supernovae are required for the explosions mentioned above. Activities near the galactic center (Morris 1993a; Genzel & Harris 1994) are manifest on various spatial scales, the dominant being the expanding molecular ring (EMR) at 200 pc, which contains over 10^{55} ergs of kinetic energy. On smaller scales, superbubble G359.1−0.5 suggests a burst of $\sim 10^{2-3}$ supernovae within the past few million years (Uchida *et al.* 1992). Shells of expanding gas in the Sgr B complex suggest energy inputs from $\sim 10^2$ supernovae in the past 10^{5-6} yrs (Tsuboi, Handa, & Ukita 1994). Within the EMR region, 6.7 keV line emission was detected by GINGA (Koyama *et al.* 1989, 1990). Ozernoy *et al.* (1993) modeled this emission with a supernova rate of 0.2 yr^{-1} over a period of 4 10^3 yrs. Stellar populations within ~ 1 pc of the center (Tamblyn & Rieke 1993) suggest a starburst of $\sim 10^2$ supernovae 5 to 9 million years ago. Observations of the central cluster of HeI emission-line stars suggests that many of the massive stars within a few parsecs of the center were born in the last $\sim 10^6$ yrs (e.g., Krabbe, *et al.* 1991; Rieke & Rieke 1994). Recent IR observations of the central 0.5 pc region suggest a small star formation burst between 3×10^6 and 7×10^6 years ago (Krabbe *et al.* 1995). Other arguments support even larger starbursts. If the total mass within the inner 1 parsec is the result of mass segregation of black holes with masses ~ 10 $M_\odot$ formed within 10–100 pc (Morris 1993b), the associated number of neutron stars would exceed $\sim 10^6$. Their high velocities and small mass would prevent buildup of a neutron star cluster in the inner region. If produced in a series of starbursts of 10^3 neutron stars, such bursts would periodically inject 10^{54} ergs of kinetic energy into the ISM. Hydrodynamic simulations (Loose, *et al.* 1982) suggest that starburst episodes producing over 10^5 supernovae could occur every $\sim 10^8$ yrs.

2 Where are the gamma rays ?

One possibility to detect the aftermath of starbursts is γ-ray line afterglow from radioactivity. A promising isotope is ^{26}Al, which has a lifetime of $\sim 10^6$ yrs. Emission from ^{60}Fe would sample even older starbursts (twice the lifetime) but fluxes would be lower (e.g., Timmes *et al.* 1995). The Al yield per supernova is estimated to be $\sim 10^{-4}$ $M_\odot$. For an instantaneous starburst at a distance of 8.5 kpc the 1.8 MeV γ-ray line flux at Earth would be

$$F_\gamma = 1.7\ 10^{-3}\ \mathrm{M}_{-4}\ \mathrm{N}_5\ \exp\left(-\mathrm{T}_6\right)\ \ \text{photons cm}^{-2}\ \text{s}^{-1}\ , \tag{1}$$

where N_5 is the number of supernovae divided by 10^5, M_{-4} is the ^{26}Al yield per supernova in units of 10^{-4} $M_\odot$, and T_6 is the age of the burst in million years. To detect the γ-rays from the burst one requires a recent event or a large number of supernovae. COMPTEL observations of the plane (Diehl, *et al.* 1994a,b) show that the γ-ray brightness is patchy, with several isolated hotspots. Some may be due to forground sources, but some may be related to the starbursts discussed here.

3 Where are the pulsars?

A different method to test the star-burst hypothesis uses another product of star formation and subsequent massive star explosions. A burst of 10^5 supernovae would generate a comparable number of pulsars. These pulsars (all of similar age) could provide a unique signature of recent star bursts. Detection of such pulsars is hampered by several factors; their distances would be larger than most of the known pulsars (e.g., Taylor, Manchester, & Lyne 1993), they would be located in a region of high dispersion measure (e.g., Taylor & Cordes 1993), and they would be born in regions with high background emission. However, a significant fraction of these pulsars may have propagated to positions at which detection is easier. The upward revision of pulsar velocities (Lyne & Lorimer 1994) improves these odds.

The pulsar birthrate in the Galaxy is $\sim 10^{-2}$ yr^{-1}, so that a typical pulsar age of $\sim 10^7$ yrs suggests that the Galaxy contains a total of $\sim 10^5$ active radio pulsars. Beaming of the radio emission and limited instrumental sensitivities cause an actual number of detected pulsars of $\sim 10^3$ (e.g., Taylor, Manchester, & Lyne

1993). The creation of 10^5 pulsars 10^7 yrs ago significantly increases the observable pulsar sample. The model discussed here assumes a burst very near the Galactic center, at D = 8.5 kpc. Since typical pulsar distances are roughly a factor 3 closer, radio fluxes from the injected pulsars are reduced by $\sim$ 10. In addition, pulsars originating from steady star formation in the disk contain many stars that are younger than those from a burst $\sim$ 15 million years ago. Detectability also depends on the intervening dispersion measure

$$DM = \int ds\, n_e(s) \quad pc\ cm^{-3} \tag{2}$$

where $n_e(s)$ is the free electron density. While most pulsars have DM $\leq$ 200, pulsars close to the Galactic center could exhibit much larger values (Taylor & Cordes 1993). Fortunately, pulsars have very high space velocities and quickly move away from regions with high obscuration. A recent evaluation of pulsar space velocities (Lyne & Lorimer 1994) suggests that typical velocities are $\sim$ 500 km s^{-1}. A linear trajectory would move such a star a distance of 5 kpc over 10^7 yrs.

A pulsar injected near the origin would be more affected by the potential than a pulsar injected in the solar neighborhood. The gravitational potential in this region can be probed with radio and IR observations (e.g., Kent 1992) and with surveys of microlensing events towards the bulge (Griest *et al.* 1991; Paczynski 1991). The OGLE and MACHO groups observed many microlensing events toward the bulge (Alcock it et al. 1994; Udalski *et al.* 1994; Kiraga & Paczynski 1994). The observations seem to suggest that the potential near the Galactic center may be less than commonly assumed. We have modified the potential of Hartmann, Woosley, & Epstein (1989) to take a reduced bulge potential into account (Hartmann 1995). Additional motivation for a reduction is derived from the detailed models of the bulge from IR observations (Kent 1992). The IR data suggest that the large circular velocities found in the HI/CO rotation curves near R $\sim$ 1 kpc are overestimated due to noncircular motions. Kent (1992) suggests circular velocities of 150 km s^{-1} instead of 250 km s^{-1}. Pulsars born in the modified potential propagate farther, causing a larger fraction to become detectable. Hartmann (1995) generates Monte Carlo samples of isotropically injected pulsars integrating orbits for a period given by the age of the star burst. Final positions are recorded and the dispersion measure for pulsar is determined from the electron model of Taylor & Cordes (1993). The radio "luminosity", L_r, of a pulsar is modeled as

$$\mathrm{Log}\ L_r\ (\mathrm{mJy\ kpc^2}) = \alpha\ \mathrm{Log}\ P(s) + \beta\ \mathrm{Log}\ \partial_t P + \gamma\ , \tag{3}$$

where P(s) is the period (in seconds), and $\partial_t P$ its derivative. The radio flux at 400 MHz derived from this relationship is $S_{400} = L_r\, D^{-2}$, where the pulsar distance, D, is measured in kpc. Surveys are assumed to detect fluxes in excess of 1 mJy.

4 Discussion

Solid evidence for star bursts at the Galactic center is missing and the best evidence in favor of a *massive* black hole may be that there is no firm argument against it (Rees 1987; but see Ozernoy *et al.* 1993). The absence of hard X-ray emission from Sgr A* also argues against the presence of an accreting super-massive black hole (Goldwurm *et al.* 1994), but it does not rule out its existence, because the cycle of active periods could be such that the present source happens to be off-duty (Grindlay 1994). Alternatively, the accretion flow may be advection dominated so that most of the energy released is carried along with the gas and lost into the black hole (Narayan, Yi, & Mahadevan 1995). Starbursts and black hole induced activities are not exclusive, accretion onto the Galactic center could be responsible in part for driving star formation episodes.

In either case, it is desirable to obtain direct observational evidence for these interesting possibilities. We discussed two methods for obtaining this in the case of star bursts. The first method uses γ-ray afterglows from radioactivties associated with the supernovae from the star bursts. The timescales involved favor ^{26}Al as a tracer. The 1.809 MeV flux from Galactic nucleosynthesis is patchy (e.g., Diehl *et al.* 1994a,b), which may be due to localized star formation events. The observational evidence is not yet convincing, but next generation detectors may change this situation. As a new method, Hartmann (1995) suggested a search for radio pulsars that should be created in starbursts. The results of his study show that starbursts near the Galactic center might be detectable through an excess of radio pulsars. The distribution of such pulsars is distinguished by age, spatial distribution, and kinematics.

Acknowledgements. This work was supported by NASA grant NAG-5 1578, a grant from URGC at Clemson, and by NSF grant PHY94-07194.

References

Alcock, C., *et al.* 1994, ApJ, in press
Diehl, R., *et al.* 1994a, ApJS, 92, 429
Diehl, R., *et al.* 1994b, A&A, in press
Genzel, R. & Harris, A. I. 1994, *The Nuclei of Normal Galaxies: Lessons from the Galactic Center*, eds. R. Genzel and A. I. Harris, (Kluwer Acad. Publ.), NATO ASI proceedings, Series C - Vol. 445
Goldwurm, A., *et al.* 1994, Nature, 371, 589
Griest, K., *et al.* 1991, ApJ, 372, L79
Grindlay, J. E. 1994, Nature, 371, 561
Hartmann, D. H., Woosley, S. E., & Epstein, R. I. 1989, ApJ, 348, 625
Hartmann, D. H. 1995, ApJ, 447, 646
Kent, S. M. 1992, ApJ, 387, 181
Kiraga, M., & Paczynski, B. 1994, ApJ, 430, L101
Koyama, K., *et al.* 1989, Nature, 339, 603
Koyama, K., *et al.* 1990, Nature, 343, 148
Krabbe, A., *et al.* 1991, ApJ, 382, L19

Krabbe, A., *et al.* 1995, ApJ, 447, L95
Leitherer, C., Robert, C., & Drissen, L. 1992, ApJ, 401, 596
Loose, H. H., Krügel, E., & Tutukov, A. 1982, A&A, 105, 342
Lyne, A. G., & Lorimer, D. R. 1994, Nature, 369, 127
Lyne, A. G. 1994a, in *Frontiers of Space and Ground-based Astronomy*, eds. W. Wansteker *et al.*, (Kluwer), in press
Morris, M. 1993a, in *Back to the Galaxy*, ed. S. Holt, AIP 278, 21
Morris, M. 1993b, ApJ, 408, 496
Narayan, R., Yi, I., & Mahadevan, R. 1995, Nature, 374, 623
Oort, J. 1977, ARA&A, 15, 295
Ozernoy, L., Titarschuk, L., & Ramaty, R. 1993, in *Back to the Galaxy*, eds. S. S. Holt and F. Verter, (AIP, N.Y.), Vol. 278, 73
Paczynski, B. 1991, ApJ, 371, L63
Rieke, G. H., & Rieke, M. J. 1994, in *The Nuclei of Normal Galaxies*, eds. R. Genzel and A. I. Harris, (Kluwer Acad. Publ.)
Rees, M. J. 1987, in *The Galactic Center*, ed. D. Backer, AIP 155, 71
Sofue, Y. 1989, in *The Center of the Galaxy*, ed. M. Morris, (Dordrecht: Kluwer), p. 213
Sofue, Y. 1994, ApJ, 431, L91
Tamblyn, P., & Rieke, G. H. 1993, ApJ, 414, 573
Taylor, J. H., & Cordes, J. M. 1993, ApJ, 411, 674
Taylor, J. H., Manchester, R. N., & Lyne, A. G. 1993, ApJS, 88, 529
Timmes, F. X., *et al.* 1995, ApJ, 449, 204
Tsuboi, M., Handa, T. & Ukita, N. 1994, in *The Nuclei of Normal Galaxies*, eds. R. Genzel and A. I. Harris, (Kluwer Acad. Publ.)
Uchida, K. I., *et al.* 1992, ApJ, 398, 128
Udalski, A. *et al.* 1994, ApJ, 426, L69
Woosley, S. E., & Weaver, T. A. 1995, ApJS, in press

When Is a Merger not a Merger?

D. L. Clements[1] and Amanda Baker[2]

[1]E.S.O., Karl-Schwarzschild-Strasse 2, D-85748 Garching b. München, Germany
[2]Institute of Astronomy, Madingley Road, Cambridge, England CB3 0HA

Abstract. Activity in ultraluminous IR galaxies (ULIRGs) is thought to be triggered by mergers or close interactions. However, optical observations by Leech et al. (1994) suggest that many ULIRGs reside in undisturbed systems. We have obtained IR observations of the galaxy 1648+5447, which Leech et al. classified as 'isolated and undisturbed'. We find that it contains two nuclei separated by 3". This result demonstrates the importance of IR observations of ULIRGs.

1 Introduction

A substantial number of ULIRGs are merging or interacting systems, many with double nuclei (Sanders et al. (1988), Armus et al. 1994). Theoretical results (e.g. Combes 1993) show that mergers can rapidly drive gas (and dust) into galaxy centres, providing the fuel for the extreme luminosities and material to reprocess optical/uv light into far-IR radiation. But this scheme is not consistent with some optical observations. R band CCD observations by Leech et al. (1994) and analysis of UK Schmidt plates by Zhenglong et al. (1991) suggest that $\sim 40\%$ of ULIRGs are non-interacting. Yet, R band CCD observations convinced Clements et al. (1995) that 91% of ULIRGs in their new sample are interacting. They suggest that Leech et al. may have missed signs of interactions in some of their galaxies.

2 Observations and Results

We present a K' image of the galaxy 1648+5447, classified by Leech et al. as 'isolated and undisturbed', obtained using MAGIC at the Calar Alto 3.5m in July 1994. There are clearly two nuclei in this galaxy, and it is thus likely to be a merger.

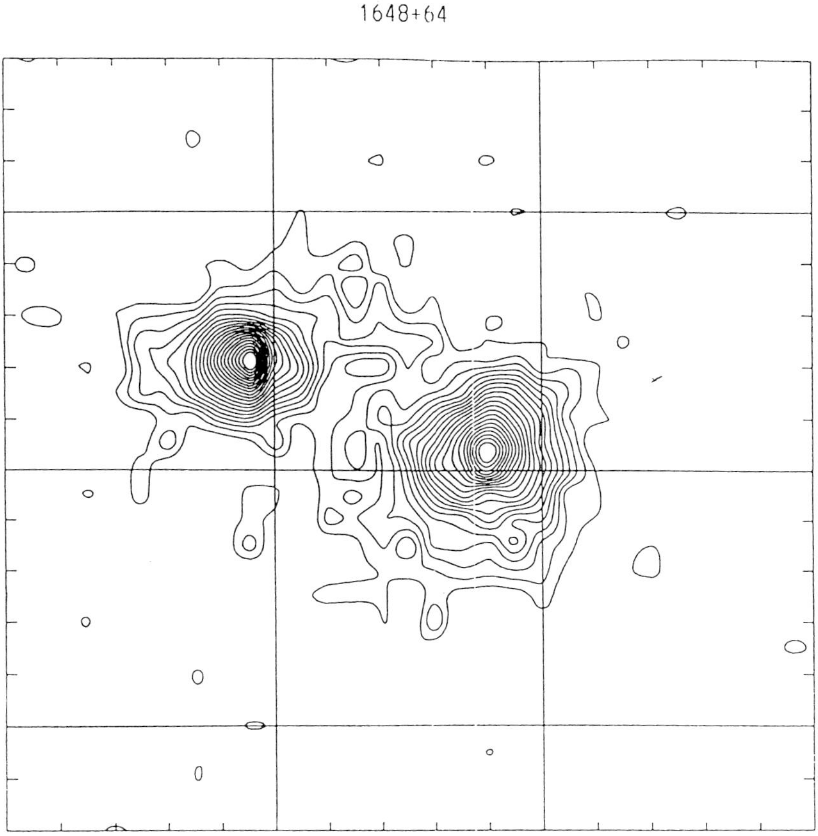

Fig. 1. A sky-subtracted, flat-fielded K' image of the ULIRG 1648+5447. The nuclei are 3" apart.

3 Discussion

A clear double nucleus galaxy has escaped identification — there may be many other misidentified galaxies in the Leech et al. and Zhenglong et al. samples. These observations show that deep IR observations are an important tool in the classification of ULIRG host galaxies.

References

Armus, L., et al., (1994), AJ, 108, 76

Clements, D.L., et al., (1995), in preparation

Combes, F., (1993), in *First Light in the Universe: Stars or QSOs?*, p. 215, ed. Rocca-Volmerange, pub. Editions Frontieres

Leech, K.J., et al. (1994), MNRAS, 267, 253

Sanders, D.B., et al., (1988), ApJ., 325, 74

Zhenglong, Z., et al., (1991), MNRAS, 252, 593

Author Index